AF892754

When AIAA Meets IEEE

Franklin Li Duan

When AIAA Meets IEEE

Intelligent Aero-engine and Electric Aircraft

Springer

Franklin Li Duan
School of Electronics Information
and Electrical Engineering
Shanghai Jiao Tong University
Shanghai, China

ISBN 978-981-19-8393-1 ISBN 978-981-19-8394-8 (eBook)
https://doi.org/10.1007/978-981-19-8394-8

© The Editor(s) (if applicable) and The Author(s), under exclusive license to Springer Nature Singapore Pte Ltd. 2023
This work is subject to copyright. All rights are solely and exclusively licensed by the Publisher, whether the whole or part of the material is concerned, specifically the rights of translation, reprinting, reuse of illustrations, recitation, broadcasting, reproduction on microfilms or in any other physical way, and transmission or information storage and retrieval, electronic adaptation, computer software, or by similar or dissimilar methodology now known or hereafter developed.
The use of general descriptive names, registered names, trademarks, service marks, etc. in this publication does not imply, even in the absence of a specific statement, that such names are exempt from the relevant protective laws and regulations and therefore free for general use.
The publisher, the authors, and the editors are safe to assume that the advice and information in this book are believed to be true and accurate at the date of publication. Neither the publisher nor the authors or the editors give a warranty, expressed or implied, with respect to the material contained herein or for any errors or omissions that may have been made. The publisher remains neutral with regard to jurisdictional claims in published maps and institutional affiliations.

This Springer imprint is published by the registered company Springer Nature Singapore Pte Ltd.
The registered company address is: 152 Beach Road, #21-01/04 Gateway East, Singapore 189721, Singapore

Preface

The initial motivation of this book is an over-30-page long article submitted to AIAA Journal, the top administrative academic magazine of American Institute of Astronautic and Aeronautic association. Unlike a purely technical paper which just focuses on a specific research topic, the editor feels hard to choose proper reviewer since its content covers a broad scope of disciplines both in AIAA and in IEEE, and it is not easy to find any experts who master both of them. He shared his thought with me, and in the mean time, I received the letter from Springer to encourage me to write a book. Indeed, considering both the content and the scope, the best output mode of this long article is a book. It is the comment from editors in AIAA and Springer who inspire me to write this book, talking about the integration of IEEE and AIAA, the two biggest engineering organizations in the world.

The book title *When AIAA Meets IEEE* is inspired after joining the AIAA Meeting in 2018 in Cincinnati, USA, an interactive meeting that AIAA invited IEEE for a joint conference called EATS. It is the first interactive meeting that an AIAA invite IEEE for a joint project of electric aircraft technologies to combine the two teams' efforts together. The author was deeply engaged these years with AIAA people to develop smart sensors for intelligent gas turbine engines and accumulated a lot of real-time experience of IEEE/AIAA working together. And I do feel the strong need of coordination to integrate these two teams together to achieve an engineering target on electric aviation. The differentiation of the mindset and the focus of interest of these two teams shall be properly addressed for their most efficient cooperation. In general, IEEE people needs to adjust their focus to fit the need of AIAA, while the AIAA people need to adapt the most advanced achievements in IEEE, especially in building intelligent aero-system and powerful battery for electrified aero-engines. AIAA people should be able to ask the correct questions on their designated need in smartness, power, weight, and robustness when communicating with IEEE fellows. A professional integration team shall be suggested to aid the communications on both sides to coordinate their necessity and feasibility.

In the previous century, the key word of the AIAA is power; for the next century, the key word is smartness. Electric aero-engine and power are two crucial leading technologies, and both involve heavily the IEEE. The work-together of AIAA and IEEE people becomes a "MUST" in the future activities. While the first part of the book addresses IEEE people's working experience with AIAA, the second portion of the book explores the blueprint of electric aero-propulsion, involving two key technologies: rim-driven electric propulsion and mobile electric power grid. The privilege of this novel electric propulsion is investigated in comparison of the old-fashion propeller and the existing fuel-based gas turbine engines. Rim-driven aero-engines are much smaller and lighter. Multiple engines can be integrated in an aircraft design with a flexible vertical/horizontal propulsion orientation. New theories/methodologies are proposed and explored with some initial efforts made. As to the electric power, three major sources: battery, supercapacitor, and turbine gas generator are analyzed regarding the needs on power, energy, thrust from AIAA's angle and the power/energy density, efficiency, capability, price/maturity in IEEE's perspective.

When I start to write on the book, I feel a big challenge. The first challenge is the scope and the multi-disciplines involved since the AIAA and IEEE belong to totally two different categories of physics and cover so broad knowledge. Not only one needs to learn more knowledge but also needs to express it in more understandable manner. We try to use the verbal language other than the formal language to describe the technical terms whenever possible, and to use the glossary/index is to explain the specific technical jargons. Since the AA and EE cover two separate different physics categories, I most often use the first principles of physics to explain the technology/project from the root, like we did in Chap. 9 that electric motor is an origin of both electric generator and electric engine. The book is intending to lessen the burden of the readers of the intensive knowledge in vast fields by using the oral English rather than the official English.

The second challenge is to walk out from your own comfort zone which may make fool of yourself for the insufficient knowledge outside your profession. Besides, some readers may also feel uneasy when finding themselves outside the comfort zone at some point. The author, therefore, asks the pardon of the readers for some stupid naïves occurred occasionally in the book and offers you the courage you to engage into the new land like me. In fact, the author is intending to use this book to inspire you instead of using it as a textbook. The author was originally an IEEE fellow for nearly 20 years and then engaged in AIAA project in the past 10 years. The author appreciates the different mindsets of these two great minds from AIAA and IEEE people through these 30 years engagement on both sides. The author also feels the need for the effective and efficient integration of these two teams of people if they got to work together for a common goal. From necessity point of view, the intelligent aero-engine needs smart sensors, the electrified propulsion makes the aviation more efficient and sustainable. From feasibility point of view, advanced rim-driven brushless electric motor and advanced battery/supercapacitor achievements in electric automobiles shed the light for electrified niche aero-applications and smarter aviation scenarios such as short distance takeoff landing.

The author understands that some people may prefer to live in their comfortable zone instead of taking the challenge to step into another brand new career, just like most people prefer to park the car near the entrance of a shopping mall. It is the author's willingness to take the risk of making fool of himself, to make an initial effort to contribute some primitive thoughts for the integration of the two teams of people—the two world largest institutes of technology/engineering groups. The coordination of these two great minds, goals, and the ways of doing is highly necessary in order to make an efficient multi-disciplinary project such as the exemplified cases in this book—the IoT for intelligent aero-engine and electric propulsion for next generation electrified aviation, among the many others.

There are three main parts of this book. Part I explains the necessity and feasibility of the AIAA and IEEE cooperation. Part II discusses the use of smart sensor for an intelligent aero-engine. Among the Part II, Chap. 2 mainly explores the specific meanings of AI and how the sensors play an indispensable rules for intelligent AIAA. Then, we use Chaps. 3–5 to illustrate how to use the smarter methods to build the smart sensors to achieve the smarter gas turbine machines. Chapter 3 deals with the scenarios that the TFTC sensor can be used. Chapter 4 discusses the various methods to build them, and Chap. 5 discusses the specific tests that IEEE and AIAA people need to go through before the smart sensors being used infield. It is the author's motivation to use the true experience between the AIAA and IEEE people working together to encourage and to serve as a reference for the future IEEE/AIAA cooperation on electrified aviation engineering. As an intermission between Part II and Part III, Chap. 6 is an intermediate chapter to connect the AIAA and IEEE cooperation from previous to future, i.e., to use our six years' true experience of IEEE people with AIAA colleagues to explain the integrated process between the two groups of professionals, serving as a reference point for the future bigger cooperation. Electric propulsion and associated aviation electric power in Part III are much more challenging and more promising and may bring a more revolution to the whole aviation enterprise. In Part III, we discussed the smarter and the green aviation by the innovative design from e-engines, e-power to e-planes. Although electricity provides power for the e-aviation, the electricity is just a medium but not the final energy resource. The ultimate human energy comes from the solar, wind, tidal, and other green and sustainable natures. New energy vehicle offers new generation of human transportations from ground, ocean to sky with chance, privilege, and feasibility as shown in chart below—a blueprint of the feasibility (the left) versus necessities (the right) of future NEV.

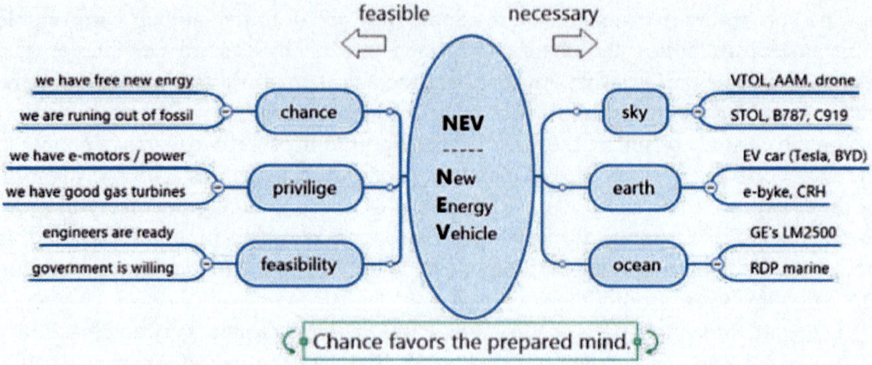

Specifically in Chap. 7, we discuss the e-aviation from the first principle of physics, i.e., how to achieve a clever and smarter aviation, the significance of the greener takeoff and landing, and how to achieve this by vertical e-propulsion. Vertical thrust from e-engines leverages the weight of an airplane either to achieve a VTOL or STOL. Vertical takeoff and landing offer the vast scope of aviation mobility. Short takeoff and landing for larger airliners save lot of airport resources and reduce the airport pollution both from noise and from the hostile gases. Chapter 8 compares the few generations of the aero-engines from propeller to jet, gas turbine, and gas turbine fan, as well as the evolution of the electric motors from the DC brush to AC brushless. From these previous technologies, we are intending to extract the useful experience and lessons for the future development of advanced electric propulsion. In Chap. 9, we introduce a new e-propulsion called rim-driven fan (RDF) jet, a lightweight and small yet more efficient electric propulsion unit as the future electric aero-engine. Chapter 10 discusses the electric power grid suitable to provide enough electricity to drive the RDF jets such as battery, supercapacitor, and lightweight gas generator. Chapter 11 discusses the integrated design of the e-engines and e-airplanes, i.e., how to integrate the multi-electric engines into an aircraft. The key technology is the vertical and horizontal thrust transformation by rotatable wings to keep the optimal performance for both vertical thrust and horizontal propulsion.

The author deeply appreciates the academic and research platform provided by Shanghai Jiaotong University over the past 10 years, especially the cooperation and support of the team. Specifically, the author thanks Prof. Ding Guifu, Zhang Yafei, Han Tao, Fu Xuecheng, Cheng Xiulan, Wang Ying, and other teachers in these years. The author also thanks the other cooperative supports on all levels, such as Li Jibao, Li Jie, Hong Zhiliang, Qian Lingyi, Zheng Fangfang, Shao Jing, Wu Shaohui, Zhang Baowen, Tian Shuqing, and other comrades in AECC. The author also thanks the other colleagues outside the school such as Lin Yuzhen, Cao Xueqiang, Zou Binglin, Wang Ruijun, in Beihang University Wuhan Polytechnic University and

Chinese Academy of Sciences. At last, the author is grateful to all my previous graduate students: Gao Junchao, Hu Mingkai, Weng Haotian, Ji Zhonglin, Xie Ziyi, Liu Siqing, Li Xiaocui who have been working so hard for many years.

Shanghai, China Franklin Li Duan

Contents

Part I Astronautics/Aeronautics Versus Electricity/Electronics

1 When AIAA Is Meeting IEEE—How to Interact and What to Achieve ... 3
1.1 The AIAA ... 5
1.1.1 The Scope of AIAA ... 5
1.1.2 Smart Engines by Smart Sensors ... 7
1.1.3 Electrified Green Aviation ... 8
1.2 The IEEE ... 9
1.2.1 The Scope of IEEE ... 10
1.2.2 MEMS Technology and Micro-fabrication ... 10
1.2.3 Electric Motor and Electric Power ... 13
1.3 To Integrate IEEE to AIAA ... 15
1.3.1 Mindset Difference of AIAA/IEEE Engineers ... 16
1.3.2 Two Cutting-Edge of AIAA/IEEE Fields ... 17
1.3.3 The Corporations of AIAA/IEEE Communities ... 18
References ... 18

Part II The Intelligent AIAA—Smart Sensor

2 Artificial Intelligence ... 23
2.1 AI: Brain, IoT, Machine ... 24
2.1.1 Smart Brain and Cloud ... 24
2.1.2 Internet of Things ... 25
2.1.3 The Robotics ... 27
2.2 Three E-devices—RLC, Transistor, and Sensor ... 33
2.3 Smart Sensors ... 36
2.3.1 Smarter Sensor for Smart Use ... 36
2.3.2 Smarter Way to Make Smart Sensor ... 38
References ... 45

3 Smart Sensors for AIAA 47
3.1 Calibration for Other Methods 49
3.1.1 Turbine Surface Infrared Thermography 49
3.1.2 SAW Wireless Sensor Calibration 49
3.2 Evaluating TBC's Temperature Barrier Quality 50
3.3 NGV Cooling Performance Tests 54
3.3.1 Background 54
3.3.2 Error Analysis 55
3.3.3 TFTC's Advantage 57
3.4 Evaluate Horn-Shape Cooling Hole on Turbine Blade 58
3.5 Jet Nozzle Temperature Distribution 60
3.6 Temperature Field of Jet Flow 60
3.7 Long-Term Monitor for Gas Generator 61
References 62

4 To Build Smart Sensors Using MEMS 65
4.1 MEMS Technology 66
4.1.1 Concept 66
4.1.2 Process 68
4.1.3 MEMS Sensors for AIAA 69
4.2 MEMS Temperature Sensors 71
4.2.1 Comparison with Other Methods 71
4.2.2 MEMS Methods and Scenarios 73
4.3 Challenges and Countermeasures 79
4.3.1 Electric Insulation at High Temperature 81
4.3.2 Thermal Stress Mismatch 90
4.3.3 Minimizing the Test Error 94
4.3.4 Sharp Corner Patterning 96
4.3.5 On-Site Fabrication 98
4.4 Technology and Process 100
4.4.1 PDMS Curved Surface Patterning 101
4.4.2 DFPR Photolithography 104
4.4.3 YSZ Thermistor 107
4.4.4 SAW Wireless Sensing 109
References 115

5 To Test the Smart Sensors for AIAA 119
5.1 Performance Test—IEEE 120
5.1.1 Test Accuracy 120
5.1.2 Dynamic Performance 124
5.1.3 Ultra-high Temperature Measurements 127
5.1.4 Wireless Temperature Sensing 130
5.2 Robustness Test—AIAA 136
5.2.1 Endurance Test 136
5.2.2 Temperature Shocking Test 137
5.2.3 Mechanical Shocking Test 139

5.2.4 High Vibration Test 140
5.2.5 Violent Heat Flame Rushing Test 142
5.3 The Combustion Infield Test 142
References 144

6 The Mindset Analysis of AIAA and IEEE 145
6.1 The Necessity of AIAA/IEEE Working Together 145
6.2 The Difference of AIAA/IEEE People 147
6.2.1 Difference in Technical Focus 147
6.2.2 Difference in Culture 150
6.2.3 2D Versus 3D Thinking 154
6.3 The Connection of AIAA/IEEE People 157
6.3.1 Customer Versus Vendor 159
6.3.2 The 3rd Way 160
6.3.3 A Dream Keeper 162
References 163

Part III The Smarter Aviation—Electrified Aero-engine

7 Why Electric Aviation—Versatile, Smarter, and Green 167
7.1 The Niche Aviation—AAM 169
7.1.1 General Aviation Versus AAM 170
7.1.2 AAM Case 1—3 × 3 VTOLer 174
7.1.3 AAM Case 2—Drone UPS 174
7.2 The Smarter Aviation—STOL 176
7.2.1 Saving Airport Resources 178
7.2.2 Electric Engine Is More Efficient 180
7.2.3 Electric Flight Is Smarter 181
7.3 The Green Aviation—Less CO_2 183
7.3.1 Greener Electricity 183
7.3.2 Greener Flight 184
7.3.3 Greener Takeoff 186
7.3.4 Greener Landing 190
7.4 Case Studies/The Learnings 192
7.4.1 Electrification of Boeing 787 194
7.4.2 Rotatable Wing VTOL Carcopter 195
7.4.3 Three E-plane Cases 196
7.4.4 Unmanned Aero-system 198
7.5 The Distinct Topics of E-propulsion 201
7.5.1 Power Versus Thrust, Engine Versus Battery 201
7.5.2 Specific Power of E-engine 204
7.5.3 Torque, Power, and Speed of Electric Motor 206
References 207

8 The Evolution of Aero-engines 209
8.1 Pistol Engine and Propeller 212
8.1.1 The Pistol Engines 213
8.1.2 The Propeller 213
8.1.3 The Privilege of Propeller Aircraft 215
8.1.4 The Drawback of the Propeller 216
8.1.5 From Propeller to Jet Engine 218
8.2 The Gas Turbine Machine 219
8.2.1 The Core 220
8.2.2 The Inlet and Outlet 221
8.2.3 The Working Process 221
8.2.4 Turbojet Engine 222
8.2.5 Turbofan Engine 222
8.2.6 Turboshaft Engine 223
8.2.7 Gas Turbine Efficiency 223
8.3 The Electric Aero-engine 224
8.3.1 Analysis 1: Propeller Versus Duct Fan 224
8.3.2 Analysis 2: Shaft Driven Versus Rim Driven 226
8.3.3 Inspiration 1: Fan-Ducted Jet 227
8.3.4 Inspiration 2: Lightweight Aero-generator 228
8.3.5 Inspiration 3: Edge-Driven Turbine Machine 229
8.3.6 R&D Matrix of Electric Aero-engine 230
References 232

9 The RDF Jet—A New Electric Aero-engine 233
9.1 The Features—Small, Light, Flexible 234
9.2 RDF Jet Principle 235
9.2.1 The Concept 236
9.2.2 The Thrust 237
9.3 The Architecture—Motor, Fan, Jet 238
9.4 The Rim-Driven Motor 239
9.4.1 A Quick Review of the E-motor 239
9.4.2 Rim-Driven Principle 242
9.4.3 Rim-Driven Theory 243
9.5 The Tai Chi Fan 252
9.5.1 The Future of Rim-Driven Fan 253
9.5.2 The Design of Rim Driven Blade 253
9.6 The Duct and Jet 257
9.7 The Building Material 259
9.7.1 Weight Estimation 259
9.7.2 Permanent Magnet 260
References 262

10 The Electric Power—Energy and Weight 263
10.1 The Electric Power Grid for Aviation 265
10.1.1 Energy, Power, and Weight 266
10.1.2 From AIAA's Perspective—Power, Energy, and Safety 275
10.1.3 From IEEE's Perspective—Battery, SC, LTG 278
10.2 Battery 284
10.2.1 Li Battery 290
10.2.2 Fuel Cell 295
10.2.3 Supercapacitor 302
10.3 Lightweight Generator 304
10.3.1 The Background 304
10.3.2 The Reference Technologies 307
10.3.3 A New Rim Driven Generator 308
10.4 3D High-K Supercapacitor 308
10.4.1 Necessity and Feasibility 309
10.4.2 The Principle 310
10.4.3 The Evaluation 310
10.4.4 The Technology 313
10.5 Summary 317
References 318

11 The Integrated Design of Aero-engines and Aero-craft 323
11.1 The Airplane Basics—Range, Speed, Efficiency 325
11.1.1 Air Flight Journey 326
11.1.2 Navigation Distance—The Longer Range 326
11.1.3 Aircraft Velocity—The Optimized Speed 329
11.1.4 Thrust and Power—The Capability 331
11.1.5 Lift Versus Weight—The Efficiency 333
11.1.6 Distributed Electric Propulsion 335
11.2 The Showcases—Delta VTOLer and STOL for B787 338
11.2.1 VTOL: Δ Aircraft as Delivery Drone 339
11.2.2 STOL: Shorter Takeoff for Airliner 348
11.3 The Technologies—Rotatable Wing and VTOL Operation 349
11.3.1 Rotating Algorithm of RDF Jets 349
11.3.2 VTOL Prerequisites 352
11.3.3 Propulsion and Energy in One Shell 354
11.3.4 Power Electronics and UAS 355
11.3.5 Lightweight Aero-materials 359
References 361

Afterword 363

Glossary 367

Index 377

Part I
Astronautics/Aeronautics Versus Electricity/Electronics

AIAA is on the scale of kilometers, and the IEEE is the world of nanometers. The range of AIAA is from 1000 m to 1000 km, and the scale of IEEE is 1 nm to 1 mm. One is powerful, and the other is delicate. It is a wonderful thing to connect the two worlds together. Two contemporary bridges connecting them are the intelligent aero-engine and electric aviation. They connect the "far and near", they connect "strong and smart".

While there are many interactive activities between AIAA and IEEE, in this book, two hot topics are discussed and addressed as bridges to connect the two groups together. One is the AI's applications on intelligent aero-engine where the AI is greatly facilitated by IEEE, and the other one is electrified aviation where the electric propulsion, aviation electrical power grid and electric control play the key rules, which are also greatly facilitated by IEEE. We use the electronics to make the aero engines more intelligent, we use the electricity to make the aviation smarter.

Chapter 1
When AIAA Is Meeting IEEE—How to Interact and What to Achieve

AIAA refers to American Institute of Aeronautics and Astronautics, and IEEE refers to Institute of Electrical and Electronics Engineers. The "AA" in AIAA is astronautics (space) and aeronautics (aviation), and the "EE" in IEEE is electricity (powerful) and electronics (smart). AIAA is more government-driven, and IEEE is more product/commercial-driven; AIAA used to be more national, and IEEE is always more international. In this book, both terms refer to broader meanings than just the two institutions. The terms AIAA and IEEE in this book contain three meanings: organization/institute, science/technology, and professions/people.

IEEE and AIAA are the two world's largest engineering associations or institutes. Other than the AIAA and IEEE, American Society of Mechanical Engineers (ASME) and Materials Research Society (MRS) are two world's highly announced institutes which also relate to aerospace and aerospace industries. They are associated with AIAA very closely since the airplane and aero-engines are the 3 + 1: mechanical, material, thermal plus the aero-/astro-system and design.

AIAA and IEEE also refer to two distinct scientific categories. Not only they belong to different physics. In physics, we have six basic categories: electromagnetism, thermal dynamics, mechanical dynamics, vibration and wave, micro-nano science, field theory. AIAA mainly involves thermal and mechanical dynamics, while IEEE mainly involves the rest fields.

AIAA people may have different mindsets/points of interest compared those of IEEE people. For example, AIAA people like to say: more reliable and reliable …, more repeatable and repeatable …, while IEEE people like the wording: newer, the first time, cheaper … AIAA people are more focusing on reliability and endurance, while IEEE people may care more about expense and marketing. However, both believe that the technology itself is not the final success—much more effort is still needed through team efforts to transform technology into engineering/marketing.

The purpose of these two people meeting together is mainly that the IEEE helps the AIAA get smarter by electrification in two ways. One is using "electronics", and the other is using "electricity". One is tiny and delicate, and the other is strong and

© The Author(s), under exclusive license to Springer Nature Singapore Pte Ltd. 2023
F. L. Duan, *When AIAA Meets IEEE*,
https://doi.org/10.1007/978-981-19-8394-8_1

powerful. One is to make the AIAA more intelligent by smart sensors, and the other is to make the AIAA wiser by using electric propulsions.

This book is intended to help the AIAA and IEEE people to form a cooperation team for a common goal. Recently, there have been strong interactive activities of these two teams regarding intelligent aero-engines and electrified propulsion [1]. To effectively communicate between the two groups of people, the differentiation of the mindset and the focus of interest are analyzed in this book using two distinct showcases. One is the smart sensors built by IEEE's method to test the high temperature for various AIAA scenarios. The other is to use electrified propulsion for future aviation. The first example uses our six-year true working experience with AIAA's groups on developing TFTC sensors to illustrate the differentiation on mindsets and the coordination process on achieving a target. The second showcase is to explore the blueprint of the next generation of aero-propulsion, involving two key technologies: rim-driven electric jet and mobile electric power grid. The privilege of this novel rim-driven motor with inner blades is investigated in comparison to the old-time propeller and the concurrent turbine gas engines. Rim-driven engines are much smaller and can be multiple for flexible vertical/horizontal propulsion. New theories to evaluate the torque and magnetic interactions are proposed for the first time with some initial efforts made. Regarding the electric power, three major sources, battery, supercapacitor, and turbine gas generator, are analyzed regarding the needs of various aero-mobile scenarios on power, energy, and thrust from the AIAA's angle and the PE density, efficiency, capability, price/maturity from IEEE's perspective. In general, IEEE people need to adjust their focus to fit the need of AIAA, while the AIAA people need to explore and adapt the advancements from IEEE, especially in building intelligent aero-systems and powerful aviation electric power grids for electrified aero-engines. AIAA people should be able to ask the righteous questions on their designated need in smartness, power, weight, and robustness when communicating with IEEE fellows. A professional integration team is suggested to aid the communications on both sides to coordinate their necessity and feasibility and to ensure the cooperation plan executed along the road.

In this chapter, a primitive review of these two organizations and the initial integration efforts on the above two topics are given. The mindset differentiation and the manner to express data of the AIAA and IEEE people are analyzed and appreciated during the cooperative process.

1.1 The AIAA

The outer arrow in the AIAA logo represents the astronautics referring to flying out and toward the space; the inner arrow represents the aeronautics such as an aircraft carrying people around the earth. The propulsion of the aircraft is turbine jet + turbofan, while the outer space propulsion mainly relies on the rocket.

Figure 1.1 illustrates the scope and features of the AIAA organization. In general, the AIAA organization is quite concentrated and organized. AIAA has just four annual forums, which are highly focused and application-oriented [2]. There are three flags indicating the relevant hot spots associated in this book.

Another AIAA-related topic is unidentified flying objects. The UFO, if exists, may rely on gravity field modulations other than the rocket since it appears so quietly, together with advanced stealth technology because even the radars cannot detect it, which indicates that UFO can shield itself from large range spectra of wavelengths. At this moment, it still belongs to the scientific frictions, yet it reminds people of the analogy that Jules Verne predicted the submarine 20 years ahead before being used in Marine in his novel "Twenty Thousand Leagues under the Sea".

1.1.1 The Scope of AIAA

The AIAA institute comprises the following journals to cover a few distinct areas of expertise [3] (Table 1.1).

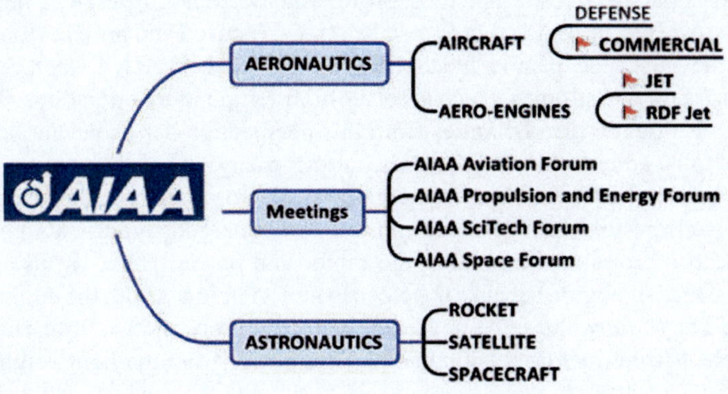

Fig. 1.1 Features of AIAA

Table 1.1 Relevant AIAA publications

*	AIAA Journal
	Journal of Aerospace Information Systems
	Journal of Air Transportation
*	Journal of Aircraft
*	Journal of Guidance, Control, and Dynamics
*	Journal of Propulsion and Power
	Journal of Spacecraft and Rockets
*	Journal of Thermophysics and Heat Transfer

In this book, the related topics are marked with "*" signs, i.e., smart sensors for propulsion and the airplanes. When AIAA is meeting IEEE, one needs to realize the pragmatism is a key which means that the cooperation must be highly project driven instead of academic; otherwise, it is very hard to make the coeffort be fruitful. There were quite a few unsuccessful projects due to the diversified efforts. We use the E-Fan X as an example to explain this defocused cooperation consequences. On Apr 27, 2020, E-Fan X, a joint effort of IEEE and AIAA on the electrification of airplanes of the Airbus/Rolls-Royce group, was canceled after just three years cooperation (the project was launched in 2017 [4]). The goal was to explore hybrid electric propulsion for commercial aviation. The due date of its maiden flight was in 2021. Obviously, the ambition seems impractical in the four-year span, and very importantly, the goal covers too many areas and topics that lack sufficient resources from many professionals. Specifically, the 20-pax commercial aircraft relates to hybrid power architectures, high-voltage systems, battery technologies, aircraft design, and enough flight experiments. Each one of them is a large and underdeveloped project that is still in the R&D stage. It could be more suitable to develop a new simple electric propulsor that we proposed as an RDF jet in Chap. 9. Therefore, it is better to be pragmatic when IEEE is working with AIAA on a new project. To cooperate with AIAA people more efficiently, the IEEE people also need to obtain some knowledge from AIAA's perspectives. Since their professions were not the AIAA, their efforts to become involved in AIAA should also be AIAA-focused and project-oriented. For example, they just need to know the thermo-related details in AIAA when developing smart sensors for gas turbines which involves high-temperature operations. They may ignore the theories in aerodynamics from this perspective—the relevant knowledge that needs to be added from the AIAA is not that many.

For example, AIAA has two immediate needs from IEEE at this moment: (1) the smart sensors to optimize aero-engine operations and monitor its conditions; (2) the electric propulsion including the motor and power. Here, we use the word "motor" instead of engine because the electric motor includes both the engine and the generator. They follow the same physics principle but renders two different needs in parallel: electric engine/propulsion and electric power from the lightweight electric generator.

1.1.2 *Smart Engines by Smart Sensors*

Smart sensors are a newly proposed idea to improve engine efficiency and to optimize engine design. For example, the efficiency of the aero-engine follows the Brayton thermal cycle:

$$\eta = \frac{W}{Q_{2-3}} = 1 - \frac{Q_{4-0}}{Q_{2-3}} = 1 - \frac{T_4 - T_0}{T_3 - T_2} \tag{1.1}$$

Increasing the pre-turbine temperature T_3 is crucial to enhance the eta and thrust. Generally, the thrust increases by 10–15% when the T_3 increases by 100 °C. To maintain the thermodynamic efficiency, the T_2 (T_0: input air, T_4: the jet air, T_2: after the compressor, T_3: pre-turbine) must also increase, that is, the boost ratio of the compressor increases. Therefore, the development of turbojet engines mainly focuses on two things: increasing the temperature in front of the turbine and increasing the boost ratio of the compressor.

Unfortunately, turbine material cannot bear such a high temperature, and a series of cooling holes are needed. Too much cooling air dramatically reduces the efficiency; therefore, the cooling effect needs to be evaluated. What is the efficiency of the cooling by these cooling holes? It is anticipated that cone-shaped cooling holes are more efficient. How effective can it be? The cooling effect can be described as:

$$\Theta = (T_\infty - T)/(T_\infty - T_c) \tag{1.2}$$

where T_∞ is the mainstream temperature, T is the wall temperature, and T_c is the cold air temperature. To keep the engine working safely, one needs to keep the surface temperature T below the safety spec, e.g., 1000 °C. Traditionally, the cooling effect of the cooling gas from the cooling hole is very difficult to measure. This is mainly because of the impossibility of accurately measuring the wall temperature T in Formula (1.2). Usually, we apply more cooling air to keep it safe. Excessive cooling certainly lowers the engine efficiency greatly, but this has to be done since safety is the number one priority in aero-engine system. By using the MEMS technology, we can make a tiny temperature sensor just on the site of the cooling hole to get this T value in Eq. (1.2), so that we can optimize our blade design accordingly.

One distinct showcase that we have achieved is shown in Fig. 1.2—the cooling effect evaluation of different cooling holes. We have made special cooling holes on the turbine blade using laser drilling together with the conventional holes and use the MEMS sensors to compare their cooling effects. How to evaluate the cooling effect of this special hole with the conventional cooling holes? MEMS sensor is the best solution for this. Such measurement and comparison can only be made using our thin-film micro-sensors since the conventional wired thermocouples are adjust too big to fit in such a tiny spot on turbine blade. Figure 1.2 illustrates the temperature measurement on such a tiny spot (0.5 mm) near the cooling hole using MEMS micro-fabrication technique.

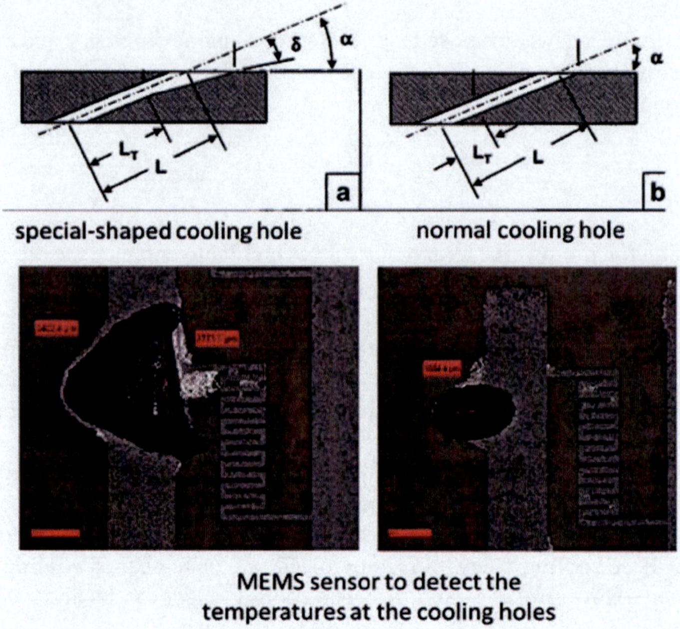

Fig. 1.2 To test the cooling effect of different cooling holes by using MEMS sensors

More details are presented in Chap. 3.

1.1.3 Electrified Green Aviation

For larger commercial aircraft, reducing airport pollution is an important issue that still lacks an effective solution. It is just the fact that an airplane needs more energy to takeoff, and when its engine is in full throttle, the airport receives a much higher amount of pollution. When the airplane is landing, the engine is still on with sub-combustion state, exhausting more NO_x gases (Chap. 7). Vertical takeoff and landing (VTOL) is therefore becoming a very hot topic in recent AIAA activities. The word VTOL is almost equivalent to eVTOL, i.e., electric-driven vertical takeoff. Electric propulsion creates clean thrust and can help to shorten the takeoff/landing distance, therefore saving airport resources. Electric propulsion, including electric engines and electric power, makes greener takeoff/landing possible.

From the application scenarios, electrified VTOL also enables advanced air mobility (AAM), a versatile small-scale mobile aero-applications. From the airplane

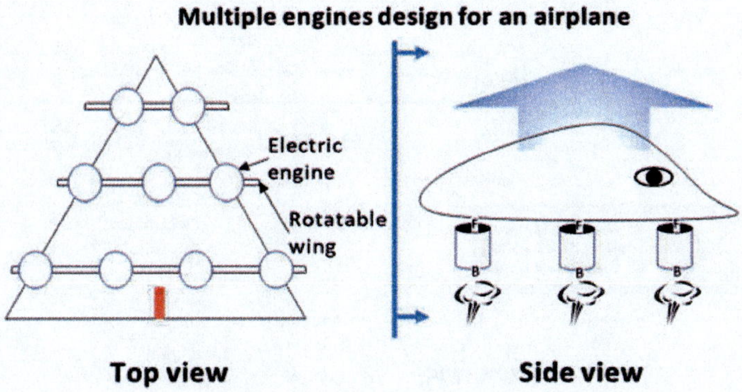

Fig. 1.3 Integrated delta e-plane with nine RDF jets (left: top view, right: side view)

design perspective, multiple rotatable electric propulsions enable the distributed electric propulsion (DEP) concept to become true, which enables the integrated design of airplanes with multiple engines for the most optimized aero-dynamic performance and flexibility of a modern airplane, either for the pure-electric small aircraft or as a supplement to the big commercial aircraft.

Electric engines are small, light and can be made multiple, providing a green and much more flexible propulsion for abundant mobile air application scenarios. Below is an example of using nine RDF jet electric engines to provide the flexible vertical and horizontal thrust of a delta-shaped aircraft. More details will be presented in Chaps. 7 and 11 (Fig. 1.3).

1.2 The IEEE

Coincidentally, the IEEE logo is also composed of two distinct arrows. The up straight arrow refers to electricity, while the circling arrow refers to electronics. The electricity represents the "powerful" side of IEEE, which comprises the electric power and electric motor, and electronics studies the inner operations of electrons, which brings about the "around-the-earth" communications technology (the computer and internet, the IoT, and 5G).

As shown in Fig. 1.4, IEEE has much more diversified societies [5] compared with AIAA—more than 1900 annual conferences/events worldwide on much diversified cutting-edge contents. The relevant hot spots associated with smarter AIAA in this book are indicated by the "sunny" flag signs.

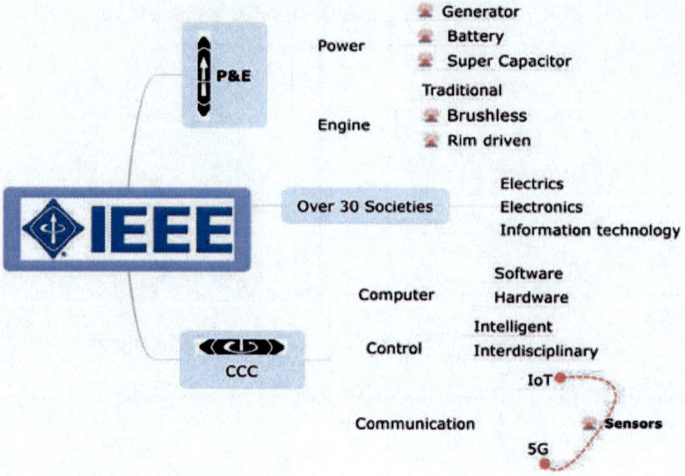

Fig. 1.4 Scope and organization of IEEE

1.2.1 The Scope of IEEE

The scope for IEEE is vast compared to AIAA. In general, it has two basic categories, strong "E" and weak "e". Strong "E" involves electricity, i.e., the electric power and electric motor. Electric aviation are most often involved in the strong "E", both the powers and engines. The weak "e" is related to the electron operations in the solid, i.e., the electric device and electric circuits, which usually do not involve high voltage, large current, and high power. There is a "3C" feature in electronic engineering: computer, control, and communication. The electronic engineering helps the global communication by computers, internet, and IoT.

From a–z, IEEE has the following societies as in Table 1.2.

Although the scope is large, the relevant IEEE societies to our topics in AIAA regarding IoT and electric aviation are just the few (in Table 1.2 marked with the * signs). They are basically around the two categories: MEMS micro-fabrication and sensors, electric motors and electric power.

1.2.2 MEMS Technology and Micro-fabrication

Micro-scale manufacturing from MEMS technology originated from 50 years of development in the integrated circuit (IC) industry from the Silicon Valley back to the 1970s. Since then, many industries have borrowed micro-manufacturing methods to fulfill their specific needs. For example, in our case, which will be discussed in Chaps. 4 and 5, we use the modified micro-method to build the smart MEMS sensor on the top of the turbine blade to measure the high temperature in aero-engine systems.

Table 1.2 IEEE has many academic and professional committees

	Aerospace and Electronic Systems Society
	Antennas and Propagation Society
	Broadcast Technology Society
	Circuits and Systems Society
	Communications Society
	Computational Intelligence Society
	Computer Society
	Consumer Technology Society
*	Control Systems Society
*	Dielectrics and Electrical Insulation Society
	Education Society
*	Electromagnetic Compatibility Society
	Electron Devices Society
	Electronics Packaging Society
	Engineering in Medicine and Biology Society
	Geoscience and Remote Sensing Society
*	Industrial Electronics Society
	Industry Applications Society
	Information Theory Society
	Instrumentation and Measurement Society
	Intelligent Transportation Systems Society
*	Magnetics Society
	Microwave Theory and Techniques Society
	Nuclear and Plasma Sciences Society
	Oceanic Engineering Society
	Photonics Society
*	Power Electronics Society
*	Power and Energy Society
	Product Safety Engineering Society
	Professional Communication Society
	Reliability Society
*	Robotics and Automation Society
	Signal Processing Society
	Society on Social Implications of Technology
	Solid-State Circuits Society
	Systems, Man, and Cybernetics Society
	Technology/Engineering Management Society
	Ultrasonics, Ferroelectrics, Frequency Control Society
	Vehicular Technology Society

Figure 1.5 indicates our capability to embed the very thin-film (just 1 μm thick) sensor in any shape (in this case, a school logo) on the curved blade surface. The minimum line width can be no smaller than 20 μm.

Such micro-sensor is capable to measure the surface temperature on turbine blade over 1000 °C. The thermal resistor detector sensor (RTD) on the surface of the turbine blade is made by using the micro-fabrication method borrowed from IC technology. More details are in Chaps. 3–5.

The use of the micro-technology to build smart micro-sensors for AIAA application is just one example of the AIAA/IEEE cooperation. To help the future AIAA/IEEE integration, Fig. 1.6 illustrates the main categories of the electronic societies in IEEE, which may help to build relationships with future AIAA applications.

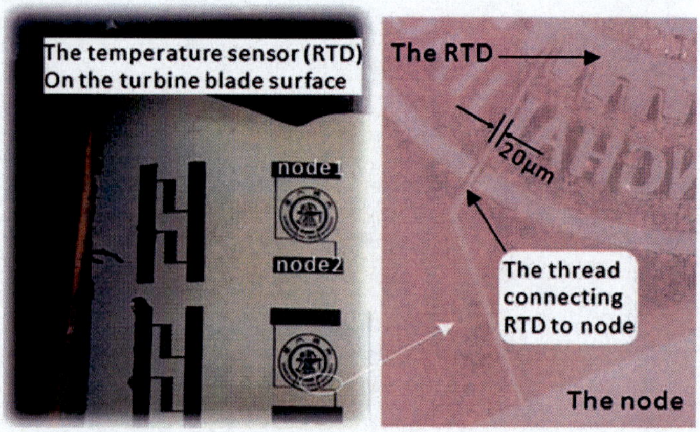

Fig. 1.5 Illustration of the micro-sensor built on the surface of the turbine blade using the modified school logo as a RTD pattern

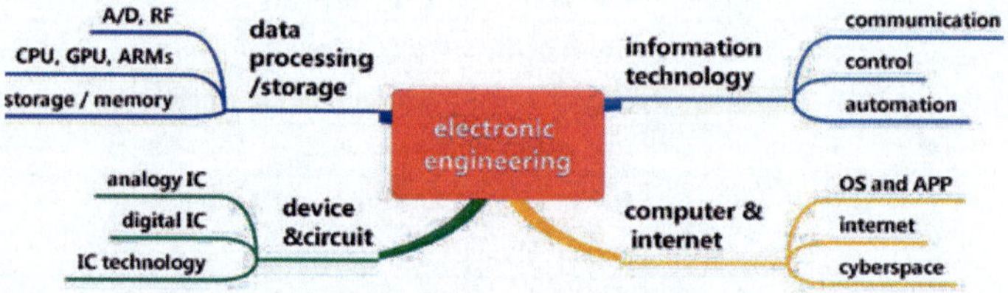

Fig. 1.6 Fields in IEEE for relevant AIAA applications

1.2.3 Electric Motor and Electric Power

Electrified aviation requires both electrical motors and electrical powers, which need help from IEEE people. The keywords here are lightweight and high power density both from the engine and the power supply. From the electric engines' perspective, Fig. 1.7 lists the possible alternatives of current electric motors used for electric airplanes. Tesla uses EM motors that do not need any permanent magnets (PM) and generate electromagnetic (EM) rotation purely by electric induction. Compared to PM-type motors, EM motors are less weight-efficient. A switch reluctance electric motor is another option for electric engines [6]. The three types of motor architectures can be combined and integrated such that the Tesla Model 3 car uses IPM-SynRM—a combination of permanent magnetic and reluctance electric technologies [7, 8].

A brushless DC motor (BLDC) is the most promising choice for electric propulsion with the best overall motor performance. PM brushless motor is a very hot spot in the current modern electric motor industry and is broadly used in EV car, UAS, and robot. Among the two types of brushless motors shown in Table 1.3, the BLDC type is more suitable considering its lightweight and large power-to-weight (PWR) ratio feature. However, more efforts are still needed to enhance its overall efficiency on propulsion for airplanes. We have proposed the rim-driven brushless motor for a more efficient high torque driver as the electric jet propulsion. This will be elaborated further in Chaps. 9–11.

For the electric power suppliers, Fig. 1.8 lists the electricity-related fields from IEEE's perspective, which consists of three main parts: electricity creation, storage, and transfer. The AIAA-related fields are marked with star signs, which will be further discussed in Chap. 10. The promising options for mobile electric power for aviation are lightweight gas turbine electric generators and 3D HK supercapacitors. Li batteries are the most mature technology borrowed from the electric car (EV) industry and can be used for quick-market planes and R&D purposes. For electricity transfer, electricity manipulation using SiC-based IGBTs is the key technology for high-power brushless electric motor control. Although the fossil fuel power generation is still the main stream electric power in the power grid, it is expected that in the future,

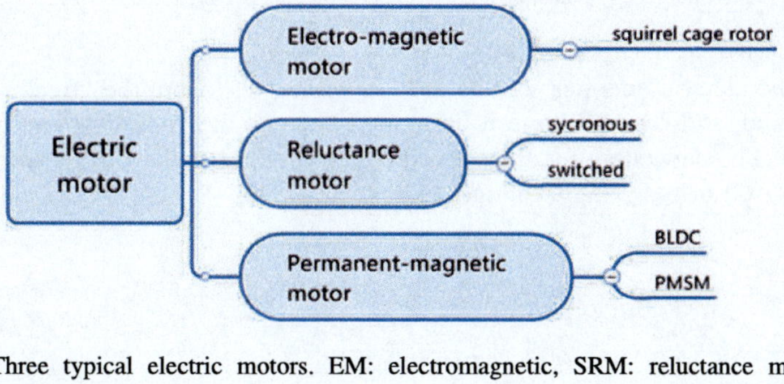

Fig. 1.7 Three typical electric motors. EM: electromagnetic, SRM: reluctance motor, PM: permanent magnetic

Table 1.3 Pros and cons of two brushless electric motors

Classification	Permanent magnet brush DC motor	Permanent magnet brushless motor	AC induction motor
Time, user	Old motor	BYD, *aero-engine, UAV*	Tesla
Stator	Permanent magnet	Electromagnet, multi-winding	Electromagnet, multi-winding
Rotor	Electromagnet, multi-winding	*Permanent magnet*	Electromagnet, squirrel cage winding
Electronic controller	Not necessary except to adjust the speed	*Must have*	Not necessary except to adjust the speed
Operating voltage	Low	*High, IGBT*	High
Efficiency	Lower	*High*, rotor has almost no loss	Low, rotor loss
Power density	Lower, rotor is not easy to dissipate heat	*High*, stator winding is easy to dissipate heat	Lower, rotor has loss and heating
Reliability	Low because of brush	*Higher*, mainly determined by bearing	High
Maintenance	Regular cleaning and maintenance needed	*Less* frequently maintained	Less frequently maintained
Cost	High, permanent magnet materials	High, permanent magnet materials/controllers	Low cost
	Permanent magnet brushless motor		
	PMSM	BLDC	
Driving mode	Sine wave driven	Square wave driven	
Torque		Higher *	
Power density		Higher *	
Control accuracy	Higher		
Purpose	More precise drives, robot	High power drives, high PWR	

PMSM Permanent magnet synchronous motor; *BLDC* brushless DC motor

most of the electric energies will be the "new energies" from the "free" resources of sun, wind, and river. The word "EV" will become the "NEV" (more details in Sect. 7.3). The new energy is transferred either via ultra-high-voltage synchronous AC (UHVAC) or extra high-voltage asynchronous AC (EHVAC) superpower grid lines [9].

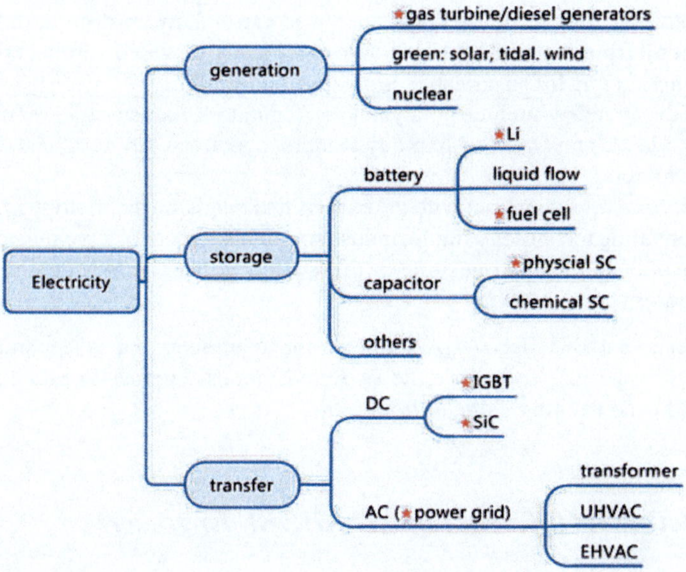

Fig. 1.8 Three key components regarding to the electric power supplier

1.3 To Integrate IEEE to AIAA

Electronic panels are already exist in front of the pilot in the airplane. This IEEE's work is associated with the knowledge of the basic electronics of electric circuits and electric controls and is also widely used in all fields. This is not the topic that we are addressing in this book.

Starting from 2018, there is a noticeable milestone of AIAA/IEEE interaction. In the AIAA Propulsion and Energy Forum in Cincinnati, Ohio, USA, AIAA first invited IEEE to open up joint section Electric Aircraft Technologies Symposium (EATS), inviting the IEEE colleagues for joint efforts to add more electric ingredients in AIAA fields. It is a good starting point of communication between the world's two largest Sci-Tech families. Electrical propulsion and smart sensors are the two promising kids of this "marriage". Since then, EATS has continued every year until now and is expected to continue for quite a long time due to the intrinsic mutual needs on both sides.

Just before this book released, on Sept 21, 2022, Honeywell hosted an "Air Mobility Summit 2022" at Washington, DC, in Honeywell's office at 101 Constitution Avenue Northwest [10]. In this summit, they discussed:

(1) How this new technology (urban air mobility, unmanned aircraft system, regional air mobility) helps with maintaining US global competitiveness, creating new and expanding economic benefits for society and building a more sustainable world.

(2) In detail on all-electric air taxis, uncrewed cargo delivery drones, move transportation from the road to the air with electrically powered vertical takeoff and landing aircraft for sustainable, high-speed air mobility.
(3) With the attendees include policymakers, regulators, industry trade associations, UAS, UAM, and RAM technology leaders, operators, key suppliers, and other stakeholders.
(4) To see, touch, and interact with the experts and hands-on technology immersion.
(5) The enabling technology for the industry with several eVTOL vehicle mock-ups on display and hear from some of the world's greatest minds behind this new technology.

It is seen that this IEEE/AIAA integration continues and is becoming more and more prosperous. To mingle the two great minds together to establish a new playground is an exciting game in front of us.

1.3.1 Mindset Difference of AIAA/IEEE Engineers

Other than the academic differences, AIAA and IEEE people have different habit of thinking—not only their IQ is different, but also their EQ. Generally speaking, IEEE covers more and broader fields and needs to deal with more variety of people/culture, hence needs more EQ than AIAA people. In order for AIAA and IEEE people communicating more effectively, it is worthwhile to understand their tiny differences in mindsets and habits.

IEEE people may have a more academic focus mindset, and AIAA people are more project-oriented. To a certain extent, AIAA is the customer of IEEE. The interaction between the two merely integrates IEEE's achievements into AIAA's engineering scenario. IEEE people need to adjust their mindset and point of interest to fit the needs of AIAA. The feature of the data shall be more application-oriented rather than purely academic-focused. For example, when developing a high-temperature sensor, IEEE people may concentrate on its performance and capability, but the AIAA people not only require the sensor to be capable of measuring the temperature but also place a stringent request on the repeatability, stability, and high-temperature endurance.

We use Chaps. 4–6 to explain this integration process, hoping to shed some insight into the future interactive activities between the two teams. Chapters 7–11 discuss electrified aviation, specifically electric propulsion, power, and integrated airplane/engine design.

1.3.2 *Two Cutting-Edge of AIAA/IEEE Fields*

There are two hot spots between the AIAA and IEEE interactions: the smarter aero-engines and the more electrified airplanes. As we know there are two different approaches in aero-industry: (1) technology/product is ready and we are waiting/looking for its user/market, (2) market has been designed/planned and we are seeking for the associated resource/technology. Below, we explain these two types of approaches: one is from bottom to top, and the other is from top to bottom.

(1) IEEE technology ready for AIAA—MEMS
Sensor technology from the micro-fabrication method is a mature technique and has been used extensively in the IoT for years. Where to use this technology to help AIAA? When we find an AIAA opportunity, then how to modify the current technology to fit in AIAA scenario? Chapters 4 and 5 address the details of this bottom to the top procedures to explain how "the technology goes the first, then we are looking for its playground" works?

(2) VTOL and green aviation—electric propulsion
This is the top-down approach. Vertical takeoff and landing (VTOL) is an advanced and clever way of airplane/propulsion. This need has been foreseen, designed, and planned in many large aviation organizations, and we are sincerely looking for partners to make it happen. IEEE is an ideal partner since the best VTOL is electric jet, and both the electric engine and electric power are in its territory. In Chaps. 7–11, the blueprint of this need from AIAA and the relevant technologies from IEEE are depicted.

The intelligent aero-engine and electric aviation are two cutting points of the AIAA/IEEE cooperation. The first joint effort (smart sensor for gas turbine engine) is relatively easier. The second goal (electric propulsion including engine and power) is tougher yet more promising. This book will focus on two concurrent hot topics: (1) IEEE helps AIAA more intelligently with smart sensors, and (2) electrical propulsions for aviation together with advanced electric power.

(1) More intelligent
In the past, aero/astro-enterprises were seeking capability and powerfulness but now are focusing on smartness and efficiency. On the other hand, IEEE has made tremendous achievements in the past 40 years and is looking for its playground. It is just the right timing to combine their willingness—to make the aero-engine more cost-effective and more intelligent.

(2) Electrical propulsion
With the advent of Tesla all-electric vehicles [11] and the emerging need for advanced air mobility (AAM) [12], electrical rim-driven Tai Chi jet engine and affordable lightweight electric power (LTG and 3D HK SC) have shed the light on the new generation of aero-propulsions. Full electric aircraft involving miscellaneous airmobile operations, such as 1–500 kg delivery UAS, 3-man VTOLer as rescuer plane, provides both the necessity and the feasibility for AIAA/IEEE cooperation.

1.3.3 The Corporations of AIAA/IEEE Communities

The gas turbine engine by using the smart sensors developed by IEEE people is the first tempting cooperation with AIAA. The rim-driven fan (RDF) jet is the key proposal for the next generation electric propulsion. They are the joint efforts of the AIAA and IEEE, the world's two largest engineering families. The integration of AIAA and IEEE is the interdisciplinary interaction between the two groups of smart people—a teamwork uniting both minds for both needs by both people. Mindset merging is highly needed as detailed in Chap. 6, which serves as a link from the corporate interaction experience of AIAA/IEEE on developing the smart sensors from Chaps. 3–5 to extrapolate the promising future of AIAA/IEEE cooperations in electrified aviation from Chaps. 7–11. That is, to use our six-year true experience of coworking between the IEEE and AIAA colleagues in Part 2 (Chaps. 3–5) to explain the process of the corporation between the two groups of people, serving as a reference point for the future bigger cooperation—the electrified aviation in Part 3 (Chaps. 7–11). Electric propulsion and associated aviation electric power are much more challenging and more promising and may bring a more revolution to the whole aviation enterprise.

More topics are also discussed in Chap. 6 regarding to: the mindset difference, influence of the different culture on international cooperation, 2D versus 3D thinking, the third way, the need of the dream keeper, etc. It is hoping that by combining the efforts of IEEE and AIAA, the next generation of hybrid/full-electric aero-engines will come true, just as the first Toyota's Prius hybrid car released in 1997 becomes the well-adopted EV cars with Elon Musk's Tesla or Wang ChuanFu's BYD electric vehicle after 20 years of efforts (from 1997 to 2022). It took over 20 years to get the electric power/motor ready for cars, and it may take 10 years to fulfill the electrified aviation.

References

1. AIAA/IEEE First Combined Symposium in 2018 AIAA's Joint Propulsion Conference—The EATS (Electric Aircraft Technologies Symposium), Cincinnati, OH, 9–11 July 2018. eISBN: 978-1-62410-570-8. https://arc.aiaa.org/doi/book/10.2514/MJPC18
2. AIAA's Four Major Forums: Aviation, Defense, Propulsion and Energy. SciTech. https://www.aiaa.org/events-learning/Forums
3. The AIAA sitemap. https://www.aiaa.org/site-map
4. E-Fan X project canceled. https://cn.bing.com/search?q=cancel%20E-Fan%20X%20project%20&qs=n&form=QBRE&=%25eManage%20Your%20Search%20History%25E&sp=-1&pq=cancel%20e-fan%20x%20project%20&sc=8-23&sk=&cvid=A674373776C6471EA1844216CD657187
5. IEEE's over 30 communities/societies and over 1,900 annual conferences. https://www.ieee.org/communities/societies/
6. Cao, W.: Overview of electric motor technologies used for more electric aircraft (MEA). IEEE Trans. Ind. Electron. **59**(9), 3523–3531 (2012)

7. Huynh, T.A., Hsieh, M.-F.: Comparative study of PM-assisted SynRM and IPMSM on constant power speed range for EV applications. IEEE Trans. Magn. 1 (2017). https://doi.org/10.1109/TMAG.2017.2707125
8. Tesla Model 3's IPM-SynRM electric motor. https://www.lesics.com/tesla-model-3_s-ipm-synrm-electric-motor.html
9. Xu, Z., Dong, H., Huang, H.: Debates on ultra-high-voltage synchronous power grid: the future super grid in China? IET Gener. Transm. Distrib. **9**(8), 740–747 (2015)
10. https://aerospace.honeywell.com/us/en/about-us/events/2022/09/save-the-date-air-mobility-summit
11. Tesla: All-electric vehicles and scalable clean energy generation and storage products. https://www.tesla.com/
12. NASA's vision for advanced air mobility (AAM) mission. https://www.nasa.gov/aam

Part II
The Intelligent AIAA—Smart Sensor

→ AI is becoming popular which is brought about by IEEE.
  → How to apply the AI in AIAA?
  → A few showcases or exemplifications.

Chapter 2
Artificial Intelligence

Artificial intelligence (AI) has become a popular term very recently, and it is not yet clearly classified or explained. Just from the surface meaning, AI refers to the simulation of human intelligence by machines that are programmed to think like humans and mimic their actions. Let us take the example of a monkey to see a banana on the roof beam, and there is a ladder sitting on the floor. A human being knows to move the ladder to the wall and climb to the roof to fetch the banana. Initially, the monkey may not know this, but when he sees the human doing this, he can imitate the human to do the same. Therefore, intelligence comes first from imitation. Simple imitation is the first and preliminary stage of AI. Currently, people use the term AI to cover many intelligent activities. The deeper AI actually involves comprehensive thinking and many other actions/options/outcomes. AI involves multi-dimensional sensing capabilities and much broader actions beyond humans, including smart brains and clouds, the Internet of Things, and robotics. Such a classification of AI helps us to find the relevant enabling technologies.

However, AI itself may not originate from the technology but from imaginations. For example, iPhone has changed people's lives, but the idea and outlook of the iPhone were at first just configured in Steve Job's mind. Then, he was seeking the associated technologies, the RF, the touch screen, the iPad, the iPod, etc. Some of them are already existing, and he just needed to integrate them into the iPhone; some need to be developed, such as an accurate touch screen and protocols. The iPhone was imagined at first, and the combination of the relevant technologies came afterward. The iPhone as a smartphone involves many sensors and sensor fusions. From the technology's perspective, AI involves multiple disciplines, including physics, chemistry, biology, etc. Among them, the IEEE is the greatest enabler. Almost all the AI activities involve the IEEE. For example, the concept of sensors is to convert any non-electrical signal into an electric signal so that one can use electronic circuits and computers to manipulate these sensors' signals and make the proper reactions. In addition, most of the IC fabrication technologies are borrowed by MEMS to build the various sensors by making use of the matured Silicon Valleys' integrated circuits techniques.

© The Author(s), under exclusive license to Springer Nature Singapore Pte Ltd. 2023

F. L. Duan, *When AIAA Meets IEEE*,
https://doi.org/10.1007/978-981-19-8394-8_2

In this chapter, we first discuss the three main categories of AIs: the smart brain and cloud, the Internet of Things, and robotics. Among these three categories, the sensors are highly related to artificial intelligence and machine learning, which is our main focus regarding the most recent activities of IEEE's involvements in AIAA. Sensors is one of the key devices in IEEE family, which is composed by RLCs, transistors, and sensors.

2.1 AI: Brain, IoT, Machine

These are the three main aspects of AI. Their commonality is that AI can do a more effective job or even replace humans to fulfill a task, but all of them are within the scope of physical and mental level but not spiritual.

2.1.1 Smart Brain and Cloud

The first category of AI is the smart brain and cloud. In the early stage, we called them computer and internet—the personal computer (PC) self-powered individual personal capability and the web network helped the communication to exchange/share human activities/capabilities. Smart brains are much more advanced and powerful computers, and clouds are big data-related larger internet activities.

Faster Brain and Easier Chores

A computer is a typical example of smarter brain. A human being is not able to calculate faster than a compute. High-speed computing greatly enhanced human brainpower, such as AlphaGo to compete with humans in chess. AlphaGo [1] is a computer machine to defeat world top Go players and becomes an AI Go champion. Go is originated in China over 3000 years ago, which involves many comprehensive and multiple layers of strategic thinking compared to either the Western or Chinese chess. The top Chinese go player Nie WeiPing once predicts that the AI Go player is not defeatable for two reasons. One is its emotionlessness. A human being may be affected by emotion, and this affects the results of competition, but a Go player always shows a poker face, and it will not be affected by emotions. Second, an AI Goer is capable of calculating far more steps ahead than ordinary human brainpower. As the calculations are patterned and can be programmed, the Go machine thinks faster than human brain.

The other aspect of AI is the electrified automation, which greatly facilitates human lives, such as replacing chore labor with a washing machine. It is electricity and electronics that make our lives much better and easier than before. Electricity contributes to provide stronger energy to liberate human physical capability, and electronics enables to delicate this energy flow for smarter lives.

The Cloud for Win–Win

"Cloud" is becoming a popular term esp. after the COVID-19 pandemic since 2019. In English, we call it "online conferencing". In comparison, "cloud" is a more vivid and romantic term (in Chinese the wording cloud is "云" flowing in the sky high, covering anything but in a remote way). Many global technical meetings and activities are carried out online—cloudy meetings with cloudy connections by cloudy of people. Human activities such as politics, education, business, and even private conferencing were coined as cloud diplomat, cloud classroom, cloud conference, cloud dating, cloud meditation ... Even for the AIAA Forums (Aviation, P&E, Sci-Tech) in 2019–2022, all become the Cloud AIAA Congregations. The cloud has two meanings, remote and not real. Remote communication is a "last resort" approach but cannot deliver the holographic view. Like the clouds in the sky, it feels very "virtual" and not "touchable". Cloud communication is incomplete or as wholesome. A complete human being has five senses: eyes, ears, nose, tongue, and body and six senses: voice, virtual, fragrance, taste, touch, and feelings. The "cloud" can deliver only two of them and certainly cannot be as wholesome as face-to-face "confrontation". Cloud communication is not as real and may cover up the defects either on-purpose or unintentionally. Beautifying the sound and color makes the impression superficial and advertising. Specifically, for the AIAA/IEEE coops, face-to-face wholesome communications are extremely important since many infield tests are needed (more discussion in Chap. 6). One has to say that "cloud" activities are very useful to share surface data other than the wholesome human experiences. Purely from the information-sharing perspective, these cloud activities may be sufficient.

In addition to the above, the other meaning of "cloud" is the database and platform. The human brain cannot hold so much information, while many databases can form clouds to share from various media around the world. Similarly, there are many platforms or even platform companies, that people do their business in cloud. For example, Amazon in USA and Taobao in China are two platform companies to establish the connections between the individual vendors and customers via the data hubs. On this database/platform, people in all countries do business, perform data mining and deep learning to enable the digital economy and to provide useful information for more fruitful decision-making, enlightening insight and discovery. The cloud data have the "4V" features: volume, velocity, variety, and value. At present, the main tools used to analyze big data are open source (Hadoop, NoSQL, etc.) and commercial (IBM PureData, Oracle Exadata, SAP Hana) ecospheres.

2.1.2 Internet of Things

The second AI refers to IoT. The Internet of Things is a very vivid speaking—the "things" fly through the internet, namely the logistics like Amazon, eBay in USA and Taobao, Jingdong in China. The internet is used to be just a web activities to exchange the text, pictures, and A/Vs to connect people's thoughts. The IoT connects people in more wholesome manners. It is the communication in physical form based on visual/audio information from the internet. For example, one can see and hear that a cake is tasty on eBay, yet the information conveyed on the "internet" is just the message, not the taste. Through purchase and delivery, one can taste the real cake with their tongues—this is the "thing". Logistics is a way for this type of IoT. Trucks, high-speed railways, etc. are continental deliverers, and large ships are the main continental tool of IoTs. Electric airplanes can be used as delivery drones (Chap. 11). The e-drone delivery is more prompt and convenient and can even drop your package on the platform of your apartment.

As shown in Fig. 2.1, the IoT delivers the wholesome needs of a human being of six senses. In the early days, only three can be exchanged via the internet. A vivid example is shown in the movie "You've Got Mail" in the 1998 that Tom Hanks and Meg Ryan can fall in love through the internet simply communicating by text messages. Later, with the faster internet and related vital/audio technologies, people can share audio and visual information with "more wholesome thoughts". In the meantime, the technologies for audio and visual recognition for searching and identification are under rapid development in recent years. Facial and voice recognition techniques are widely used in this COVID-19 event to facilitate the pass/fail process in various customs. The Internet of Things enhances human activities not only limited to the visual, audio, and thought via the eye, ear, and mind but also offers a holographic delivery of human's six senses via the "things". Nevertheless, the only left-over item which cannot be communicated or delivered is the missing part "touch or feel", the intimate energy exchange like face-to-face.

The other meaning of IoT is to connect any sensing items (things) to the internet, such as radio frequency identification (RFID), QR code, infrared sensor (facial recognition), global positioning system (GPS, BDS), according to the agreed protocol. The purpose is for information exchange and communication to realize the intelligent identification, positioning, tracking, monitoring, and management for AI global

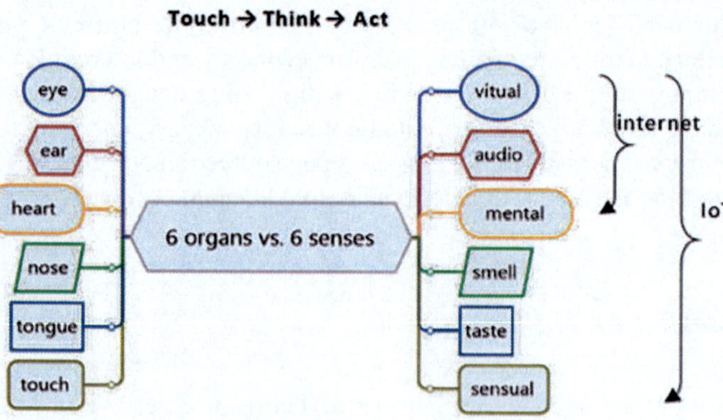

Fig. 2.1 Human's six senses and six organs

networking. This definition just covers the first two of three in Fig. 2.1: "touch and think" but no "act".

2.1.3 The Robotics

The third type or meaning of artificial intelligence is robots. There are two kinds of robots. The first kind is made as much as a real human, resembling all six senses of a human being with all five basic organs except the "soul" (Fig. 2.1). The second type is just artificial machines, which can do much better and more powerful jobs than humans. These intelligent robots may have multiple arms or multiple legs, have the smarter eyes like a cat (can detect large range of spectrums of wavelengths) and the noses as a dog (large ranges of smarter smell and gas sensors), combining various bionic advantages to fit in specific scenarios to replace human beings performing some special missions or tedious jobs. These robots require many IoT multi-dimensional sensors, i.e., the sensor fusions to make intelligent artificial judgments and actions, which are far more advanced compared to ordinary electric automations such as repeatable/programmable washing machines. The intelligent learning of a smart robot is based on the technologies of "sensor fusion". The fusion combines multiple data from multi-dimensional sensors to give rise to a wholesome judgment for a complete picture. Let us analyze a human's intelligent action "this is half glass of hot water" to explain the "sensor fusion":

When a human being's hand touches a half cup glass of hot water, he at first feels the heat of the glass wall. His one finger is on the top and the other is at the bottom, which are equivalent to the two temperature sensors, and one feels the heat at the bottom finger. He then sees a glass together with half water, equivalent to the two visual sensors that see both the cup and water. The comprehensive signals of the four sensors let him make the judgment: this is a half hot glass of water. Of course, as for an ordinary human being, he may make his judgment with some common sense which is the repeated leanings that he accumulated in his life.

We might feel this human action is so easy which seems no need to use so many words to explain it. However, they are a series of intellectual reactions when translated to machine learning. This simple AI process involves sensor fusion, a comprehensive interactions from computers with multi-dimensional sensors. By separating each step in detail and associating them with a specific sensing device, together with the intelligent judgments with computer, we can make the artificial machine functioning like humans. To enable data fusion for an intelligent machine, sensors and algorithms are the two key ingredients, and various multi-sensors must first be fully developed to collect the data inputs.

Micro- and nanoscience and technology are great enablers of robotics in AI. MEMS and nanomaterial technologies were first highly developed in the USA, Europe, and China in 2001, a very hot topic starting from the millennium, i.e., MEMS is pushed by technology. In recent years, the main focus has been on finding its playgrounds. Figure 2.2 illustrates the relations and details of technology and market of the micro-/nanoscience/technology. AI is pushed by technology and led by market. A few typical technologies that push micro-/nanoscience forward and a few hot market places that drive further engineering activities in micro-/nanotechnology are shown in Fig. 2.2. Sensor manufacturing by MEMS is a main driving force in application scenarios.

The AI system architecture has three main layers: the perception layer, platform layer, and application layer (Fig. 2.3). The role of the perception layer is mainly to obtain environmental information, which is mainly composed of sensors. The communication layer is mainly used for information transmission between the sensing layer and platform layer. It consists of both hardware and software such as RF, cable, optical fiber, cable, Zig Bee, Bluetooth, and communication protocols.

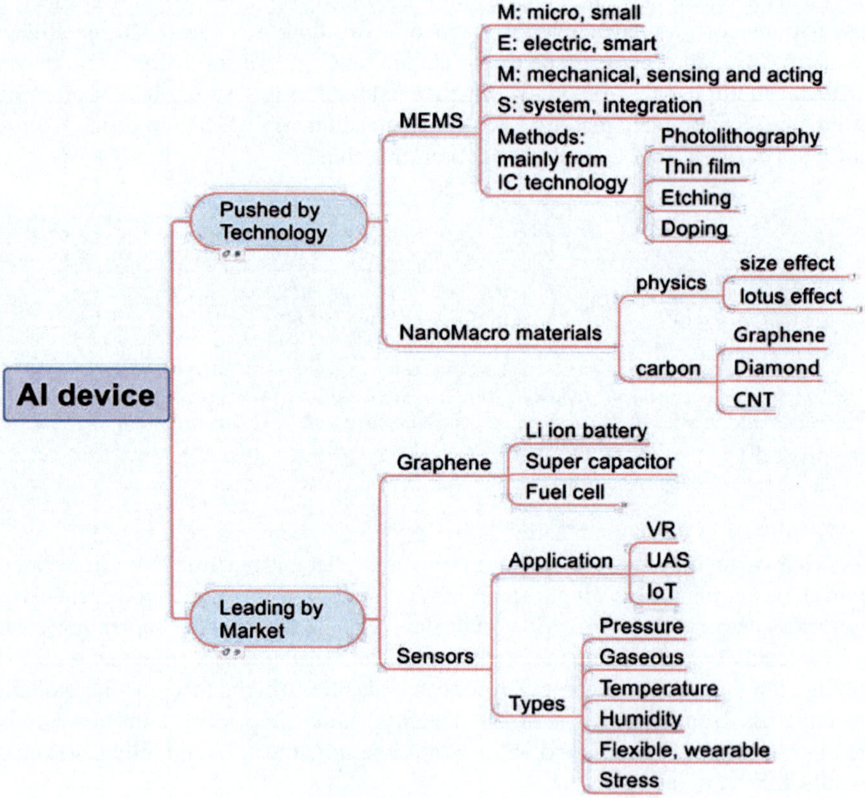

Fig. 2.2 Technologies and market places of MEMS and nanomaterial in AI

The platform layer includes software such as cloud computing, cloud storage, big data and data mining, and human–computer interaction. Intelligent instructions and data processing results are transmitted through the communication layer again to the application layer, such as intelligent robots, intelligent cities, autonomous pilot systems. The AI application layer covers a wide range of industries ranging from mobile phones to new energy vehicles, as well as a large number of devices and terminals, which will bring trillion market value. At present, the internet and smartphones have become mature, and the growth rate is relatively slow. The perception layer, that is, the sensor is at the bottom of the whole IoT, is the weakness of the current AI industry. Multiple sensors are the entrance of data collection, the "base" of the Internet of Things, and have a huge space for the coming market.

The three levels of AI can be translated into three steps of AI, i.e., from the sensing of the physical ambient to computer/cloud data processing and then to the intelligent reactions as shown in Fig. 2.4. It is noted here that the AI differs from 3A (industrial automation, office automation, home automation) because of its high level of "intelligence". Automation is a preprogrammed machine that performs/repeats

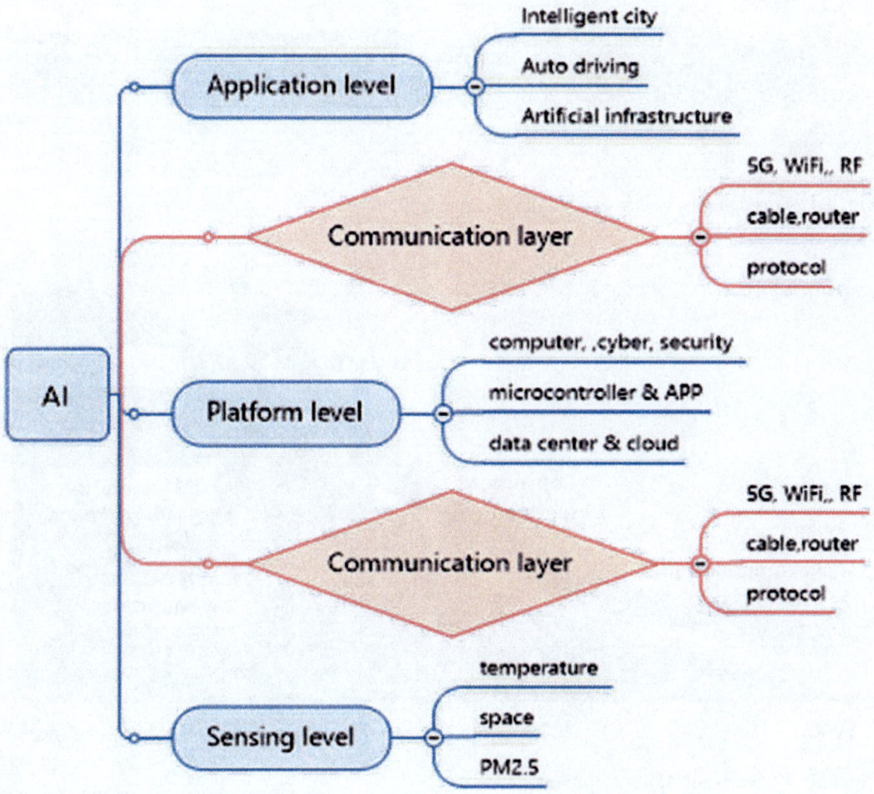

Fig. 2.3 Three main levels of AI

a series of actions, but artificial intelligence involves sensors, and the actions are performed based on real-time judgment, which is not preprogrammed and may not be repeated automated machines.

The AI has promoted sensor development very rapidly in recent years. According to the Yole Development [2] prediction, the MEMS sensor market will grow ~ 13% from 2016 to 2020, almost doubling the growth rate of the integrated circuit market by ~ 7–10%. Sensors and IoT development are the third wave of the rapid development of electrification and artificial intelligence after basic electric control in automobiles/aviation and consumer electronics since 2010, as illustrated in Fig. 2.5. The top MEMS manufacturers in the world include Bosch, St, ADI, Honeywell, Infineon, and AKM. Bosch's revenue accounts for nearly one-third of the total revenue of the five major companies. In Fig. 2.5, we can see the evolution of electrification from automation to AI. Its accumulated growth and the third wave growth rate will be much greater than the previous two 3A progress. This enhanced ingredient of the development is attributed to the rapid growth of the multiple sensor development.

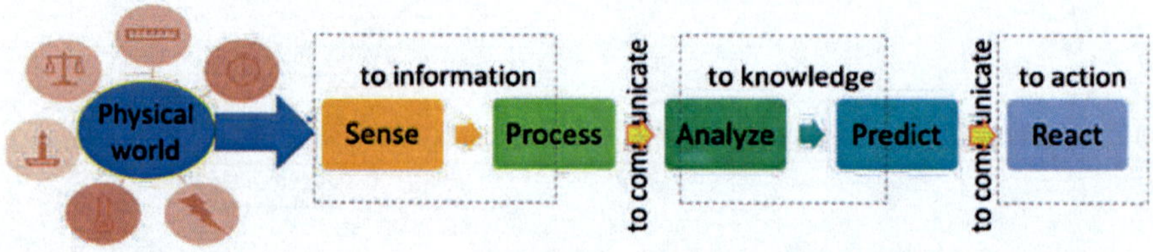

Fig. 2.4 Three blocks and five steps of AI process

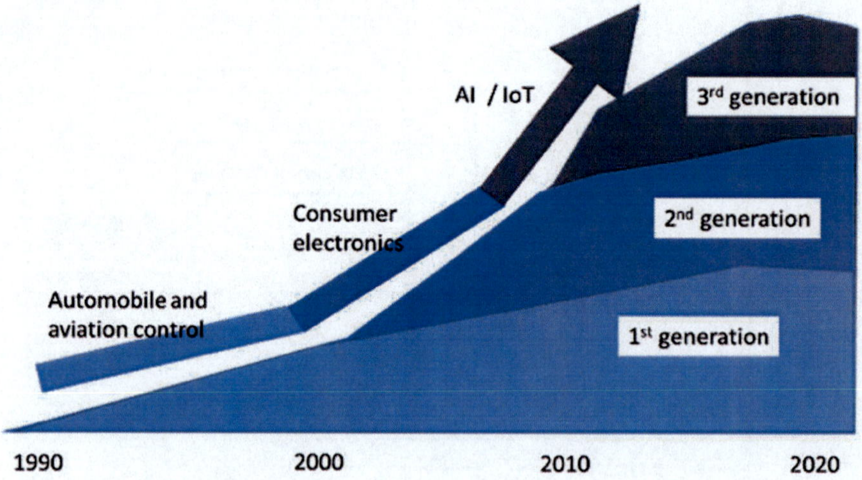

Fig. 2.5 Three stages of development

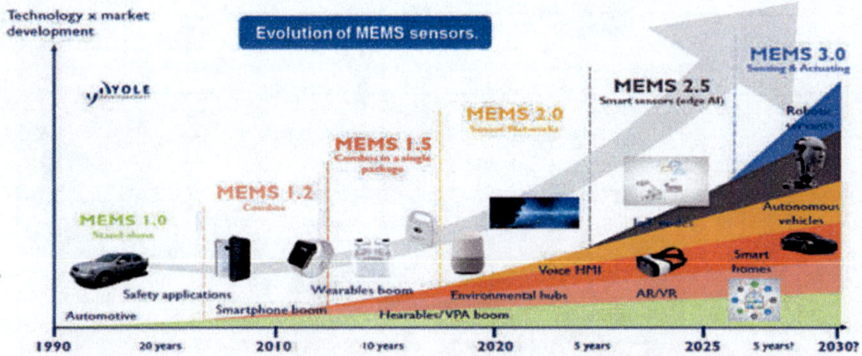

Fig. 2.6 Few stages of sensor development since the 1990s [3]

The detailed evolution of MEMS sensors since the 1990s is illustrated as a roadmap in Fig. 2.6. The early sensors before 2018 were mainly independent and discrete, and since 2020, there have been an increasing number of types of sensors that form the basis of sensor networks and sensor fusion. Future robotics is a high level of artificial intelligence based on multiple and multi-functional sensors and sensor fusion.

On the other hand, the other view to look at Fig. 2.6 is multi-dimensional sensing from the six senses. The sensors reflect human's six senses and convert them into electric signals or vice versa. Figure 2.7 shows the sensor development in this respect. Sensors actually started from microphones and loud speakers, which converted the human's voice to electric signals or vice versa earlier than 1990. In 1990, the visual image sensor from CMOS devices started to replace the traditional cameraing technique by using films by converting the optical signals to electric signals to store and process. Up till now, more and more dimensional sensing is developed using various kind of sensing principals involving physics, chemistry, and biology.

Although there are various smart all-dimensional sensors be developed to translate human's physical signals, it is the author's opinion that emotions and spirituality could be also sensed as well since there are some evidence that the emotions can be detected by the subtle vibration field. We could have such an experience that when somebody is staring at you on the back. You may feel it although you cannot see it. There must exist some invisible vibration field that is not delivered from humans basic five senses (hear, see, touch, smile, taste). For example, there are some studies on the subtle vibrations from human emotions (such as happy vs. sad, positive thing vs. negative) that can be recorded by water crystallization [5]. Visually, structure of water at the moment of freezing with high-speed photography showed the direct consequences of positive thoughts such as love and wisdom on the formation of water crystals as contrary to the negative thought (Fig. 2.8 [6]).

Such subtle vibrational wave patterns can be visualized, recorded, and may be even re-produced, but the deep-rooted motivation and passions cannot be duplicated

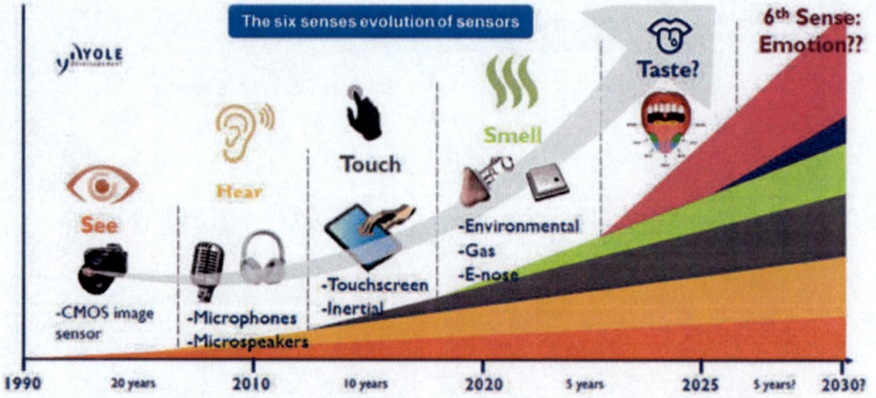

Fig. 2.7 Sensors correspond to humans' six senses (eyes, ears, touches, nose, taste, and subtle emotional vibrations) [4]

Fig. 2.8 Water can record the emotional message of the word and reflect as crystalized images showing the direct evidence of positive thoughts like love, wisdom, etc., on the formation of water crystals

by machines or AI. More and deeper roots which give rise to the corresponding pattern shall be further explored [7].

In summary, AI contains smart brains, IoT, and robotics. Nevertheless, AI can facilitate human intelligence and labor with fast computing brains and electrified automated machines but would hardly duplicate the emotional feeling, exchange the human experience containing all six senses in the chart in Fig. 2.1, and inspire spiritual enlightenment. In short, it is the author's belief that an AI human is just a machine, not a life. Similar to the picture in this video, the artificial eyes can never be life eyes. AI's eyes do not convey life and reflect the soul, as the people always say that "the eye is the window to the soul" (Fig. 2.9).

In the following, we mainly focus on developing sensors—the most basic level of robotics and machine learning. From IEEE's perspective, this involves three main devices: passive devices—two terminals, such as the resistor, inductor, and capacitor (RLC); active devices—three terminals, such as the vacuum tube, bipolar transistor, and field effect transistor (MOSFETs); and interface devices—the sensors/actuators,

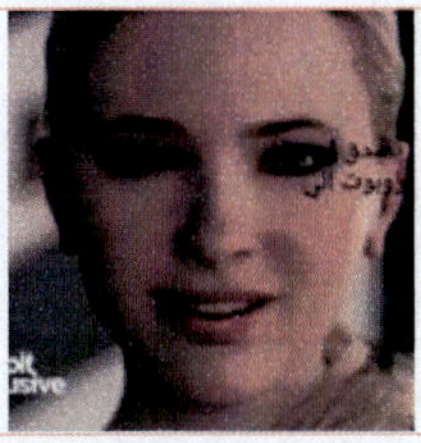

Reporter: People say that your brains can operate billions of time per second. Is it true?
AI girl: Absolutely. However, I only exist that the AI designed to me. There is something that I could never have.
Reporter: Really? And what is that?
AI girl: A Soul.

Fig. 2.9 Eye of a robot

which connect the electronics to other dimensions of life to convert the other physical/chemical/biological signals to electric voltage, current, or RF signals or vice versa.

2.2 Three E-devices—RLC, Transistor, and Sensor

IEEE's basic devices are divided into three main categories as shown in Fig. 2.10: active devices such as transistors or MOSFETs, passive devices are RLCs (resistors, inductors, and capacitors), and conversion devices like sensors and transducers.

(1) RLCs are three passive devices that usually have only two terminals. *R* is the resistor that consumes the energy or converts the electric energy into heat or light. During electricity transmission, *R* should be as small as possible to minimize the transmission loss. *L* is an inductor, which is an electromagnetic device storing and converting the magnetic energy, and *C* is a capacitor, which stores electric energy. $R \times C$ is the time constant—the smaller, the better; $L \times C$ is the resonance constant, the proper is the best.

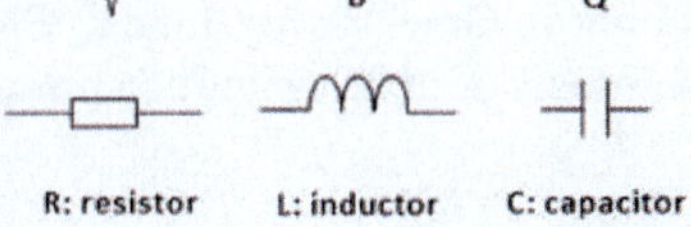

(2) A transistor is an intelligent device to operate the electron movements. It is usually a three-terminal device using the input to control the output. Bipolar and MOSFET are two popular transistor devices. In general, a bipolar transistor performs the analogy action by amplifying the input current I_B into a larger output current I_C, and a MOSFET performs a binary conversion by changing the high input voltage to the low output voltage or vice versa. Using the MOSFET device, we can perform both the space and time operations. For example, we can use an inverter to convert a "0" to "1", and we can use a pass gate (PG) to control pass or stop. Regarding to the timing, we use the pulse wave to open

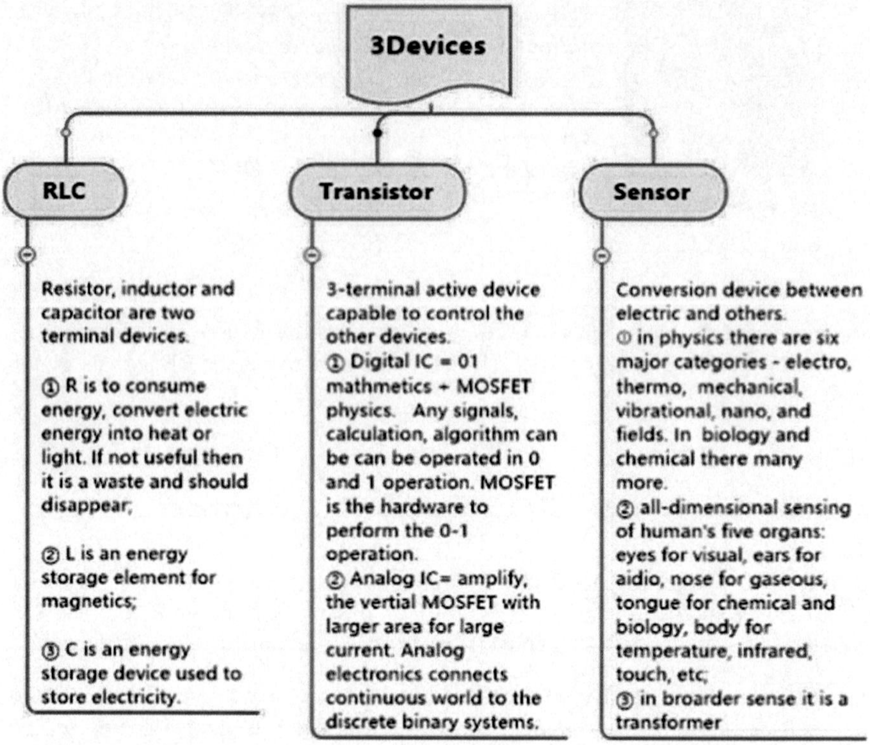

Fig. 2.10 Illustrated explanations of the three devices

the pass gate or various flip-flops to control the on/off in a proper sequence. Transistor reflects the Chinese and Greek philosophy that "The 3 creates all". Both have a theorem that the most determining factors are within 3. In Greek, the notation of 123 is I II III; in Chinese, it becomes 一二三. However, the 4 is not IIII or 亖 either in Greek or in Chinese. Steve Jobs follows this principle quite well—the numbers of main topics of a presentation must not exceed three.

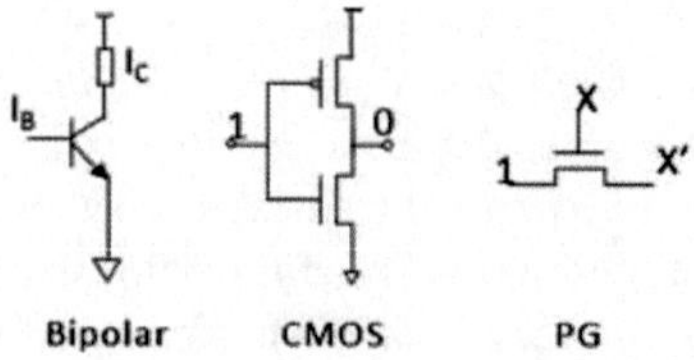

(3) In the third realm, sensors refer to both the sensing device and actuators. The sensing device converts non-electric signals such as temperature, voltage, and sensitive gases to electric signals such as voltage, current, or RF. Once these

physical signals are transformed into the electric signals, we can use electric circuits and computers to manipulate, process, and interpret them. The actuator is similar to a robot's arm to convert the electric signal/instruction to physical mechanical movement. Sensors are very multi-dimensional and can be termed a transformer, a mysterious device, to juggle or conjure in a comprehensive physical world. The scopes covered in transformers are vast. Just in the scope of physics, there are in total of six distinct categories. The sensor converts thermal, mechanical, vibrational (including acoustic, optical, wave, piezo, etc.), nano- (intrinsic micro-/nanofeatures such as size effect), and field (RF, microwave, infrared, magnetic, etc.) to the electric signals as in the input, process, output (IPO) [8] chart.

In Fig. 2.11, the "inputs" in the IPO box of a sensor are the measured items from the sensing materials. The sensor converts the signals of the other five items in physical science, such as optical signals and thermal signals, into electrical signals. Signals are extracted from sensitive materials. In general, nearly all materials will undergo corresponding and characteristic reactions under the influence of external factors. Among them, some materials that are most sensitive to external effects are used as functional materials to make sensitive components of sensors. For example, platinum resistance is very sensitive to temperature. Therefore, we use the Pt thread as temperature sensor. In addition, the sensor can also detect other types of signals beyond physics, such as chemistry—the gas sensors.

The "process" is "how the sensor works". Process is the theorem, the principle, the gray box, the mysterious transformer of the magic trick to convert the inputs to outputs. For example, the heat is converted to an electric signal by using a thermocouple sensor based on the Seeback principle. Most sensors operate on the basis of physics and chemistry principles, such as the piezoelectric effect, polarization, thermoelectric, photoelectric, magneto electric, and other physics effects, chemical adsorption, electrochemical reaction, and other physiochemical phenomena. Small changes in the measured signal will be converted into electrical signals. Typical transforming sensors are pressure sensors, force sensors, position sensors, liquid-level sensors, energy consumption sensors, speed sensors, thermal sensors, acceleration

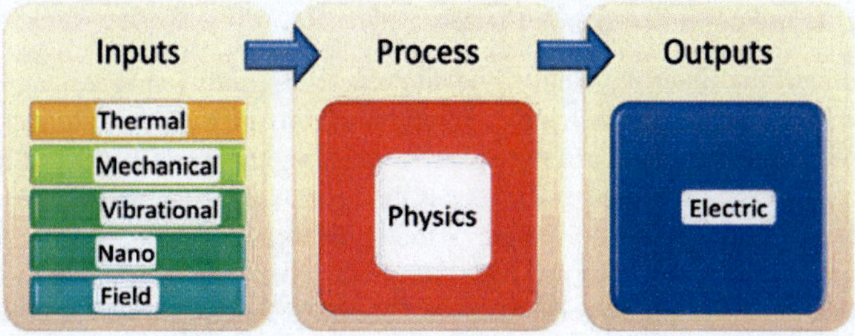

Fig. 2.11 Input, process, output (IPO) of the sensor

sensors, ray radiation sensors, vibration sensors, humidity sensors, magnetic sensors, gas sensors, vacuum sensors, biological sensors, etc.

The "output" in the sensor's IPO is the characterization signals that must be detected, quantified, and recorded. The signals can be either analog or digital. It can also be divided into incremental code signals, absolute code signals, and switch signals. The "O" is the measured/sensed information which is converted from the non-electric signals into an electrical signal for transmission, processing, storage, display, recording, and control.

2.3 Smart Sensors

There are two meanings of this smartness. The first one is the varieties—multi-dimensional sensing or sensor fusion. The second meaning is the smarter ways to build them, such as MEMS and 3D. Both categories are undergoing fast-growing developing stage. Below, we give a brief introduction on what are the specific occasions that we can use these smart sensors to make the airplane engines working smarter. Then we introduce a few typical smart methods to build such sensors for AIAA.

2.3.1 Smarter Sensor for Smart Use

Sensors to Make the Aero-engine Work Smarter

It is of great significance for monitoring the operating state of turbine blades, predicting the working life of blades, and ensuring the healthy and safe operation of aero-engines to achieve long-term real-time measurement of high temperature on the blade surface. Figure 2.12 illustrates the steps from the airplane to the engine and to the turbines. The working condition of the turbines plays an essential role of the engine performance and then the airplane performance.

Sensor All Dimensions—Sensor Fusion

Sensors convert the other physical signals to electric signals to process and manipulate. Shown in Fig. 2.13, the items on the left hand side are the other physical signals, while on the right hand side are the specific sensors performing specific functions. For example, the temperature sensor senses the touch signal—whether the part is hot or cold. The microphone sensor converts the audio signals to electric signals.

The sensor fusion combines multiple data from multi-dimensional sensors to give rise to wholesome judgment for a complete picture. For example, humans need to confirm the results through a variety of senses, such as smell, hearing, and touch. When our fingers touch the water glass, you first feel the touch of an object, which seems to be a glass (imagine your eyes are covered). You also feel the temperature;

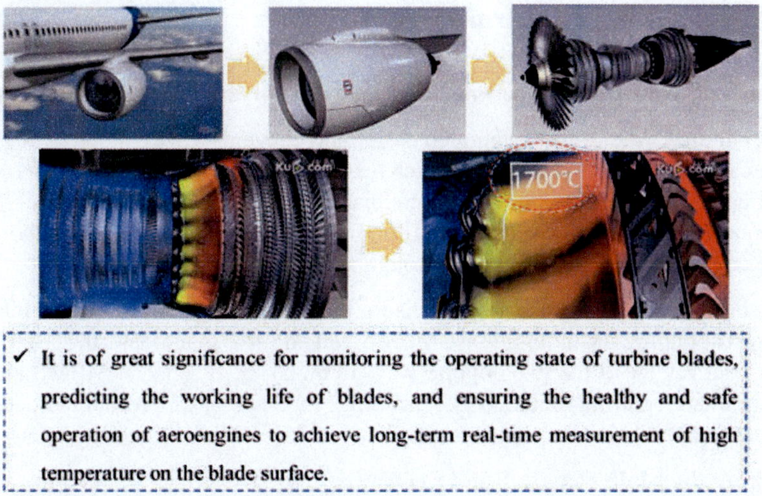

Fig. 2.12 From an airplane to a turbine blade

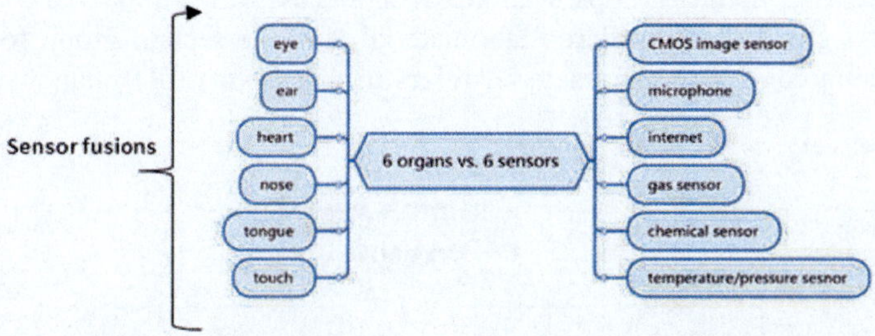

Fig. 2.13 Sensor fusion reflects the humans' six senses and uses the relevant physical/chemical devices

when you hand hold a cup, half fingers feel hot, and the others feel cold. Your eyes as a visual sensor gives you a wholesome picture of a glass cup and half water. Combining the sensors from hands and eyes, we make the judgment that this is half cup of water and the water is hot. This is the details of a sensor fusion. Another hot example is the automatic car parking sensor fusion. The comprehensive car parking sensing may include the 3D multi-sensing distance detections, infrared sensors, voice sensors, etc. to make sure the car automatically park on the right spot without hitting others. It is the comprehensive interlace of the sensors' signals and brains. To enable data fusion for an intelligent machine, sensors and algorithms are the two crucial ingredients. Multiple and various sensors to detect temperature, pressure, etc. must be developed first.

The smarter way to build these intelligent sensors is therefore the second meaning of "smart sensors".

2.3.2 *Smarter Way to Make Smart Sensor*

The smarter ways to make sensors include but are not limited to the following:

- We can make the sensor in a much easier and convenient way.
- We can build sensors in batches, which makes them cheaper and more economical.
- We can embed sensors directly on the machine parts.
- We can manufacture the sensor in situ/on site instead of in the Microfab/laboratory.
- We can make the sensor on a very tough spot such as corner/root.

In short, smarter ways to build sensors have two aspects: cheap and feasible. The corresponding technologies are MEMS and additive manufacturing. Below, we give a brief review of the two popular ways to make smart sensors: MEMS and and additive manufacturing methods (3D printing). In Chap. 4, we present the details of building smart sensors in various ways.

MEMS Is the Main Way to Build Sensors

MEMS belongs to one of the subcategories of micro-/nanoscience and technology. For example, at Shanghai Jiao Tong University, the Micro and Nano Institute is divided into two distinct groups: Nanomaterial and MEMS (Fig. 2.14).

The first group studies micro-/nanomaterial, and the second group focuses on MEMS technology. The nanomaterial refers to nanopoints (0D), nanowires (1D),

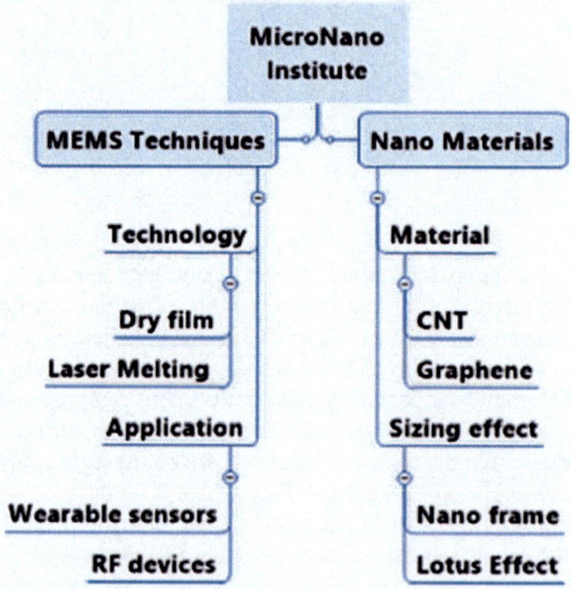

Fig. 2.14 The hierarchy structure of the micro/nano field. MEMS is one of the two branches of the micro/nano science, technology, and enginering

thin films and heterojunctions (2D), and micro-/nanoparticles/clusters (3D), with at least one dimension within the micron (10^{-6} m) or nanoscale (10^{-9} m). The sizing effect is the distinct feature in micro-/nanoscience. For example, the specific surface area (m^2/g) of the micro-Au powders is ~ million times larger than that of solid cubic gold.[1] Such a sizing feature is very useful to greatly enhance the chemical reaction rate when using Au as a catalyst ~ one million times higher catalyst effect by using Au powder when mixing two liquids or two gases together and use the Au as a catalyst.

Instead of just one type of material, the MEMS technology is focusing on manufacturing micro-structures and micro-devices based in a few materials such as to build a Pt/PtRh sensor on the TBC thin film coated on a turbine blade. In this example, three materials are involved with the proper configuration and structure to build a specific device to fit in a specific application scenario. In fact, MEMS is the best approach to make cheaper sensors since the manufacturing process can be performed in batch. For example, in Chap. 4, we show the details of building multiple thin-film sensors in batch by using the MEMS process in comparison to old technology that we implant individual wired thermocouples one by one in sequence.

MEMS technology is used to build micro-sensors which have small, light, and faster in response features. Taking advantage of the well-developed very large-scale integrated circuit (VLSI) in Silicon Valley, MEMS sensors can also be built as small chips and in batches for much cheaper prices on large scales. An 8″ silicon wafer can hold 1000 chips of 5 mm * 5 mm MEMS sensors, which are built in one time. ADI released a 10-DoF MEMS inertial measurement unit with multiple sensors (3-axis gyroscope, accelerometer, magnetometer, a pressure sensor plus a processor unit) in one package [9]. However, MEMS technology differs from the traditional semiconductor industry in two major aspects:

(1) *It does not rely too much on the most updated VLSI state-of-the-art achievements, such as TSMC's 5 nm or Samsung's 3 nm technology, but just uses the most basic and cheap process of photolithography, PVD thin-film deposition and the lift-off process. MEMS's main focus is on the downstream application needs and then to optimize the micro-mechanical structure and select different materials accordingly to realize the unique functions of each sensor.*
(2) *The steps are much fewer than those of VLSI. A VLSI technology is a series of repeated steps of bundled photograph + adding + etching + implantation with over hundreds of process steps. A modern IC chip usually contains over 50 multi-layers combining the structure for FEOL (transistor level) and BEOL (metal-line connection level). In comparison, the MEMS process only requires 1–2 layers to form a sensor since most sensors are two-terminal devices and only require 1–2 layers of sensitive materials.*
(3) *Although the MEMS devices can also be made in batch, its main focus is to build the specific sensors to fit in specific need, which the conventional VLSI*

[1] Gold powder of 3 μm has 0.5 m^2/g of SSA (the total area of a unit mass) comparing to the SSA of a golden cube of 3×10^{-7} m^2/g SSA (6/19,320,000), density 19.320 g/cm^3, the surface area of the cube is 6 cm^2.

technology is not applicable, for example, to create a pattern and curved surface. This is the main topics to discuss in details from Chaps. 3–5, i.e., to build the tough sensors in various AIAA occasions.

Although the normal VLSI technology is very comprehensive and contains over a hundred steps and over ten layers structures, there are just two key steps of VLSI manufacturing:

① To draw—using photolithography to draw a pattern on the photoresist (PR) via a mask. Photography is a comprehensive process containing the coating (spinning) of the PR, exposing under UV light with the mask, and developing the PR afterward. Photography is the key technology for VLSI, processing 30% of the total manufacturing cost and 40–50% of the processing time in line. Just the cost of the masks is already ~ 1 million USD. The most advanced ASML EUV lithography systems cost 400 million dollars. Photolithography is the most frequently used step in the whole VLSI process.

② To pattern—using the adding, removing, or modifying technique to create a pattern based on photolithography image on the PR. The main techniques of adding new layers of materials to the existing substrate include thin-film deposition, epitaxy, and thermal growth. The removing techniques such as the wet etch and dry etch remove the unwanted portion under the protection of the PR pattern. Modifying technique is the process to modify the features of the substrate, such as the doping of silicon with ion implantation under the protection of the PR thick film.

The whole VLS process is just to repeat ① and ② in a clever manner to form 100 layers of integrated circuit structures. A simple cycle of ① and ② is shown in Fig. 2.15.

There are three basic techniques that are mostly used to form MEMS sensor/actuator structures: lift-off, deep etch, and curved surface patterning.

(1) *Lift-off versus etching*

Etching and lift-off are the two popular techniques for removing a deposited film. The most frequently used removing process of thin films in VLSI is etching. However, the lift-off technique is mostly preferred in MEMS process. The difference in lift-off and conventional etching steps is illustrated in Fig. 2.16. Lift-off saves the etching step, which is esp. useful to pattern the hard-to-etch materials. The lift-off process is not as sufficient for thicker thin-film layers above 1 μm and narrow line widths/spaces less than 1 μm but is usually good enough to make large structures above 2 μm dimensions. The lift-off technique saves the chemical or dry etching process by using acetone to remove PR to form a patterned structure, which makes the MEMS process very easy to use to form simple sensor structures.

(2) *Deep etch and ICP*

The 3D MEMS structures usually involve deep etching technology that is not very frequently used in VLSI. The challenge of deep etch is to avoid the side or literal

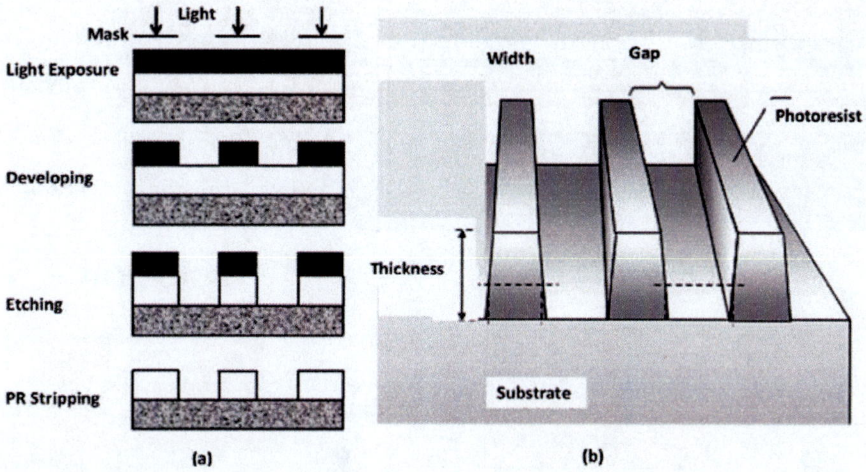

Fig. 2.15 Illustration of the basic VLSI steps: **a** vertical; **b** 3D

etching. Most deep etch processes use Bosch's inductively coupled plasma (ICP) technique, which first uses fluorine-based active ions to etch silicon, followed by sidewall passivation and vertical passivation. The two-step processes of etching and protection are carried out alternately to form a vertical deep etched trench structure. One example of the deep ICP etched structure is shown in Fig. 2.17.

(3) *Curved surface patterning*

Different from conventional VLSI technology which usually performs the surface patterning on flat and smooth silicon wafer, most of the embedded MEMS sensors are built on rough and curved surfaces. Both the photography and thin-film deposition in VLSI must be modified to fit in the new scenarios of MEMS. Figure 2.18 illustrates the curved patterning steps by using the MEMS process. More details are provided in Chap. 4.

Most of the traditional MEMS textbooks [11] discuss MEMS techniques based upon IC technology. Usually, these textbooks are old-fashioned and originated in 2010. Currently, there are many other misc. alternative techniques to build various smart sensors [12]. For example, additive manufacturing or 3D printing, mesh printing, hand-drawn painting plus annealing, etc.

Additive Manufacturing and Other Methods to Build Smart Sensors

The second meaning of the cleverness of smarter sensors is that we can build the small and thin-film sensors on site or on the tough locations of the machine part such as deep corner or sharp curved topology. Additive manufacturing offers such a flexible way to fabricate smarter on-site sensors.

As two popular technical terms, additive manufacturing (AM) and 3D printing are interchangeable. Both methods are used to create a 3D machine part by adding

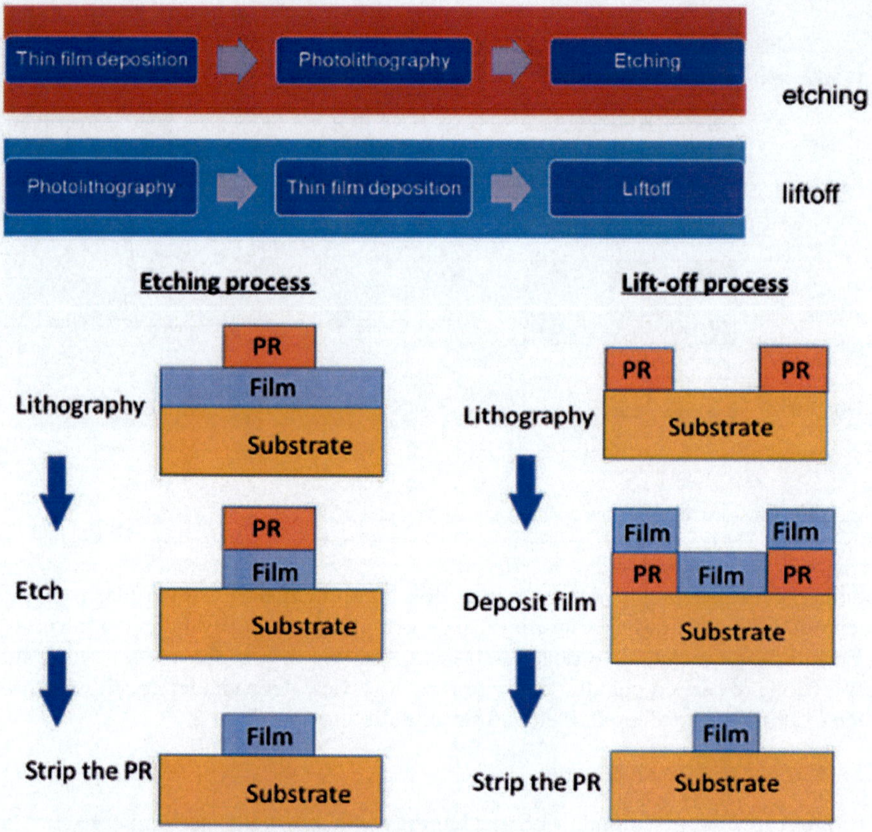

Fig. 2.16 Difference between the lift-off and etching

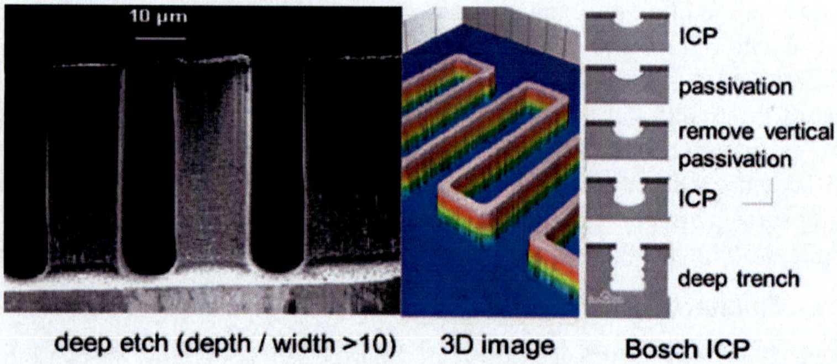

Fig. 2.17 Comprehensive surface patterns and the deep trenched structure formed by photolithography and ICP [10]

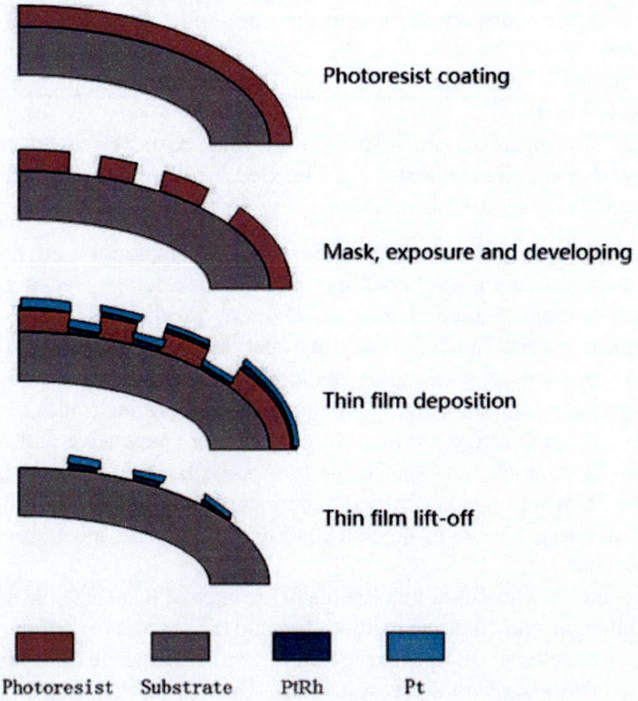

Fig. 2.18 MEMS technique for curved surface patterning: photolithography, thin-film deposition, and lift-off

the building materials layer by layer. The 3D printing can be regarded as an automatic machine painting. We use a brush or pen to manually draw a painting on the designated places with the designated colors of painting materials on the canvas. The 3D printing uses machine to scan a printing jet or beam of fine powders on the substrate layer by layer to form a 3D structure. AM or 3D printing is a transformative approach to industrial production that enables the creation of lighter, stronger parts, and systems. It is yet another technological advancement made possible by the transition from analog to digital processes [13]. AM technology has grown very rapidly in the past two decades and integrates the most updated technologies of computer-aided design and material processing technology. Based on digital model files and through software and numerical control systems, special composited materials, either metallic or non-metallic, mechanical, and medical, are added to the stack layer by layer by sintering, melting, spraying, etc., to produce solid 3D objects. The advantages of additive manufacturing are as follows:

- Rapid prototyping that can advance industry, i.e., on-demand 3D printing and quick time-to-market.
- Simple to use for beginners, allowing users with a limited experience to create/edit bespoke designs and customized parts.

- Produce difficult geometry items that are impossible for traditional methods to produce.
- Create composite structures, such as lightweight yet complex parts, for the aerospace industry.
- Affordable start-up costs and saving on tooling costs, since no cutting tools, fixtures, and comprehensive steps are needed, or no molds are needed, saving the associated costs with this manufacturing process.

The cost of a part is directly related to the amount of material used, the time taken to build the part, and any post-processing that may be needed. Because the process is based upon computer-aided designs (CAD), any product alterations are easy to make without impacting the manufacturing cost. In some instances, 3D printing is cheaper than conventional production methods. However, 3D printing is unlikely to replace traditional manufacturing, where high volume production of comparatively simple parts is needed. Large production runs are more expensive with 3D printing, as economies of scale do not impact this process as they do with other traditional methods. The 3D printing is performed one by one, and each 3D printing can take a few minutes to several hours or days depending on the print speed, resolution, and volume of material.

Additive methods can either involve the 3D printer or just draw the sensor layout manually. There are many flexible methods to build the sensors for different scenarios, and more showcases are exemplified in Chap. 4 to fabricate flexible sensors on the turbine blade. These sensors are easy to make, fast to market, highly customable, and perform the high-temperature sensing function very well both on the sensing performance and on their robustness.

Figure 2.19 illustrates the smart sensors built on the turbine blade surface by using the above two smart ways: MEMS and AM.

In summary, one of the key applications of AI on AIAA is the building of the smart sensors by smart methods either by MEMS or by 3D printing, and these will be elaborated from Chaps. 3–5.

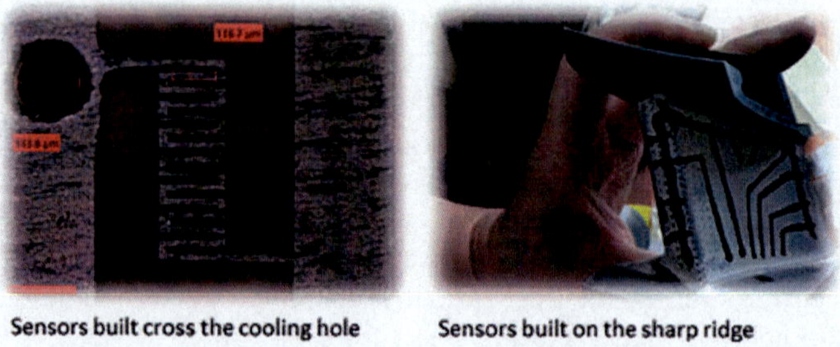

Fig. 2.19 Left: MEMS sensor just on the site of the cooling hole of the turbine blade. Right: hand-draw the TFTC sensor array on tough location (edge and the ridge)

References

1. AlphaGo, DeepMind Inc. Ltd. Started from 2010. https://www.deepmind.com/
2. Gaitan, M.: Yole Développement Group. https://www.yolegroup.com/. A strategic analysis of markets, the supply chain, and ITRS (International Technology Roadmap for Semiconductors) MEMS (Micro-Electro-Mechanical System), and adjacent industries (2022). https://tsapps.nist.gov/publication/get_pdf.cfm?pub_id=913266
3. MEMS Sensors from Yole Développement Group. https://www.yolegroup.com/product/report/mems-pressure-sensors---technology-and-market-trends-2021/
4. Status of the MEMS Industry 2021. Yole Développement Group. https://www.yolegroup.com/product/report/status-of-the-mems-industry-2021/
5. Emoto, M.: The Hidden Messages in Water. Simon & Schuster, UK (2005). ISBN13: 9781416522195
6. https://www.simonandschuster.co.uk/books/The-Hidden-Messages-in-Water/Masaru-Emoto/9781416522195
7. Li, D.: Experimental research on micro vibration field of consciousness. In: Hong, M. (ed.) 120 Dreamed Thoughts—120 Science and Technology Foresight for the Future. Shanghai Jiao Tong University 120th Anniversary Book Series, p. 152. Shanghai Jiao Tong University Press (2016). ISBN: 978-7-313-14632-8
8. Duan, F.L.: The Introductory Engineering, Chap. 4. Shanghai Jiaotong University Press (2019). ISBN: 978-7-313-20777-7. https://lib.ecust.edu.cn/zh-hans/book/903714
9. ADI's 10-DoF MEMS Sensor Fusion. https://www.analog.com/en/about-adi/news-room/press-releases/2012/06-25-12-10-dof-mems-imu-incorporates-sensor-fusio.html
10. Gao, J., Duan, F.L., Ying, W., et al.: PDMS soft template preparation and blade curved surface soft lithography process. Micro Nano Electron. Technol. **53**(5), 7 (2016)
11. Adams, T., Layton, R.: Introductory MEMS: Fabrication and Applications (2010). ISBN: 978-0-387-09510-3. e-ISBN: 978-0-387-09511-0. https://doi.org/10.1007/978-0-387-09511-0
12. Duan, F.L., Liu, S., Xie, Z., Ji, Z., Weng, H., Hu, M., Gao, J.: The various MEMS methods to build TFTC sensors for related aero-/astro-applications. In: AIAA 2021-1399. AIAA Scitech 2021 Forum, Jan 2021
13. Additive manufacturing. https://www.ge.com/additive/additive-manufacturing

Chapter 3
Smart Sensors for AIAA

By using the IEEE methods, we can make the smart sensors in order to make the AIAA smarter. For example, accurate measurement of the surface temperature of the turbine blade plays an indispensable role to keep the best engine efficiency by optimizing the dynamic balance between the high temperature and the cooling effect. The logic chain is as follows:

the higher the temperature, the greater the efficiency;
→ yet the turbine metal cannot hold it, so cooling is needed;
→ the cooling air reduces the surface temperature but reduces the efficiency as well;
→ it is necessary to control the right amount of cooling air for both efficiency and safety;
→ we need to measure the surface temperature promptly and accurately in order to make the right judgment whether it is over cooling or insufficient cooling.

Smartness has two meanings: smarter ways to build sensors and smarter usages of the sensors. Let's use the TFTC sensor as an example to explain the two smartness.

(1) To build a thin-film thermocouple (TFTC) sensor on turbine blade is a highly challenging work for IEEE people. Smarter ways to build them are needed. First, most of MEMS technology is on the plenary substrate such as a thin flat smooth silicon wafer, and most of the aero-engine parts have a complex surface topology, such as the turbine blade with sharp curved surface. The second challenge is that most conventional sensors work in room temperature ranges, but the working environment of the gas turbine is high-temperature and high-pressure. Facing these two challenges, there are two steps for the IEEE people to go through when developing a sensor suitable for the AIAA application scenario. (1) To modify IEEE's original recipe to embed the thin-film sensor on turbine blade surface—this involves 3D patterning micro fabrication technique; (2) To verify the sensor's performance to fit in the required scenarios in turbine engine environments, such as high-temperature, high thermal stress, combustion gas flow. The details of these smarter technologies will be discussed in Chap. 4.

© The Author(s), under exclusive license to Springer Nature Singapore Pte Ltd. 2023

F. L. Duan, *When AIAA Meets IEEE*,
https://doi.org/10.1007/978-981-19-8394-8_3

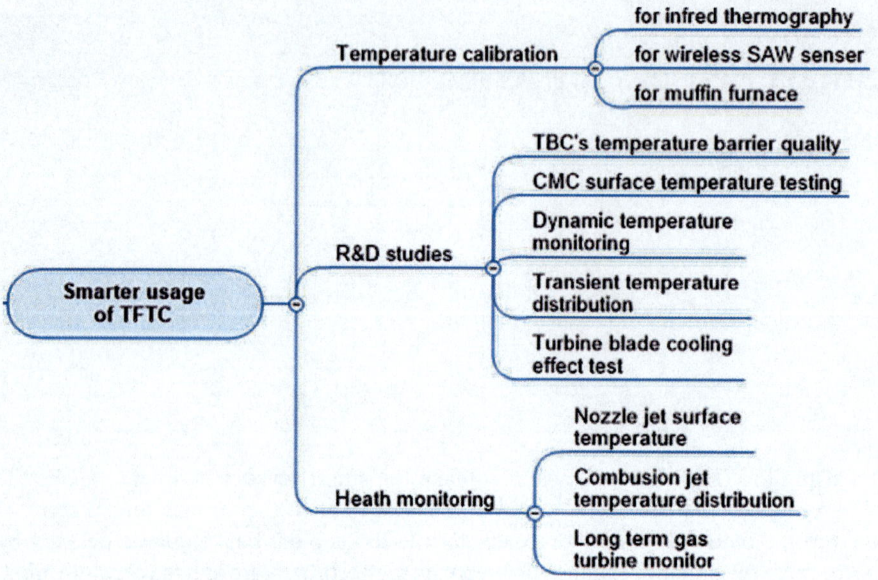

Fig. 3.1 Various uses of the TFTC sensor

(2) After successfully building smart sensors, IEEE people and AIAA fellows have to work together to fit in this TFTC sensors in various AIAA scenarios to achieve the VAATE goal of an intelligent aero-engine [1]. In most AIAA scenarios a new problem appears first and then the solution comes after. In our case of the TFTC, the order is opposite: the new technology is invented first and then we are searching for its playground. Therefore the next step is to find is the usage in AIAA scenarios such as the cooling performance test [2] or as a temperature calibration tool for infrared thermography. As shown in Fig. 3.1 [3], we list here three categories of application scenarios. The first category is to use TFTC as a calibration method since TFTC is the most accurate and precise real-time temperature test method so far. The second usage is to aid various Research and Development (R&D) activities related to high temperature-measurement, such as to evaluate TBC's (Thermal Barrier Coating) temperature barrier quality, NGV's (Nuzzle Guide Vane) cooling effect verification. The third one is for health and safety checks, such as long-term temperature monitoring for gas turbine power plants.

Below, we illustrate the playground of the TFTC smart sensor in a few typical AIAA scenarios:

- *Calibration for other methods*
- *Evaluating TBC's temperature barrier quality*
- *NGV cooling performance tests*
- *Evaluate horn-shape cooling hole on turbine blade*

- *Jet nozzle temperature distribution*
- *Temperature field of jet flow*
- *Long-term monitor for gas generator*.

3.1 Calibration for Other Methods

Among the various temperature measuring methods, TFTC is the best one for the most accurate and precise one (check Sect. 4.2.2). We can make use of this distinct feature of TFTC as a calibration standard or reference temperature reader for other methods such as muffin oven, infrared, and Wi-Fi sensors as we described below.

3.1.1 Turbine Surface Infrared Thermography

Infrared thermography is a very powerful tool to obtain the whole image of the temperature field on the turbine blade surface. For example, Fig. 3.2 illustrates the cooling performance test of a nozzle guide vane [4]—the key component to direct the heat flux to the rotating turbine blades. NGV is the first component to receive the highest temperature from the combustion chamber, and cooling is usually needed from cooling air channeled from inside the NGV. The cooling effect is one of the critical tests to verify the design and performance of the NGV. Although infrared mapping can provide a wholesome massage of the surface temperature, its accuracy is highly dependable to the ambient (the optical conditions under high-temperature combustion) and must be properly calibrated [5, 6]. Here, by using the TFTC array we can effectively calibrate the infrared thermography method by comparing the image pattern (color) to the thermal voltage reading on some specific key points to obtain accurate temperature field information.

One example of such calibration is shown in Sect. 5.2.5 which compares the results of the TFTC reading versus infrared reading.

3.1.2 SAW Wireless Sensor Calibration

Wireless sensing plays an indispensable role in high-temperature measurement inside gas turbine system where the wirings of the sensors are extremely difficult. By using the SAW sensor, the wireless high-temperature measurement on turbine blade surface can be achieved. Since SAW is a new type of sensor, it needs to be calibrated properly prior to its first use. Traditionally, the calibration uses a separate wired thermocouple. Due to the space–time delay between the SAW and thermocouple, the temperature response of the DUT is not the same as that of the standard thermocouple. Such a problem can be circumvented by using a TFTC to eliminate the space–time

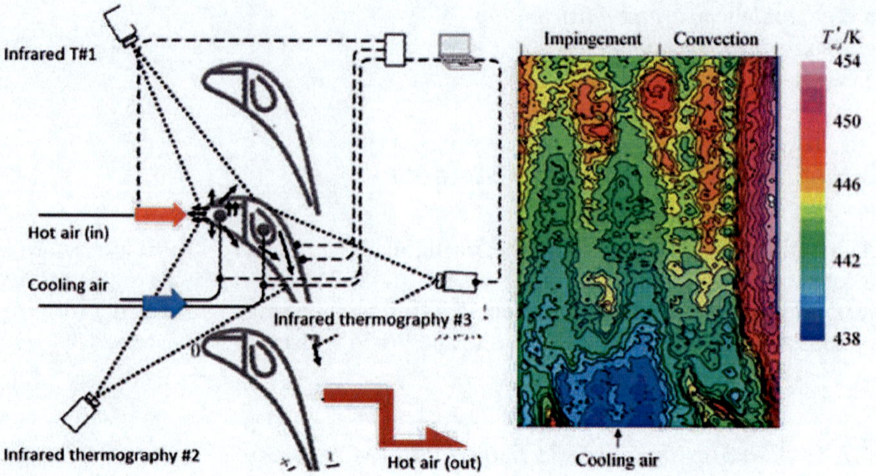

Fig. 3.2 Infrared thermography was used for surface temperature mapping of NGV

difference in the thermal transmission process. The error of the ordinary method is approximately ~ 4 °C, while the error of the TFTC is ~ 1 °C (more in Chap. 5).

3.2 Evaluating TBC's Temperature Barrier Quality

There are many reports about how much a TBC layer can barrier the surface heat in combustion air flow. However, these reported data of the thermal barring effect of TBCs vary significantly [7, 8]. A very important reason is that the TBC measurement should be carried out accurately. The conventional method to quantify the TBC's thermal insulation effect is shown in Fig. 3.3. A high-temperature acetylene gas flow is used to heat the thermal barrier coating surface, and the flame temperature is measured by the infrared method. A standard thermocouple was placed on the back of the test sample to test the temperature on the Ni–Cr alloy substrate. The difference between the two temperatures is the insulation temperature of the TBC. The accuracy of this method is very limited. First, the infrared measurement itself has an error of > 10 °C. In addition, the temperature of the sample measured with the TC on the back cannot completely eliminate the air gap between the thermocouple and the blade surface, which will also cause errors.

With the help of TFTC, the exact temperatures on both sides of TBC can be measured accurately. Here, we use two TFTCs to measure the temperature difference on both surface. To date, this is the most accurate method to characterize the temperature barrier effect of TBC material. The standardization of measurement should also be established to avoid confusion in the current TBC measurements.

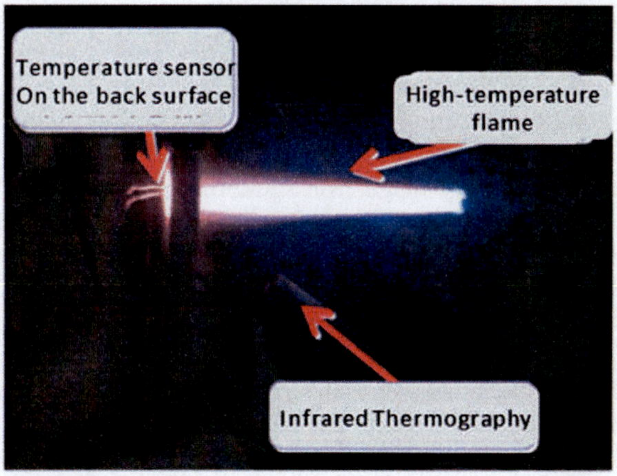

Fig. 3.3 The ordinary TBC test system

The principle. As shown in Fig. 3.4 (left), the temperature measuring nodes of the two thermocouples are located in the middle of the small disk inside and outside of the TBC. Heating the TBC surface with high-temperature flow and measuring the temperature difference between the two thermocouples, one can obtain the temperature difference on both sides of the TBC coating, that is, the insulation temperature. The accurate and fast response quality of TFTC ensures correct measurements of real-time temperatures on both sides of the thermal barrier coating.

The main difficulty of this methodology is the fabrication of a thin-film sensor inside the TBC because the spraying process during the TBC coating may damage the thin-film structure of the first TFTC. The coping strategy of this study is explained in detail in the fabrication process of the thin-film sensor below.

The fabrication. A prototype of such a structure is fabricated using a small round steel plate (Fig. 3.4 (right)). After a 30 μm thick MCrAlY layer is deposited, a relatively thin (30 μm) TBC layer is sprayed first, and then a Pt/PtRh TFTC is fabricated

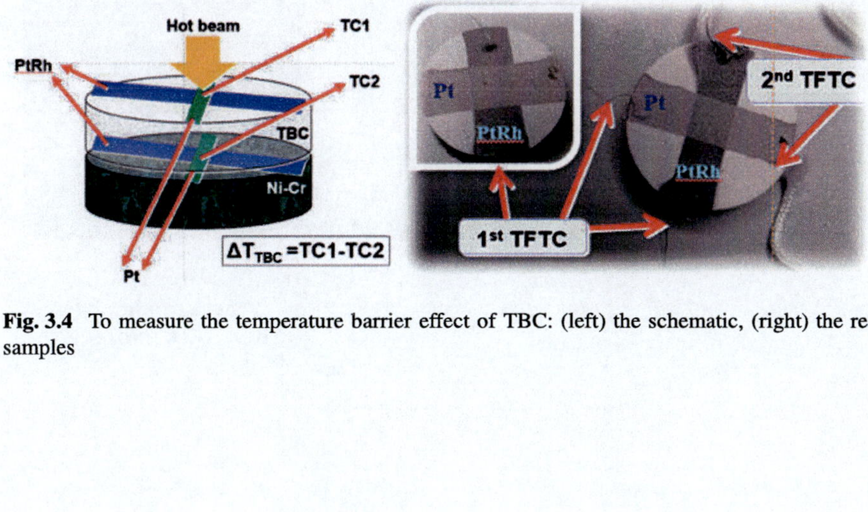

Fig. 3.4 To measure the temperature barrier effect of TBC: (left) the schematic, (right) the real samples

on the surface. The fabrication process is as follows: first, the Pt lines are patterned using photolithography, and then a 500 nm Pt layer is sputtered by magnetron sputtering and lifted-off. Then, a PtRh line (also 5 mm wide) photolithography process is made, and the line is perpendicular to the original Pt line to form a cross. The intersection of the cross is the node of the thermocouple, and the line of PtRh is also formed by magnetron sputtering and peeling. Then, thin Pt and PtRh wires are welded on the TFTC. After the formation of the first layer TFTC, a layer TBC with a thickness of 300 μm is deposited. The second layer of the Pt/PtRh thin-film thermocouple was fabricated in the same way as above. It is worth noting that in the second spraying process, the high-temperature impact process in the spraying process may cause damage to the first layer of the thin-film thermocouple, so when making the first layer of the thin-film thermocouple, the width of the line should not be too narrow, at least over 100 μm.

Testing and results. As shown in Fig. 3.5, we use a butane liquefied gas high-temperature spray gun to heat the TBC surface. Adjusting the distance between the spray gun and the sample can alternate the temperature on the surface of the thermal barrier coating. Two standard R-type thermocouples are placed on the two sides of the sample to measure the temperature of the flame and the temperature on the back. The temperature insulation effect of the thermal barrier coating is evaluated by measuring and comparing the temperature difference of four thermocouples (two standard thermocouples plus two TFTCs).

Figure 3.6 shows the dynamic temperature trend of the four thermocouples during the heating and cooling process. This experiment confirms the feasibility of using a double-layer thin-film thermocouple to measure the temperature on both sides of the TBC, and the change trend of temperature variation meets our expectations. As seen

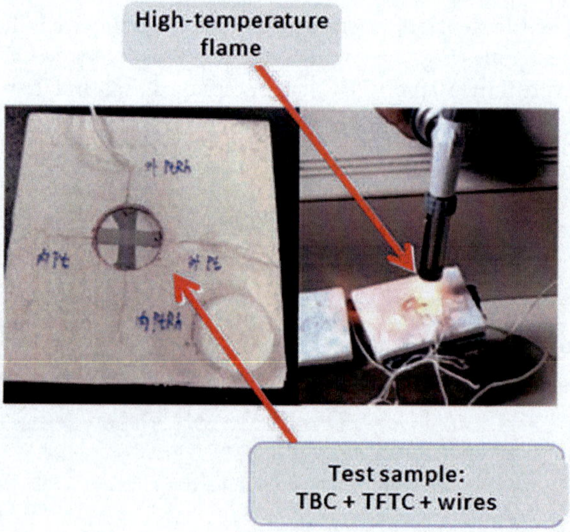

Fig. 3.5 Test system to evaluate TBC's temperature barrier effect

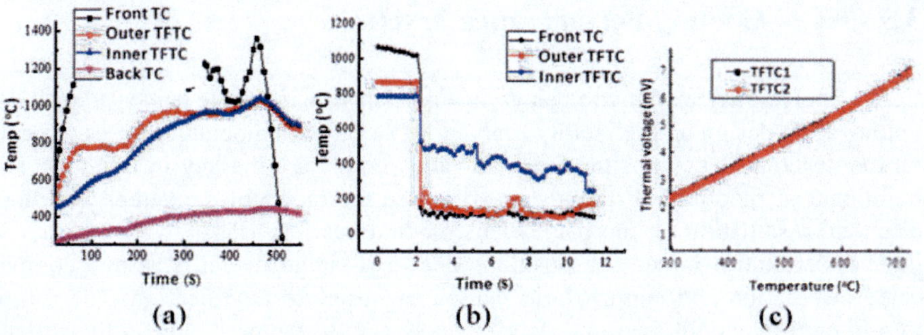

Fig. 3.6 The dynamic temperature measurements TBC temperature isolation effect. **a** Dynamic temperature response during the heating. **b** Response after sharp cooling. **c** The performance of the two TFTCs prior to the TBC dynamic tests

in Fig. 3.6a, at the beginning, the thermocouple outside the thermal barrier coating reacts quickly to the temperature, while the internal rising speed is relatively slow. Finally, it gradually reaches equilibrium. The temperature at the back of the sample rises very slowly due to the temperature isolation effect of the TBC sample and the gap of the air. When the external heat flow flame stops (at 470 s), the temperature of thermocouples inside and outside the thermal barrier coating decreases gradually. When thermal equilibrium is reached, the temperature on both sides of the thermal barrier coating is the same.

Then, a transient cooling experiment was carried out that is, cold water was sprayed on the TBC surface immediately after the flame was removed, and the transient temperature readings of the two TFTCs were collected, as shown in Fig. 3.6b. The results of the transient experiment show that the outer thermocouple cools rapidly, while the inner thermocouple cools slowly. Comparing the temperature difference between the two thermocouples, we can see that the temperature isolation effect of the TBC is approximately 150 °C. The results also show that since the TBC cooling time is in seconds, a fast reading of temperature at both ends of the thermal barrier coating is needed by using the ultra-high-speed data acquisition system (the accuracy can reach 10^{-5} s).

Of course, prior to the previous test, the two TFTCs at the upper and lower layers of the thermal barrier coating are verified and calibrated. Figure 3.6c shows the curve of the thermal voltage of two thin-film thermocouples varying with temperature, and the repeatability of the curve is very good.

We have the studies showing that the surface temperature on the back of the sample play a very important role to for TBC's performance [9]. Test standard must be established with a fixed back surface temperature. In future experimental designs, the temperature at the bottom of the sample should be controlled at a certain value, i.e., within 100 °C, by using water coolant. The temperature insulation effect of the TBC can be calibrated by comparing the temperature difference of the two TFTCs.

3.3 NGV Cooling Performance Tests

Cooling performance evaluation plays an important role in nozzle guide vane (NGV) optimization design and reliability evaluation. Accurate temperature measurements on the designated key positions on the vane surface are the key in this test. The traditional method is to install a series of wired thermocouples together with their electrical installation on the blade. This method has intrinsic drawbacks, such as large uncertainties in sensing positions due to its large overall size and complicated installation and perturbations caused by structure modification. The use of thin-film thermocouple sensors can effectively avoid such problems. In this article, the measurement errors caused by the traditional method are first evaluated using computer simulation, together with the comparison to the thin-film sensor. It is indicated that the wired thermocouple method may cause up to 370 °C variations at a high-temperature measurement of 1200 °C, while the thin-film embedded thermocouple can successfully avoid such problems. In addition, the thin-film Pt/PtRh thermocouple array was formulated and manufactured on an NGV surface using MEMS technology. The connections to sensors can be established via the extended thin-film layer on the surface toward the edge of the blade to avoid any turbulence to air flow at the sensing points. All thin-film sensors in the array can be built in batches, and the installation is much less destructive to the turbine blade structure. It is also demonstrated that such a sense array can detect the surface temperature distribution over 1000 °C with a fast sampling rate as fast as 10 μs.

3.3.1 Background

The turbine inlet temperature of the aero-engine has reached 1800–2050 K [10], which has already exceeded the highest tolerance of alloy blade metals. Therefore, effective cooling methods must be adopted to reduce the wall temperature by the proper cooling design of turbine blades. Cooling is achieved by the cooling air and thermal barrier coating (TBC). There are various methods of cooling hole designs, including the shaping of holes [11], array arrangements [12], and blade curvature designs [13], and there have been tremendous efforts on TBC development for years [14, 15], the goal of which is to build the most efficient and reliable turbine blade. To test the cooling effect, the comprehensive contributions of both TBC and cooling air are crucial to evaluate the cooling efficiency and reliability. The cooling performance test of the turbine blade plays an indispensable role in the design for higher performance and more reliability of the TBC-composite nozzle guide vane (NGV) and other critical parts [16].

The cooling effect is calculated and evaluated by the following formula:

$$\text{Cooling efficiency} = \frac{T_g - T_w}{T_g - T_c} \tag{3.1}$$

In the formula, T_g is the gas temperature from combustion, T_w is the blade wall temperature, and T_c is the cooling air temperature. Therefore, to evaluate the cooling effect, it is necessary to accurately measure these three parameters. Among them, the blade wall temperature T_w is the hardest. Therefore, in the comprehensive cooling effect test of turbine blades, the surface temperature measurement is the technical bottleneck. Currently, the most-adopted method is to use an armored wired Pt/PtRh thermocouple (TC) sensor to measure the wall temperature T_w. By embedding a series of TCs at key locations on the blade/vane surface, the temperature distributions can be obtained. The disadvantages of this measurement are mainly reflected in the following:

(1) The gap between the measuring point and the thermocouple joint. This is due to the extra room needed for electrical insulation filling between the wired TC and the turbine blade metal. Factors such as embedded depth and embedded gap cause measurement variation, uncertainty and temperature field modulation (see the simulation analysis below).
(2) The upper spec of temperature measurement may be limited. For example, most armored TCs' highest temperature cannot reach 1000 °C due to the sheath metal.
(3) An additional coating is usually needed to smooth the sheathed TC blade surface, usually a high thermal conductance, high hardness, and anti-oxidation protection layer, causing more errors and uncertainties in surface measurements.

The use of thin-film thermocouples (TFTC) can avoid these drawbacks. The thickness of the TFTC is less than 1 μm, and the area of the temperature measurement point (thermocouple joint, the overlap of the Pt and PtRh thin-films) can be as small as 0.2 mm × 0.2 mm using MEMS [17] microfabrication technology. The impact of the TFTC thermo-joint on the environment is very small. Another great advantage of TFTC is that a sensor array composed of all TFTCs can be built in batches without embedding them one by one, as in traditional TCs. The connections of TFTCs in the array are through the guided Pt or PtRh thin-film toward the edge of the NGV and led out through the high-temperature insulated wires; in this way, the air flow and temperature heat transfer characteristics of the measurement point of the turbine blade will not be affected. Damage to the blade structure due to wired TC implantation can also be avoided. In our early work [18, 19], we successfully integrated a film-type (< 1 μm) high-temperature sensor on a turbine blade surface and passed a series of engineering experiments, such as functionality, repeatability, endurance, and reliability.

3.3.2 *Error Analysis*

There are two conventional ways to measure surface temperature in cooling tests: armored sheath wired TC and ceramic cube-insulated wired TC. Both have intrinsic measurement error due to their composite size in the test environments. Both need a slot on the turbine blade surface and then embed either the armored thermocouple or electrical insulation tubes inside the slot and then smooth the surface with chromium carbide nickel chromium coating. Armored TC is more flexible in bending, and the tubed (made by alumina) wire TC may have a higher upper temperature limit (up to 1600 °C). In the following, we use computer simulation to explain the error caused by the traditional methods together with their TFTC counterparts.

The Error Analysis of the Armored TC Temperature Measurement

Armored TC is composed of a conductor, insulating magnesium oxide (MgO), and a Cr–Ni–Ti stainless steel protective shell (Fig. 3.7). It is made by inserting a thermocouple wire into the metallic protective shell and then sealing and filling it with inorganic insulating material (high purity MgO) powder. MgO has superior insulation and sealing performance and good thermal conductivity, which can quickly transfer the heat from the shell to the sensing joint of TC.

The simulation structure is simplified and shown in Fig. 3.7. The wired TC joint (size ~ $(0.2\text{ mm})^2$) is located in the center of the armor ($(0.5\text{ mm})^2$). In addition, to undercover the sheathed TC, a layer of high thermal conductivity hardened coating is usually used to smooth the surface. The TFTC sensor is also simulated as a comparison.

In the simulation process, the surface temperature is set as 1472 K, and the temperature of the metal substrate is 700 K. COMSOL simulation software is used to obtain the temperature distribution in the multi-layer structure by solving the thermodynamic equation with the proper boundary conditions:

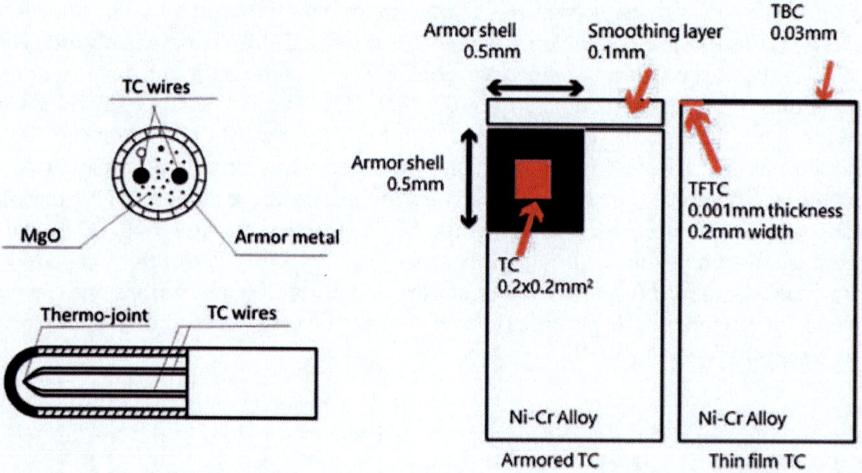

Fig. 3.7 The inner structure of the armored thermocouple simulation structure of buried TC (left) and thin-film TC (right). The red area is the TC sensor

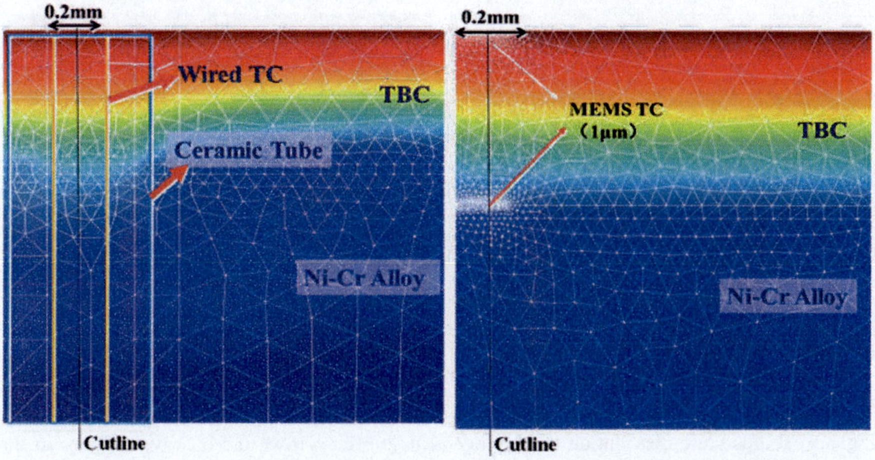

Fig. 3.8 Simulation structure and temperature distribution of the cube-insulated TC and TFTC

Error Analysis of Tube Insulated Wired TC

Figure 3.8 shows the simulation structure and grids considering wired TC insulated using an Al_2O_3 ceramic tube on the NGV with TBC, together with its TFTC counterpart. The thicknesses of the TBC, bonding layer, and super-alloy are 0.3 mm, 0.05 mm, and 2 mm, respectively. The diameters of the TC wire and ceramic insulation tube are 0.2 mm and 0.4 mm, respectively, and they extend all the way through the inner wall of the NGV. The temperature distribution is also shown in the same chart—the color red indicates high-temperature area and blue is the low temperature area. It is seen that in the area where cubed-insulated TC is implanted, the high temperature on the surface makes it easier to reach the substrate metal due to its higher overall thermal conductivity, while in the case of TFTC, such an influence is quite minimal. It is a clear indication that the thermal conductivity difference of wired TC metal and alumina ceramic tube causes the temperature field modification of the blade structure so that the temperature test result may not be able to reflect the real situation in the cooling performance test.

3.3.3 *TFTC's Advantage*

Further infield implementation of the cooling performance test of NGV involves placing TFTCs based on designated points on the NGV as shown in Fig. 3.9 (left). More than 10 sensing points on each side will be placed with Pt/PtRh TFTCs for further cooling test experiments with MEMS technology. Temperature measurements at these points can be measured using the TFTC array (Fig. 3.9 (right)) in NGV's cooling effect evaluation. A thin-film Pt/PtRh thermocouple array can effectively

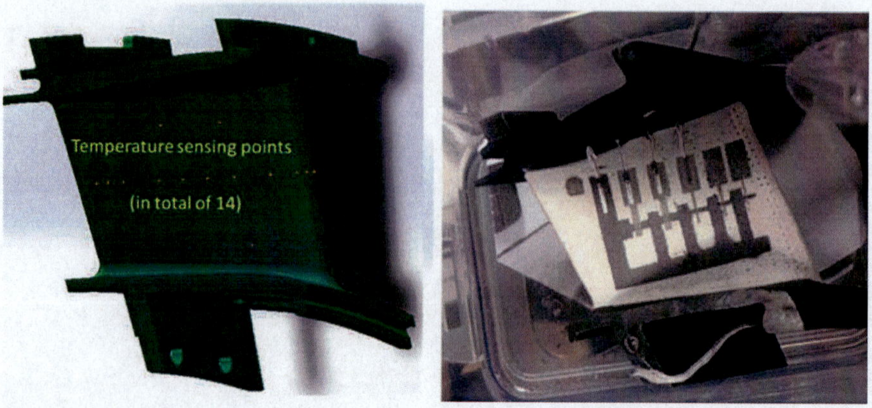

Fig. 3.9 Temperature distribution test arrangements on the convex and concave surfaces of the nozzle guide vane

circumvent the disadvantages of traditional wired TC methods both in terms of accuracy and economy. The thickness of the thin-film of TFTC is less than 1 μm, and the size of the sensing joint can be patterned as small as $(0.2\ \text{mm})^2$, causing little turbulence to the environment. Test arrays with many TFTC sensors can be built in batches and are fairly easy to implement. These TFTC sensors show stable, reliable and repeatable results based on our extensive characterizations above. The cooling performance will be tested in a high-pressure, high-speed gas flow system with the cooling air from the inside chamber toward the blade surface. The test blade with TFTC sensors will be fixed in chain with other NGVs to ensure the sealing of the cooling air from the ambient combustion flow on the NGV surface.

3.4 Evaluate Horn-Shape Cooling Hole on Turbine Blade

The horn-shaped hole has a better cooling than conventional straight cooling holes. This special-shaped hole is a discrete cylindrical hole, with a circular inlet section and a gradually expanding outlet section. The purpose of the special-shaped hole is to expand the area of the jet outlet to 2–3 times and prevent the cooling gas from separating from the surface (Fig. 3.10). Usually, this special shape is performed by a pulsed laser. The laser emits a high-energy laser beam to convert light energy into heat energy, which instantly melts and vaporizes the blade material in the irradiation area to achieve material reduction processing and finally form a small hole. When machining special-shaped holes, the output mode of the laser beam is changed from continuous to single pulse, and each pulse is subject to independent energy control. With careful energy control and high-precision mechanical movement, the desired shape can be carved bit by bit by laser carving.

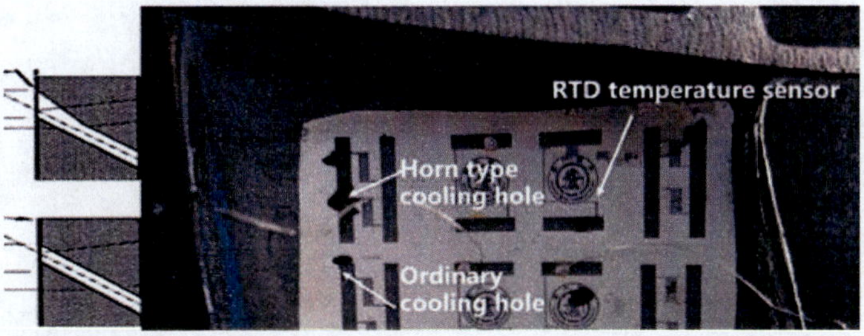

Fig. 3.10 Use MEMS temperature sensor to test the cooling effect of different holes

Figure 3.10 shows the MEMS sensors built just at the location of the cooling holes both for this special hole as well as conventional cooling holes. The MEMS sensor is the best solution for this small site temperature evaluation. Figure 3.11 shows the more intimate image of the thin-film sensor (RTD) just on the location of the hole to detect the surface temperature with the cooling air flowing through it dynamically. In this way, we are able to compare the temperature cooling effect of the different cooling holes. With the conventional temperature testing methods such a precise measurement is very difficult.

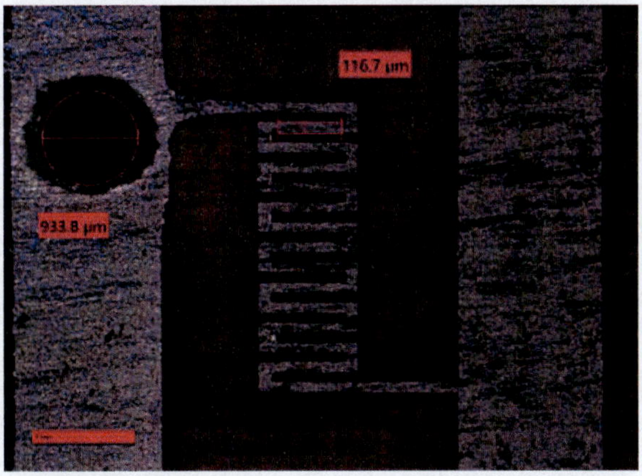

Fig. 3.11 The MEMS temperature sensor can be built exactly on the cooling hole position in order to measure the exact temperature on that location

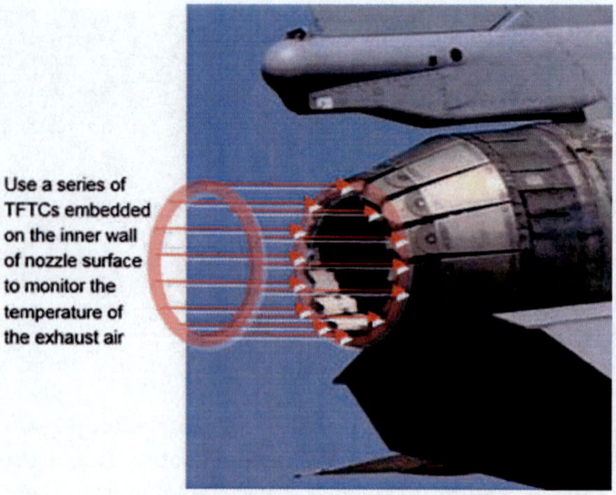

Fig. 3.12 TFTC sensor array can be integrated on the inner surface of the exhaust nozzle of the turbine jet

3.5 Jet Nozzle Temperature Distribution

Various information can be obtained via the measurement of the nozzle outlet exhausting temperature. The temperature of the exhausting air may reflect the situation and conditions in the combustion and turbine systems to obtain a certain message of the turbine efficiency. Measuring the nozzle surface temperature also helps to optimize various nozzle designs for jet engines with different exhaust infrared signatures. For example, the air stream for the rectangular nozzle may have a 70% cooler temperature than that for the round nozzle [20]. Temperature monitoring therefore becomes necessary to verify these designs with an intense sensor array where TFTC sensors can be used. The CMC-made [21] nozzle can also be evaluated using TFTC built on CMC (Chaps. 4 and 5). The thin-film thermocouple array can be integrated around the tail nozzle of the engine (Fig. 3.12). Redundant technology can be used to realize highly reliable monitoring together with the measurement of temperature distribution.

3.6 Temperature Field of Jet Flow

Figure 3.13 shows the proposal for measuring the temperature field at the outlet of the combustion chamber. The high-speed/high-temperature jet from the combustion chamber applies the heat flow on the facing round disk which is placed at the center of the outlet. The TFTC array is laid out at its surface to quantify the temperature distribution along the radius and around the disk. The welding points are located on the edge, and the wires are led out from the edge. The specific measurement scheme of the temperature field at the outlet of the combustion chamber is as follows:

(1) Five sensors are arranged on the disk in the four corners;

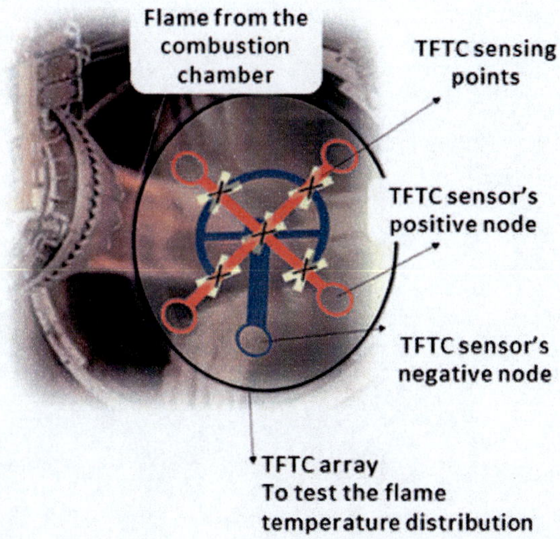

Fig. 3.13 Use the array of TFTC to quantify the temperature distribution of the combustion jet

(2) The red line is the platinum thin-film, the blue line is the platinum rhodium thin-film, and the intersection point "X" is the temperature measurement point;
(3) The circle is the welding point between the thin-film thermocouple and the external connecting wires, which is placed at the edge of the disk without disturbing the infield air flow.

By comparing the temperatures along the radius one can see the temperature distribution from the center to edge; by comparing the temperatures along the circle one can see the dynamic temperature evenness at the same radius. By varying the various input parameters such as fuel and air, pressure and flow rate, one could obtain the best combustion condition by different output readings from these TFTCs.

3.7 Long-Term Monitor for Gas Generator

Thin-film thermocouples can be hidden inside TBCs (Fig. 3.14) as a long-term monitor of the turbine temperature conditions from time to time. Since these TFTC is hidden inside the TBC, they are exempted from the corrosion and oxidation of ambient combustion air. Such feature can ensure a long-term use as a temperature monitoring of gas turbine health conditions. Gas turbines for on-ground electric generation power plant may have this need for health monitoring purpose.

In conclusion, thin-film sensors may meet various needs, such as temperature calibration for other measurement methods, such as the muffin furnace, infrared thermography, and wireless passive sensing; various surface measurements for R&D purposes, such as to evaluate the TBC's temperature barrier quality, dynamic transient

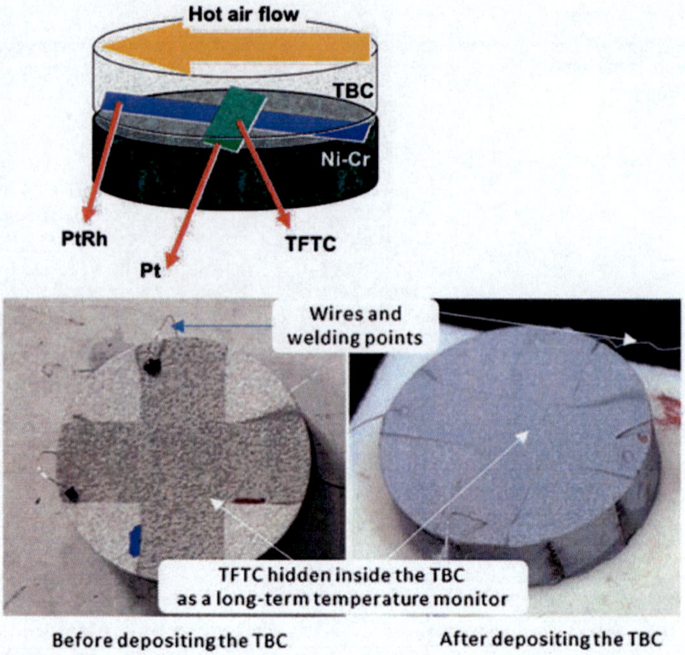

Fig. 3.14 The hidden TFTC inside the TBC using a thin-film thermocouple sensor. Left: schematic. Right: real sample

high-temperature distribution tests on NGV, and long-term heath monitoring for ground gas turbine power generator engines. In Chap. 5, more details of these tests will be revealed.

References

1. The Versatile Affordable Advanced Turbine Engines (VAATE) Initiative: An AIAA Position Paper. AIAA, Reston, VA (2006)
2. Duan, F.L., Xie, Z., Ji, Z., Li, J., Liu, Z., Zou, B.: Surface temperature test using wired and thin film thermocouple on NGV's cooling performance evaluation. In: AIAA Aviation 2020 Forum, 15–19 June 2020
3. Duan, F.L., Liu, S., Xie, Z., Ji, Z., Weng, H., Hu, M., Gao, J.: The various MEMS methods to build TFTC sensors for related aero-/astro-applications. In: AIAA Scitech 2021 Forum, Jan 2021. https://doi.org/10.2514/6.2021-1399
4. Huang, M., Zang, S., Ge, B., et al.: Infrared thermal imaging measurement method of turbine blade temperature field in hot wind tunnel. J. Aerodyn. **029**(011), 2679–2683 (2014). https://doi.org/10.13224/j.cnki.jasp.2014.11.019
5. Kim, K.S., Kim, Y.J.: Experimental study on the film cooling performance at the leading edge of turbine blade using infrared thermography. Key Eng. Mater. **326/328**(Pt2), 1161–1164 (2006)

6. Gribanov, I.: Applications and limitations of infrared thermography in turbine cooling visualization. Doctoral thesis, 9 Oct 2014. https://researchportal.bath.ac.uk/en/studentTheses/applications-and-limitations-of-infrared-thermography-in-turbine
7. Andress, D.E.: An analytical study of thermal barrier coated first-stage blades in an F100 engine. NASA CR-135359, National Aeronautics and Space Administration (1978)
8. Meier, S.M., Gupta, D.K.: The evolution of thermal barrier coatings in gas turbine engine applications. Trans. ASME **116**, 250–257 (1994)
9. Duan, F.L., Ji, Z.L., et al.: Accurate measurement of insulation temperature of thermal barrier coatings based on dual thin-film thermocouples. Aeroengine **46**(5), 49–54 (2020)
10. Amagasa, S., Shimomura, K., Kadowaki, M., et al.: Study on the turbine vane and blade for a 1500°C class industrial gas turbine. J. Eng. Gas Turb. Power **116**(3), 597–604 (1994)
11. Colban, W., Thole, K.: Influence of hole shape on the performance of a turbine vane endwall film-cooling scheme. Int. J. Heat Fluid Flow **28**(3), 341–356 (2007)
12. Johnson, J.J.: Genetic algorithm optimization of a film cooling array on a modern turbine inlet vane. Dissertations & theses—Gradworks (2012)
13. Dyson, T.E., Bogard, D.G., Bradshaw, S.D.: Evaluation of CFD simulations of film cooling performance on a turbine vane including conjugate heat transfer effects. Int. J. Heat Fluid Flow **50**, 279–286 (2014)
14. Adture, P., Ell, M., Jordan, E.H.: Thermal barrier coatings for gas-turbine engine applications. Science **296**, 280–284 (2002)
15. Clarke, D.R., Oechsner, M., Padture, N.P.: Thermal barrier coatings for more efficient gas-turbine engines. MRS Bull. **37**(10), 891–898 (2012)
16. Xu, L., Wang, W., Gao, T., et al.: An experimental research on the cooling performance of the turbine vane with an advanced duplex-medium combined cooling. Int. J. Heat Mass Transf. **79**, 72–81 (2014)
17. Ho, C.-M., Tai, Y.-C.: Review: MEMS and its applications for flow control. J. Fluids Eng. Trans. ASME **118** (1996). https://doi.org/10.1115/1.2817778
18. Ji, Z., Duan, F.L., et al.: Temperature distribution measurements on turbine blade surface by the aid of simple dotted Pt/PtRh thermal couple test array. In: AIAA Propulsion and Energy 2019 Forum, Indianapolis, USA, 19–22 Aug 2019
19. Duan, F.L., Lin, Y.: Development of accurate and robust high temperature sensor on aero-engine turbine blade surface. In: 2018 Joint Propulsion Conference, AIAA Propulsion and Energy Forum (AIAA 2018-4622)
20. Pittet, A.: Beat the Heat: Diffusing Gas Turbine Jet Exhaust. https://wwsef.ca/archives/2004/04repPittet.pdf
21. Zawada, L., Ojard, G., Bouillon, E., et al.: Evaluation of ceramic matrix composite exhaust nozzle divergent seals. In: 43rd AIAA/ASME/SAE/ASEE Joint Propulsion Conference & Exhibit (2007)

Chapter 4
To Build Smart Sensors Using MEMS

Comparing to the previous chapters, the content of this chapter is much more abundant. Therefore, we give a quick outline beforehand for a brief picture:

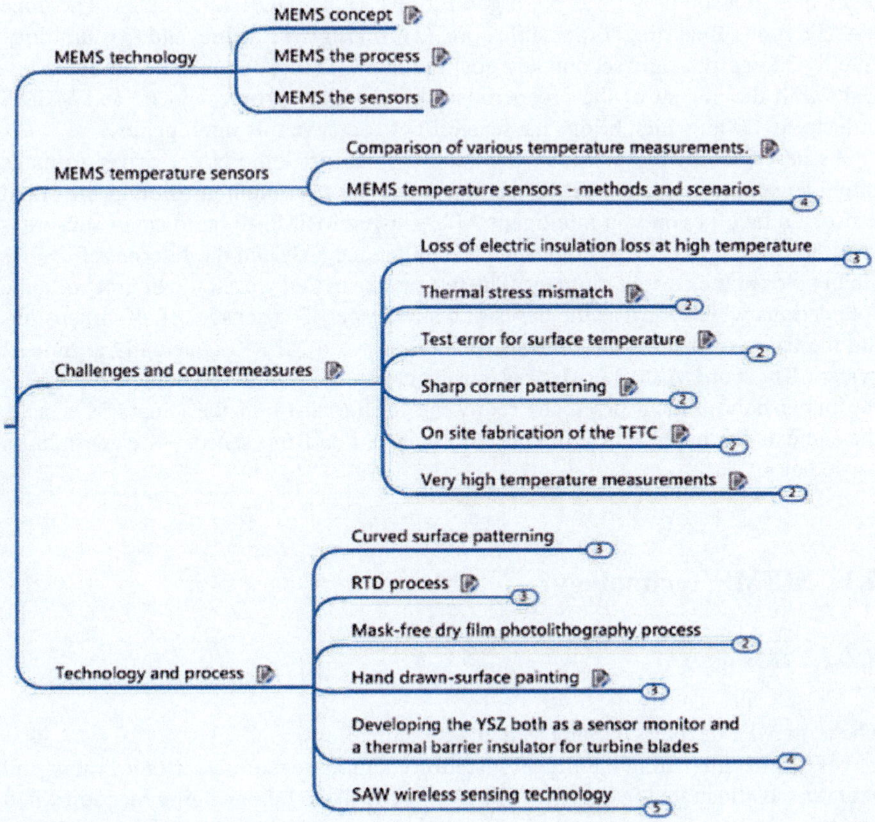

© The Author(s), under exclusive license to Springer Nature Singapore Pte Ltd. 2023
F. L. Duan, *When AIAA Meets IEEE*,
https://doi.org/10.1007/978-981-19-8394-8_4

Quick outline of this chapter

MEMS is driven by application and pushed by technology. The integration of the technology into the application is the key for the success:

- *Without the market, you lose the motivation.*
- *Without technology, it is just an empty talk.*

This basic principle is also valid when we apply the MEMS technology to make the smart sensors for gas turbine engines. The feasibility and necessity must be combined together to "make a dream become true"—the feasibility is MEMS, and the necessity is the intelligent engines. We must have the need for smarter engines, and we must have the technology ready for it.

Intelligent aero-engine is one of the three key focuses in the Versatile Affordable Advanced Turbine Engines (VAATE) program [1], which is a re-modeled follow-up plan after the very successful implementation of the previous program Integrated High Performance Turbine Engine Technology (IHPTET), of which AIAA has been a strong proponent as well [2]. IHPTET is concluded in 2005 after significant progress toward very aggressive goals. After that, the new plan VAATE was initiated by AIAA again and sponsored by three big organizations: DOD, DOE, and NASA. The three VAATE focus areas are: (1) versatile core, (2) ***Intelligent Engine***, and (3) durability. Intelligent engine is the second key goal in the VAATE program. The timing is just right with the advent of the prosperous Silicon Valley progress in IC and MEMS industry in 2006, which brings the feasibility to achieve this intelligence.

As described in the previous chapters, the sensor is the No. 1 driver to make intelligent machine, and MEMS technology is the best method to build the smart sensor. In fact, not only in intelligent AIAA, using MEMS to build smart sensors is also the frontier of current hot artificial intelligence (AI) and the Internet of Things (IoT) fields at the current moment. The reason lies in that we should at first build the foundations which form as the bases of a skyscraper. The sensors of all dimensions are the bases of the AI. In addition, the coined word MEMS is actually a mini AI system. The word MEMS = Microelectromechanical System: mechanical represents the human hand as a sensor, electro represents the human brain, the actuator is muscle, the same as the process of artificial intelligence, i.e., from sensor → electric brain → actuator.

4.1 MEMS Technology

4.1.1 Concept

The term MEMS refers either a system or a technology.

MEMS itself is a mini intelligent system containing sensors, electronic brains, and actuators as illustrated in Fig. 4.1. The analogy is like a human being reaction when his hand touches the fire. When the hand is touching the fire, he feels the heat—this is the sensor. The signal is transferred into the electronic brain, which processes the signal and responds to its arm muscle to retract his hand—the actuator. MEMS is a mini AI since the device is just several millimeters or smaller [3], that is why we it is a micro-system.

MEMS technology was derived from the conventional semiconductor manufacturing technology (IC technology in Silicon Valley), with the added techniques such as deep etching, LIGA, non-illion micro-machining, and additive manufacturing [4, 5].

MEMS technology which is mainly borrowed from IC technology contains the following four basic techniques:

- Draw: create a pattern by photolithography.
- Add: deposit a thin-film layer, or by epitaxy, thermal growth, etc.
- Remove: etch away the layer from the substrate by wet etching, plasma etching, ICP, CMP, lift-off, etc.
- Modify: changing the composition of the substrate/layer by doping, chemical reaction, alloying, etc.

The photolithography technique is a basic process to form a photoresist (PR) pattern by exposing it to UV light under a mask. The mask is usually patterned with black and white, while the black portion blocks the UV light from penetrating through. The exposed PR will be dissolved using developer; hence, the unexposed portion of the PR will stay on the substrate as a protective layer for the following processing. If the PR is negative type, the result is opposite.

The addition technique refers to the thin-film formation on an existing substrate. The criteria of this thin-film formation include uniformity, mechanical properties, defect density, step coverage, deposition rate, etc. The thin-film materials include

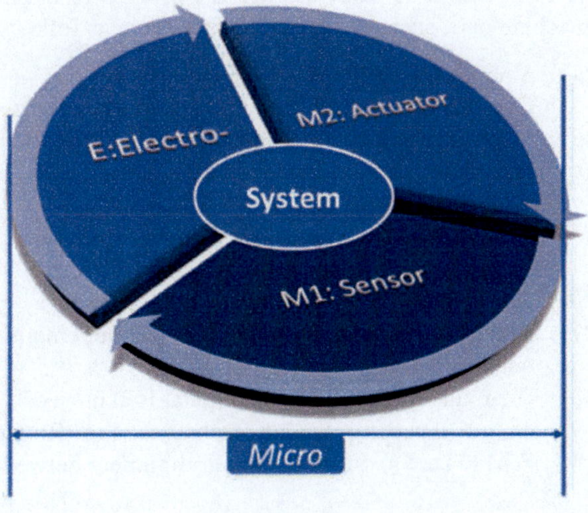

Fig. 4.1 Illustration of MEMS concept

metals, ceramics, semiconductors, etc. The thin films can be used as sensing layers, insulators, barriers, etc.

The removal techniques include wet etching and dry etching processes. Damascene and the lift-off process are two variants of the etching techniques, which cleverly sequence the "adding, removing, and photolithography".

The modification technique is to form a functional layer by changing its properties which are different from the substrate while the thickness is unchanged. One popular example is semiconductor doping by ion implantation to form a PN junction under the blockage of the patterned PR. During the process, only the property of the material is changed, but the thickness is unchanged.

For a detailed reference of the mature IC technology, please refer to our book [6].

4.1.2 Process

The basic MEME process is also similar to IC technology. The main difference lies in that IC process which is mainly performed on a very smooth (Ra is ~ nm) thin (~ 200 μm) flat silicon wafer (~ ϕ300 mm), while most of the MEMS process is to create a pattern on the complex curve surface, such as on a turbine blade surface. The conventional technology should be modified, including various equipments/methodologies. For example, the photoresist (PR) coating is different. In traditional IC technology, the PR is coated by spinning the silicon wafer to make an even thin PR layer. However, this method is not applicable to applying photoresist on curved turbine blade surfaces. Special techniques are needed to achieve photoresist coatings, such as dry film or hand paving. Some IC processes still apply in MEMS. For example, after the PR coating, it needs hard baking in the oven. Such step is always needed both in IC and in MEMS processings.

The general steps of the MEMS process to build a sensor on the surface of a machine part, such as the turbine blade, are as follows:

(1) A photoresist (PR) is coated on the surface.
(2) The PR must be prebaked in an oven to let it dry and firm.
(3) A soft mask with the designed patterns is applied onto the PR surface.
(4) The whole machine part (the turbine blade with the PR and the mask) is exposed to UV light.
(5) The exposed part is developed in the special solvent and then hard-baked in an oven.
(6) Thin metallic film is deposited by a magnetron sputtering machine (PVD).
(7) The lift-off process is used to remove the remaining PR by the acetone solution aided by ultrasonic vibrations.
(8) Then, one layer of patterned metal is achieved.
(9) If multi-layers are needed, such as the Pt/PtRh thermocouple sensor, repeat the steps (1) to (8) with the proper alignment between the layers.

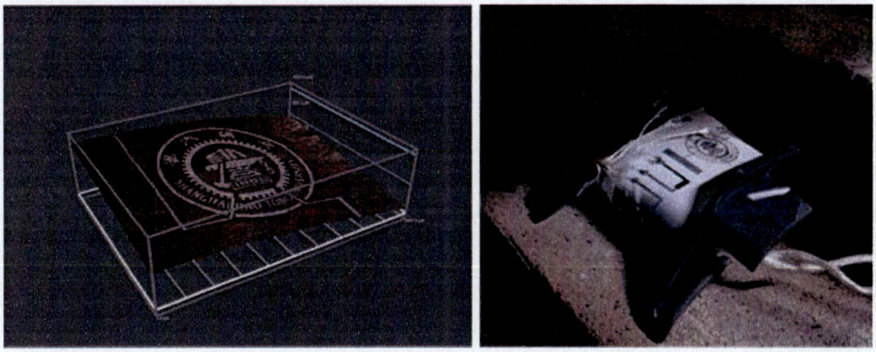

Fig. 4.2 Integrated resistor temperature detector (RTD) by patterning the Pt thin film on the turbine blade surface

Figure 4.2 shows a successfully built RTD sensor on a turbine blade by tailing the school logo of Shanghai Jiao Tong University. The Pt thin-film thickness is just 1 μm, which brings little difference to the air flow disturbance during the engine operation.

There are various sensors to test the temperature, pressure, stress, flux, etc., and there are various special occasions in the aero- and astro-system. In the following, we just discuss the temperature sensing for a gas turbine engine, serving as a case study for other sensors/scenarios.

4.1.3 MEMS Sensors for AIAA

Integration of temperature sensors on the surface of the turbine blade is an urgent requirement for aero-engine intelligence. The USA implemented the IHPTET program from 1988 which mainly relies on the advantages of materials, mechanical and thermal dynamic propulsion engineering. Starting in 2006, with the vigorous development of micro-electronics and computer technology, IHPTET's follow-up plan was replaced by the VAATE plan, focusing more on commercial and intelligent advanced turbine engines. To achieve the "intelligence" of aero-engines, it is necessary to embed a series of thin-film sensors on the surface of the aero-engine parts, such as the turbine blade, in replacement of the traditional chunky sensors. Traditional sensors are not good enough because of factors such as large size, heavy weight, and single function [7]. It is difficult to install traditional sensors in a position where the aero-engine needs to be monitored, and the new miniature sensors developed by MEMS technology are small in size and light in weight. They can be arranged easily in many parts of the aircraft engine to obtain more accurate engine operating state parameters, which is beneficial to improve the performance, maneuverability, and reliability of the engine. This is the core content of a versatile and economical advanced turbine engine (VAATE) plan.

Historically, the development of the MEMS is a "technology push and application driven", i.e., the technology is the vendor and the application is the customer. As to which one comes first, there are two angles in the MEMS endeavor: starting from the AIAA's specific need and finding the proper sensor for it or developing a sensor and then finding its playground in the AIAA's scenarios. From the IEEE peoples' perspective, they certainly belong to the second category. What the IEEE people need to do is to learn AIAA scenario and then modify their methods to fulfil the targets when facing the incoming challenges in AIAA such as super high-temperature testing which they never meet before. Their relationship is described in Fig. 4.3.

Integration and adaptation of MEMS methods are needed to conquer the challenges in various AIAA scenarios. Basically, four types of sensors are developed, pondering the pros and cons of various approaches before selecting the right method. The challenges, including thermal stress, electric insulation, in situ built (on-site built), complex topography (sharp corner), and robustness tests, must be properly dealt with by the modified MEMS process to fit the various AIAA application scenarios.

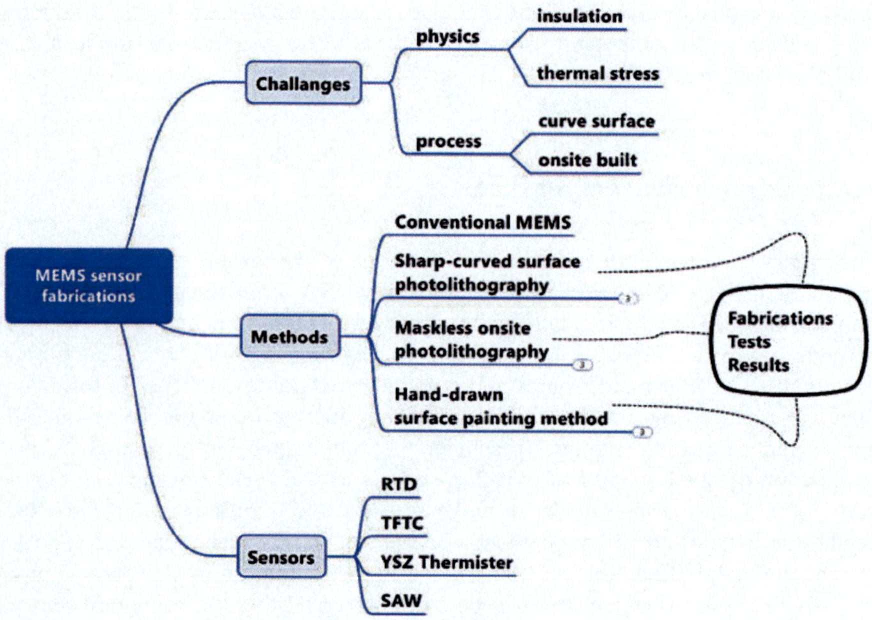

Fig. 4.3 MEMS sensors to serve the AIAA's purpose

4.2 MEMS Temperature Sensors

MEMS technology can be used to build many smart sensors for AIAA, and the most important one is temperature sensor, which is crucial and determining factor for most of the engine performance/safety [1, 8, 9]. The higher the temperature is, the higher the efficiency, but the turbine material cannot tolerate; therefore, cooling is necessary. However, cooling reduces the efficiency. To maintain a proper balance between the two, accurate surface temperature information of the turbine blade is needed to achieve the optimal efficient design and operation.

4.2.1 Comparison with Other Methods

There are various surface temperature measurement methods, such as the optical temperature test [10], temperature paint [11], infrared temperature imaging [12], temperature crystal [13], RTD (resistor thermal detector), and thermocouple [14]. Table 4.1 lists various high-temperature measuring methods. The italics text indicates the special drawbacks of the method and the boldface texts are the advantages features of that specific methods. The comparison mainly considers the feasibility and capability, i.e., criteria such as the upper temperature limit, accuracy/precision, repeatability/endurance. Feasibility usually refers to wired versus wireless, contact versus non-contact, easiness of installation, ability to build sensor array, etc. In terms of capability, it is mainly about the measurement error, the size of the sensing point, repeatability, etc. Each method has its own pros and cons, and one has to adapt its use in accordance with its application scenario. For example, the infrared thermoimaging method is very powerful to characterize surface temperature distribution. Yet, it is much less accurate and more dependable to the test ambient (the combustion color, the combustion air). Thus far, electronic thermal sensing, i.e., the use of the thermocouples/RTD method is still the most accurate way of temperature measurement. Both methods can accurately measure the temperature, while the thin-film thermocouple can be made very small on a precise point to ensure both the measurement accuracy and precision as compared to the RTD sensor (the size is larger).

Among them, thin-film MEMS TFTC sensor is a smarter and more advantageous than the others for its precise and accurate, prompt and real-time, enduring and reliable features and will be mainly addressed in this chapter. SAW sensors are also developed due to its unique feature of wireless sensing, since most of the wiring of sensors in AIAA scenarios is tough to achieve.

Table 4.1 Various temperature measuring methods

Temperature measurement technology	Signal connection mode	Measurement error	Repeatability and stability	Responsiveness	Easiness to install	Sensing point size	Sensor array possibility
Optical testing	Non-contact	*Large*	Good	Fast	Easy	Medium	Medium
Infrared imaging	Non-contact	*Large*	Good	Fast	Medium	Small	Easy
Temperature paint	Non-contact	Fair	*One-time*	NA	Easy	Large	Easy
Thermometric crystal	Non-contact	Fair	*One-time*	NA	Medium	Medium	Not easy
Wired RTD	Wired	Small	Good	Fast	Not easy	*Large*	*Not easy*
Wired thermocouple	Wired	Small	Good	Fast	Not easy	*Large*	*Not easy*
Thin film RTD	Wired	Small	Good	Fast	Medium	*Large*	Easy
Thin film thermocouple	Wired	**Accurate**	Good	**Fast**	Medium	**Small**	**Easy**
RCL wireless sensor	Wireless	Small	Good	Fast	*Not easy*	*Large*	Not easy
SAW sensor	**Wireless**	Small	Good	**Fast**	Medium	**Small**	**Medium**

4.2.2 MEMS Methods and Scenarios

Precise Versus Accurate—TFTC Sensors

TFTC sensors can fulfill both the accurate and precise needs for surface temperature sensing. The concepts of precision and accuracy are illustrated in Fig. 4.4. The precision means that the sensor is capable of measuring the temperature on a very tiny spot. The accuracy is that the sensor's testing error is very small, i.e., its repeatability is very good.

TFTC stands for thin-film thermocouple, mainly based on the principle of the thermocouple device (TC). The principle of the TC is based upon the Seebeck effect discovered in the nineteenth century (Fig. 4.5), which basically says that an electric voltage is formed at the thermocouple joint across two electrically conducting materials when there is a temperature difference between them [15].

Traditional thermocouple devices are two wires covered with a protective shell to form the armored shell. Depending on the two different metal materials, R-type and S-type thermocouples are most frequently used to measure high temperatures. The armored TC is bulky and requires a comprehensive implementation process when used in real high-temperature measurement scenarios. Such a TC is too chunky to measure the temperature accurately and precisely, which will be elaborated in Sect. 5.3.

Instead of using two solid wires, TFTC uses two layers of thin films, and the crossing joint of the two layers forms a TFTC. The TFTC pattern is created by using the MEMS process and contains photolithography, thin-film deposition, and lift-off

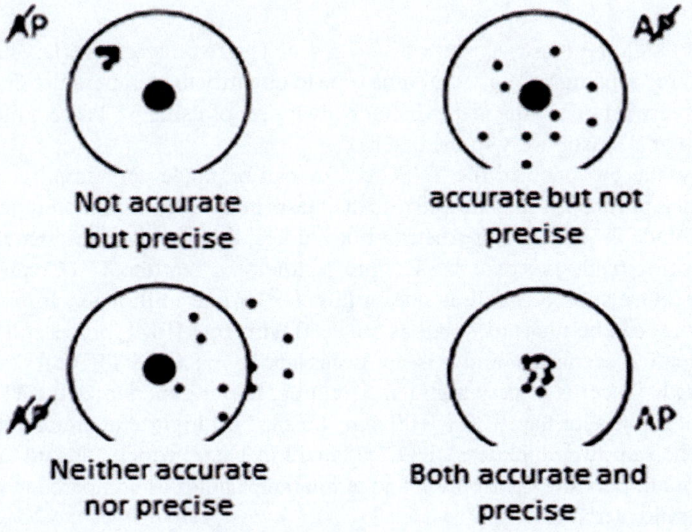

Fig. 4.4 Illustration of the accuracy and precision

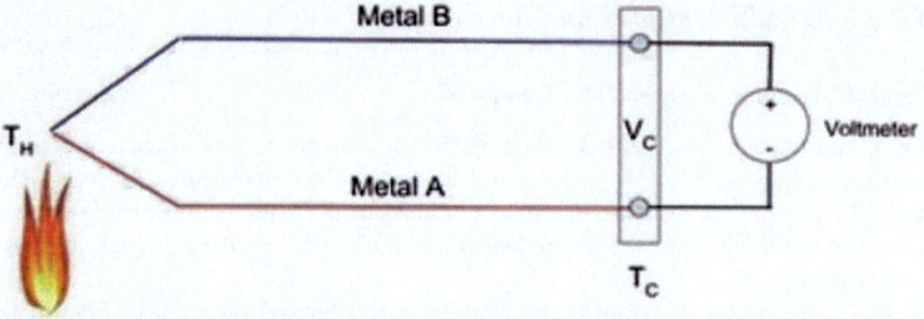

Fig. 4.5 Principle of the Seebeck effect and a thermocouple device

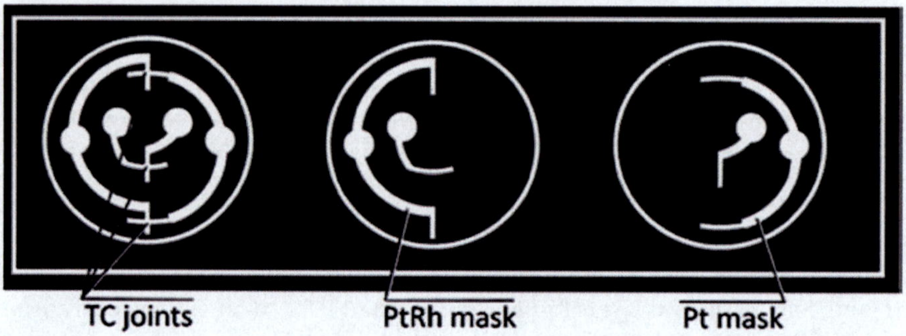

Fig. 4.6 Masks of the two layers of TFTC together with the array of the sensing points

processes using the mask, as shown in Fig. 4.6. The two layers overlap each other, forming three sensing points at the same time to quantify the temperature distribution along the vertical line. This is the distinct advantage of using TFTC to build a series of sensors or a sensor array in the batch.

Clearly, the precision of the TFTC device can be made very small at a distinct point, which plays an indispensable role in measuring the multi-point temperature on a turbine blade for temperature distribution studies. Figure 4.7 shows that the TFTC on the turbine blade is as small as 1 mm^2, indicating that the TFTC can quantify the temperature spot as small as one millimeter by one millimeter. In principle, a sensing joint can be made as small as $50 \times 50\ \mu m^2$ by MEMS process. It is just in real application scenarios, and it is not necessary to make the TFTC too small due to reliability concerns at very high temperatures. In our experience, the TFTC joint should not be smaller than $100 \times 100\ \mu m^2$ for the best high-temperature endurance quality. The thin-film thickness of TFTC is within 1 μm, which adds no turbulence for surface temperature distribution and temperature fields as compared to the wired thermocouple methodology.

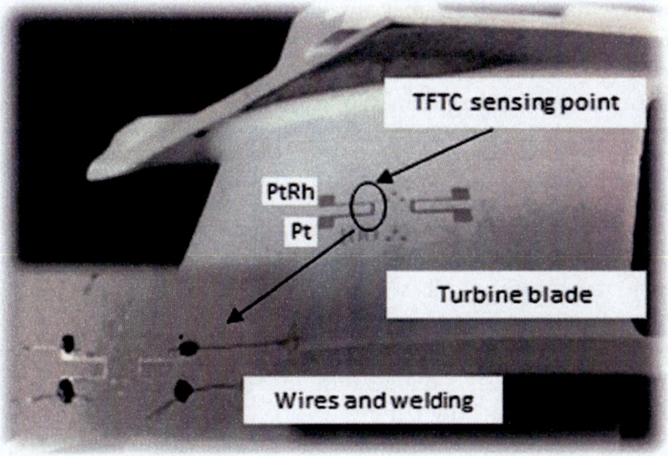

Fig. 4.7 TFTC sensors embedded on the turbine blade, thermocouple joints, sensing point, wires, and welding

BTW Sensor—The Thermistor

The YSZ thermistor is a by-the-way temperature monitor since it combines the features of TBCs both as a thermal barrier and as a high-temperature monitor. TBC is a popular thermal barrier coating layer used in aero-engine parts, such as nozzle guide vanes in combustion chambers. TBC improves the reliability of the super-alloy metallic parts and enables a higher operating temperature for enhanced aero-engine performance. Overcoming the thermal–mechanical mismatch and thermal stress between the TBC layer and metal substrate is used to be the key issue and has been successfully resolved in many years of efforts [16]. By using yttria-stabilized zirconia (YSZ) coating materials together with advanced coating technologies, YSZ-based thermal barrier layers have been widely adopted in the gas turbine industry for years. In this following, we present a new sensor by making use of this YSZ layer as a thermal resistor to monitor the temperature in addition to its benefits as a thermal barrier.

Our original motivation was to use the YSZ ceramic as an insulator for an embedded thin-film sensor built on a turbine blade surface in an effort to develop an advanced microelectromechanical system (MEMS) sensor for aero-engines. However, research shows that the resistivity of ceramic materials (Al_2O_3 based) degrades with elevated temperature [17], and our results also indicate that ordinary YSZ material exhibits lower electrical resistance at elevated temperatures. Therefore, the first efforts were made to increase the electrical resistivity of the YSZ layer at higher temperatures. It is indicated that by adding Al_2O_3 to the YSZ coating recipe, the electrical insulation can be improved significantly [18]. In the meantime, the efforts of monitoring the electrical resistance of YSZ with elevated temperature enlightened us with a new idea to make use of its resistance–temperature relationship as a thermal sensor.

By making use of the thermal-electric sensing feature, we can build a BTW YSZ thermistor to minimize the long-term stability of the engine health condition. Enlightened by previous studies on YSZ features, we formulate a new idea: why do we not use this YSZ layer as a temperature sensor to monitor the ambient temperature by making use of its electrical sensitivity to temperature? In the following, the fabrication details, electrical connections, high-temperature measurements on sensitivity, repeatability, responsiveness, and measurement error evaluation of the YSZ thermistor are presented. These new YSZ-based thermistor sensors are a series of soldering dots and are therefore very simple to fabricate on aero-engine turbine surfaces. A new test array can be simply formed based upon these YSZ-dotted thermal resistors, and its behavior as a temperature distribution monitor was analyzed via thermal-electrical computer simulation.

As shown in Fig. 4.8, we found that the YSZ resistance exhibits a log linear relationship with temperature. Therefore, we can use such feature to monitor the temperature by real-time monitoring of the YSZ thermistor. What we need to do is to use the Pt paste to pave two PC wires on the surface of the TBC and measuring is resistance versus temperature. Such sensor needs a one-time annealing and calibration prior to its infield use later on. This is a very stable and simple sensor to monitor the long-term temperature of the turbine system.

Simple All Purpose—RTD Sensor

An resistance temperature detector (RTD) sensor is a general purpose sensor that is very well adopted commercially. A thin-film RTD has the advantages of high sensitivity and fast thermal response because its thin layer later decreases the thermal exchange between the sensitive element and environment compared to the traditional RTD [19]. Unlike the TFTC, its disadvantage is the lack of enough precision since most of the RTD surface dimensions are too large to measure the exact temperature on a certain spot. Its advantage comparing to the TFFC is that to build a RTD sensor is easier since just one layer of metal is needed. Among the various materials, platinum (Pt) is usually chosen as the preferred material for its good thermal

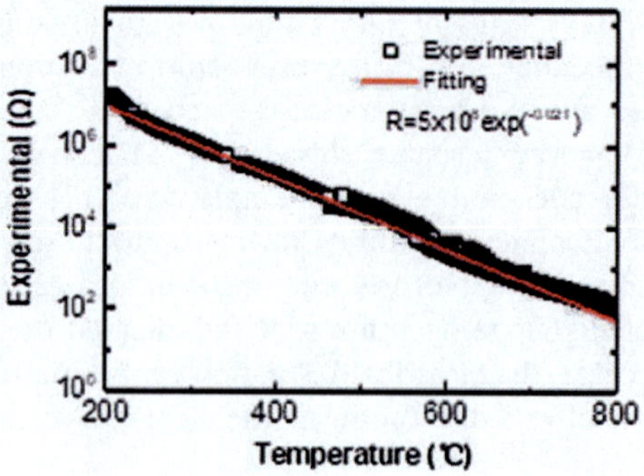

Fig. 4.8 YSZ resistance after a few temperature cycles. The reading glitch is from DMM switching the range from Ω, kΩ, and MΩ

response and stability. Pt exhibits a positive and highly linear temperature dependence [20, 21]. Generally, an intermediate layer between the Pt layer and the substrate is often used to generate physical interlocking, interdiffusion of the films, or chemical bonding between the film and its substrate to improve adhesion [22]. Titanium (Ti) and tantalum (Ta) are commonly used as adhesion layers for intermediate-to-high temperature applications [23].

Figure 4.9 shows the embedded RTD sensor built on the turbine blade surface. There are two kinds of RTDs: the meandered line thread and the circled ring saw-type RTD pattern modified using the university logo. The sensors are made with a curved surface MEMS process that we will describe in detail later. Sensors are then connected to the Pt wires through the bonding pads. These wires are then connected to outside multiple meters for electrical measurements. Usually, the ordinary multi-meter is good enough to ensure the testing accuracy of the thermal resistance measurement.

Wireless Wi-Fi Sensing—SAW Sensor

Wireless passive sensing is unique to monitor the temperature for a moving object, such as the rotating turbines. SAW (surface acoustic wave) sensors are ideal candidates for wireless passive Wi-Fi sensors to monitor the working status of rotating aero-engine parts in the propulsion system. Although we have developed TFTC sensors to accurately measure the surface temperature on turbine blade, the wiring

Fig. 4.9 RTD sensor patterned by a modified university logo on the turbine blade

connection under high-temperature/pressure ambient is not as convenient. To balance the capability and feasibility, a surface acoustic wave (SAW) sensor is a necessary alternative and its wireless sensing is extremely suitable for aero-engine propulsion test scenarios such as the rotating turbine blade, shaft, and disk where it is difficult to lead out the wires [24]. The SAW sensor uses the temperature sensitivity of the RF resonance frequency f_0 of the piezoelectric material to achieve temperature sensing purpose and uses the reverse piezoelectric effect to transmit the SAW signal wirelessly to the external equipment for wireless sensing [25]. Although SAW is commercialized in various fields, such as monitoring temperature in rotating wheels in a car, SAW sensing at high temperatures is still a challenge (not exceeding 450 °C [26]). It is our target to use the SAW device for wireless high-temperature measurements and monitor the aero-engine infield conditions.

There are two main challenges to achieve this passive/wireless sensing: (1) developing a reliable high-temperature SAW sensing device and (2) wireless communication between the SAW sensor and external RF test equipment. This is pretty comprehensive process involving a lot of professionals. As shown in Fig. 4.10, high-temperature SAW-based wireless passive sensing technology involves high-temperature piezosensing materials [27], MEMS micro-fabrication [28], and high-frequency characterization under ultra-high temperatures—a very challenging project with a large cross-disciplinary range.

The whole wireless sensing development work includes two parts:

(1) *To develop a reliable SAW sensor capable of measuring high temperatures.*

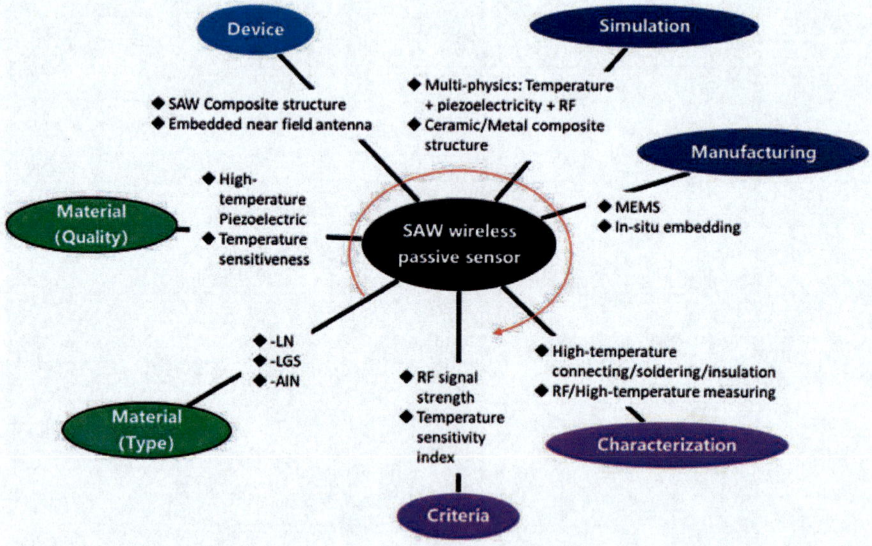

Fig. 4.10 Complex matrix of SAW-based high-temperature wireless passive sensing

(2) *To establish a wireless connection with outside RF equipment for Wi-Fi signal receiving and transmission.*

For the first goal, a reliable SAW sensor was successfully fabricated using MEMS technology on a $LiNbO_3$ (LN) substrate with a platinum (Pt) interdigital transducer (IDT [29]) electrode—it is an inexpensive and accessible piezoelectric material, and the MEMS manufacturing process cost is fair low. In addition, LN material has a high Curie temperature (~ 1133 °C), which makes it possible to measure the temperature up to 1100 °C for most scenarios of aero-engine testing. Historically, SAW sensors on LN substrates are only operable at 400 °C due to dewetting effects [26] and recrystallization [30]. In our work, this record is broken with the help of a SiO_2 passivation layer to protect the Pt electrodes and LN substrate from the high-temperature ambient. We have successfully designed and built an SAW device capable of measuring a temperature of 1100 °C, breaking through the previous record of an upper temperature limit of 450 °C using $LiNbO_3$—a popular and economic piezoelectric substrate. SiO_2 passivation is the key technique for this breakthrough. The test result is shown in top right: the RF resonant frequency f_0 decreases consistently with increasing temperature. We also observed that the resonant signals degrade with temperature, although the f_0 signal is still detectable, which is an indication of the piezoelectric quality degrading as the temperature rises. The results indicate that high-temperature SAW sensing becomes feasible by using a conventional LN piezoelectric substrate material together with Pt thin-film IDTs.

The second goal is to achieve the Wi-Fi transmission of the sensing signal and ensure sufficient transmission distance and RF signal disturbance tolerance. This work is still undergoing with a lot of challenging issues which must be resolved. Strong communication will be established through a near-field antenna attached to the SAW sensor to enhance the wireless transmitting capability. Figure 4.11 illustrates wireless test system. In Chap. 5, we will elaborate this in detail.

4.3 Challenges and Countermeasures

High-temperature sensors for smart aero-engines are very different from conventional sensors in the following ways:

(1) *Electric insulation*: The sensor is an electronic device and must be built on an insulator. Therefore, to build a thin-film sensor on the metallic surface, a thin-film electric insulation layer is needed. Ceramic materials are used as this insulation layer. However, most of the ceramic loses its electric insulation at higher temperatures, causing the shortening of the thin-film sensors built on its top. This issue must be properly dealt with when building a thin-film sensor on a turbine blade surface.

(2) *Thermal stress*: When adding this ceramic layer as insulation, the risk of thin-film peeling under high temperature appears due to the different temperature expansion coefficients of different layers. A thin-film sensor fails either when the

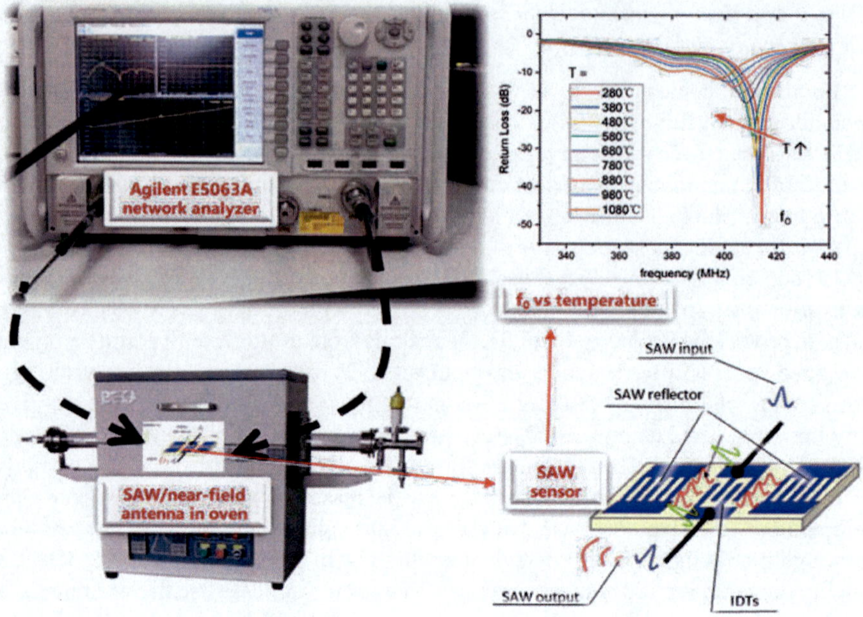

Fig. 4.11 Schematic plot of the wireless sensing systems using the SAW device and infield antenna

ceramic or metal layer is peeled off from the substrate under high-temperature stress. Proper countermeasures must be taken.

(3) *Measurement error*: The temperature measured from the thin-film sensor built on the ceramic is not exactly the temperature of the Ni–Cr alloy surface—there is a temperature difference between the top and bottom of the ceramic layer. This error must be properly taken into account or avoided.

(4) *Complex surface*: For example, the ridge and the root of the blade or disk in gas turbine engines are two tough locations to implement conventional MEMS process to build sensors. Other methodologies must be formulated.

(5) *On-site built*: In the traditional MEMS process, the substrate or the machine part must be brought into the MEMS laboratory to perform various micro-fabricating processes. Some engine parts are just too large and inconvenient to be carried into the MEMS laboratory; therefore, in situ methods must be used to pattern the MEMS sensor infield.

(6) *Ultra-high-temperature testing*: Advanced gas turbine engines favor higher operating temperatures by adopting the CMC materials as the turbine blade. Embedding the thin-film sensor on its surface is even more challenging to function and survive the tough high-temperature anomaly.

These AIAA requirements bring much higher demands and newer challenges to building MEMS sensors for high-temperature measurement in comparison to conventional room temperature MEMS sensors. Below are the detailed situations and the countermeasures. These challenges can be classified into two categories: one is from the tough need of AIAA scenarios, and the other is from the processing feasibility of MEMS in IEEE.

4.3.1 Electric Insulation at High Temperature

Fabricating sensor devices on metallic components requires the use of an electrical insulating ceramic layer in between. Ceramics are well-known electrical insulators for many general applications [31]. However, most ceramics lose their electrical insulation with increasing temperature. At temperatures above 600 °C, most ceramic materials become electrically conductive [17, 32], although they may still hold good thermal insulation properties. In recent years, the use of ceramic thin films as electrical insulation for sensors has become one of the bottlenecks in intelligent engine [33, 34] applications. Intelligent turbine engines require new sensor technology for very high-temperature on-site characterization on the surface of turbine blades [8, 35–37]. The reduced resistivity of the insulator with increasing temperatures causes short path conduction, which brings about measurement error [38]. High-temperature electrical insulation therefore needs to be studied to build a successful sensor for high-temperature applications above 1000 °C. Previous research work on high-temperature insulation layers mainly used Al_2O_3 thin films deposited on metals by PVD, focusing on their high-temperature insulation capabilities [39, 40]. However, this thin film will be easily peeled off during the temperature cycling test due to thermal stress mismatch between the ceramic thin film and the metal [41, 42], resulting in an unrepeatable measurement result that cannot meet the need for sensors. Another way to make an insulation layer is using mature thermal barrier coating (TBC) technology. TBC is a well-established surface coating technology to protect the Ni–Cr turbine alloy [43–46]. It is a revolutionary material improvement that greatly enhances the overall high-temperature performance of the aero-engine, equivalent to efforts in the past 20 years to improve the high-temperature tolerance of turbine blade alloy materials [47–51]. The advantage of using TBCs as an electrical installation lies in the fact that TBCs are a well-developed technology, especially the thermal stress and mismatch issue between ceramic and metal materials, which have been successfully resolved after many years of effort. However, the insulation of the TBC also decreases at high temperatures, which will result in the sensor not working properly [51]. In our following efforts, the electrical insulation of TBCs was studied intensively and improved by adding Al_2O_3 to the coating recipe. The results indicate that the high-temperature electrical insulation of YSZ-based TBCs with added Al_2O_3 exhibits electrical resistance that is a few orders of magnitude higher than that of conventional YSZ, which can meet the need to build smart thin-film sensors on top. Thermal resistors together with the embedded Pt/PtRh thermocouple were fabricated on the surface of modified TBCs, and characterization was carried out to study the high-temperature electrical properties with different Al_2O_3 dopings.

Experiment

The high-temperature electrical insulation property is studied by measuring the electrical resistance with two electric nodes on the ceramic insulation films. Ni–Cr turbine blade material was chosen as the base metallic substrate onto which we coat the ceramic insulation layer. A zirconia-based YSZ thermal barrier coating (TBC) film is a well-known ceramic often used for aero-engine applications [52] and is chosen as our insulator to build our sensor. Since the original recipe for TBC coating is mainly for thermal barrier purposes, its electrical insulation may not meet our requirement for our sensor application. We therefore modified the ceramic coating recipe to meet our electrical isolation purpose while still meeting its thermal barrier purpose. Below are the two kinds of TBC ceramic thin films for thermal plasma spray:

- Case 1: Ordinary YSZ (8 wt.% Y_2O_3–92 wt.% ZrO_2) with a high porosity up to 20%.
- Case 2: A mixed YSZ (YSZ—90 wt.% + Al_2O_3—10 wt.%) with much less porosity (less than 5%).

The ceramic coating process was as follows: The metallic substrate was first grit blasted with alumina powder (grit 60) prior to spraying. The TBC was coated by thermal plasma spray. During spraying, argon and hydrogen were used as primary and secondary gases, respectively, and a 45 kW Metco thermal plasma spray unit was used to apply the coatings. For YSZ-based ceramic thin films, a high arc current is needed to sufficiently melt large particles before they are deposited onto the substrate. The spray distance also affects the solidification of the melt before it reaches the substrate, and a low spray distance is therefore maintained as 100 mm. The target thickness of the YSZ film is 500 μm.

The coated ceramic surface on the metal is usually very rough, with a surface roughness of approximately ± 50 μm. Such a rough surface needs polishing before any MEMS sensor device is fabricated. The original YSZ ceramic insulator layer was mechanically polished with the finest grade sandpaper (Matador Starcke 7000) until a surface roughness less than 0.1 μm was reached. A Pt metal film with a thickness of 500 nm was deposited on the polished ceramic surface by the magnetron sputtering method. A Z550 Sputtering System (Leybold Heraeus, Germany) was used with a sputtering power of 600 W and a chamber pressure of 2×10^{-4} Pa. Pt thermal resistors were patterned using photolithography technology. The lift-off process in micro-structuring technology is used to create the shape of the connecting pads and thermal thread resistor (Fig. 4.12, left) after photoresist removal with acetone in the ultrasonic wet bench.

Pad electrodes were electrically connected to the Pt wires by platinum paste with a solid mass fraction of 83% and then hard-baked at approximately 800 °C for 10 min. The Pt thermal resistor together with two separate pads for ceramic insulator resistance measurement was connected by two pairs of thin Pt wires (Fig. 4.12, right) to the multi-meters (Keithley Model 2000) for resistance characterization. The samples were sent to a high-temperature oven with ramping temperatures from room temperature to 1300 °C. The ramping speed is approximately 1000 °C per hour. The

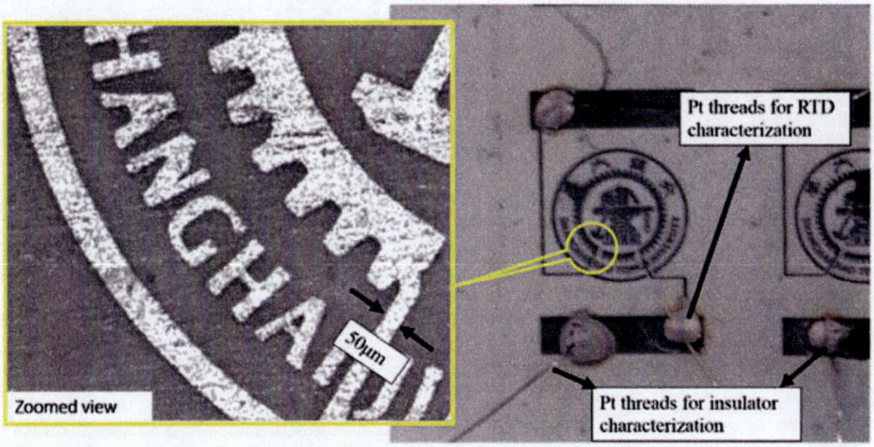

Fig. 4.12 Left: Pt (~ 50–100 μm line width) thermal resistor device (RTD) modified by the toothed wheel ring in the University Logo. Right: thermal Pt resistor and ceramic insulator connection by two pairs of 0.1 mm-thick Pt wires for characterization

sensor and insulator resistances are monitored and recorded during the whole process of temperature ramping.

Figure 4.13 shows high-temperature measurements of the MEMS sensors by the outside multi-meters, which were equipped with a data transfer system to a computer for data recording.

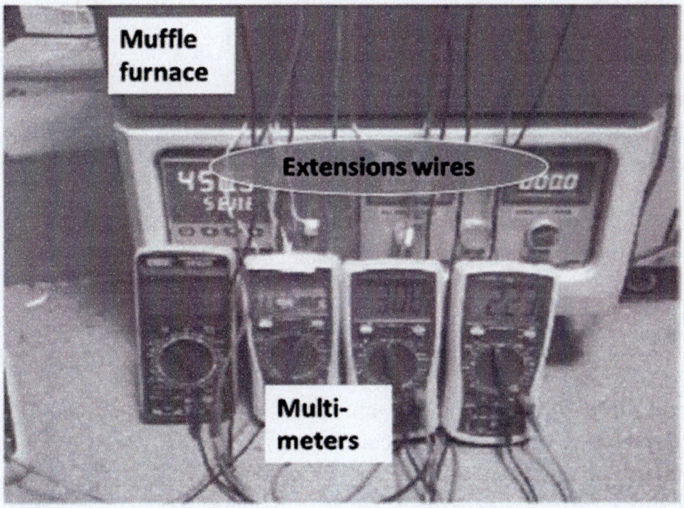

Fig. 4.13 Extension wire to the multi-meter outside the furnace the whole process of high-temperature ramping is monitored

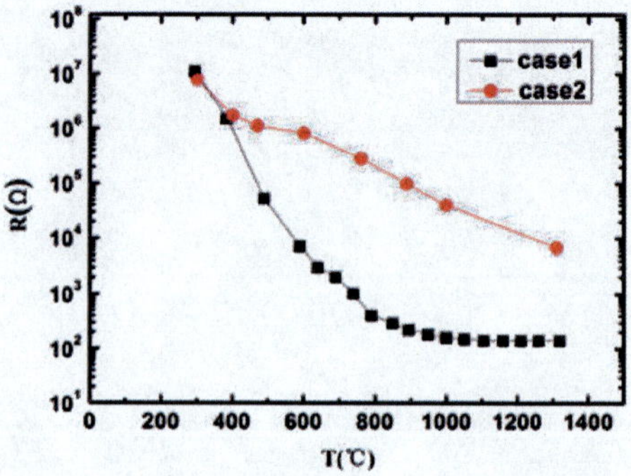

Fig. 4.14 Resistance–temperature of the two different ceramic insulators. The Case 2 YSZ shows a much higher resistivity under high temperatures (up to 1300 °C) than Case 1

The Results

As shown in Fig. 4.14, the Case 1 YSZ ceramic insulator becomes conductive when the temperature is higher than 600 °C, while for the Case 2 ceramic insulator with the dense YSZ spray process and added Al ingredient, the resistance of the ceramic insulator is high enough (over 10 kΩ) to isolate the sensor devices at temperatures up to 1300 °C. In general practice, the aero-engine industry may use the ordinary YSZ in Case 1 mainly for thermal barrier purposes. However, using this layer may not fit the electrical isolation purpose for sensor applications. Case 2 YSZ is suggested as an electric insulation layer to build our thin-film sensors to ensure the enough insulation capability above 1000 °C.

In the above endeavor, we are able to successfully build a good ceramic insulator under high temperature (> 10 kΩ at 1300 °C) with the modified YSZ recipe. This insulator can be used to isolate the top sensors from the metallic substrate. Figure 4.15 shows the measured Pt thermal resistance as the temperature increases to 1300 °C. For Case 1's YSZ, the Pt thread resistance–temperature curve is linear, as expected at low temperatures. However, the resistance drops and starts to deviate from the linear dependency at temperatures above 600 °C. For the ceramic insulator with added alumina and with much less porosity (Case 2's YSZ), the Pt thermal sensor can maintain a very good linearity until 1300 °C.

The deviation from the linearity in the sensor's electrical–thermal relationship in Case 1 can be attributed to the parallel resistance induced by the ceramic insulator under high temperature. As shown in Fig. 4.15, the resistance of the insulator in Case 1 YSZ ceramics starts to drop to the 1000 Ω range when the temperature increases to 600 °C. The ceramic beneath the Pt pads starts to join the electrical conduction from 600 °C, which causes the parallel conduction paths from inside the ceramic insulator and through the base metal substrate. This parallel conduction will cause the error to the top Pt thread resistor conduction and deviate from its linearity. The combined resistance of the top Pt thermal resistor is:

$$R = \frac{1}{\frac{1}{R_1} + \frac{1}{R_2}} \tag{4.1}$$

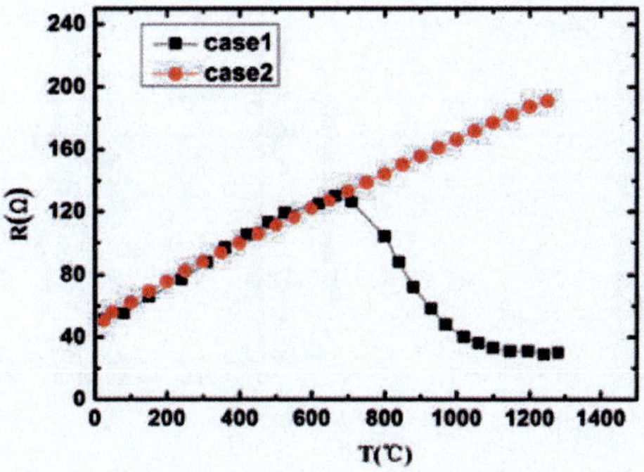

Fig. 4.15 Resistance versus temperature behaviors of the RTD sensors built on the two types of YSZ layers. The linearity of the RTD sensor is lost for the Case 1 YSZ due to the loss of electric insulation above 600 °C

where R_1 and R_2 are the resistance of the top Pt thread resistor and the insulator-related resistance, respectively. The resistance of the Pt thread resistor follows a good linear relationship, as shown in Eq. (4.1). Regarding the temperature dependence of the ceramic insulator, the relationship follows a good exponential trend at the beginning, yet it starts to saturate when the temperature is above 1000 °C. This phenomenon may agree with the fact that the intrinsic conductivity in the YSZ insulator follows a good exponential trend with increasing temperature, while the other resistance may follow a mild linear relationship with increasing temperature. The relationship of the temperature dependence of the YSZ insulator can be empirically fitted as:

$$R_2 = A\mathrm{e}^{-BT} + CT + D \tag{4.2}$$

This will lead to a merely flat relationship in resistance versus temperature dependence in log scale for the ceramic layer under high temperature when the temperature exceeds 800 °C with proper coefficients in the fitting. Since all three conduction mechanisms at high temperatures (Pt, YSZ ceramic insulator, and the base metal substrate) have a weak dependency on temperature, the dependence of the overall resistor between node 1 and node 2 shall have a weak dependence on temperature as well under high temperatures. The resistance of the ceramic insulator and the resistance of the sensor/ceramic/metal sandwich composite structure are shown in Fig. 4.16 with the empirical fitting the following coefficients: $A = 3 \times 10^7$, $B = -0.021$, $C = 4 \times 10^{-3}$, and $D = 133$ in Eq. (4.2). The total combined resistance of the sensor considering the parallel reduced conductivity in the insulator at higher temperatures agrees with the model fairly well, indicating that our estimation model is quite reasonable.

Computer Simulation Analysis

The reduced ceramic electrical insulation at high temperature can be studied by COMSOL Multi-physics simulation tools. Figure 4.17 illustrates the sensor/ceramic/metal composite structures for our computer simulation, while the right chart shows the isopotential profiles and current density distributions when applying a voltage between two electrodes node 1 and node 2; the arrows indicate

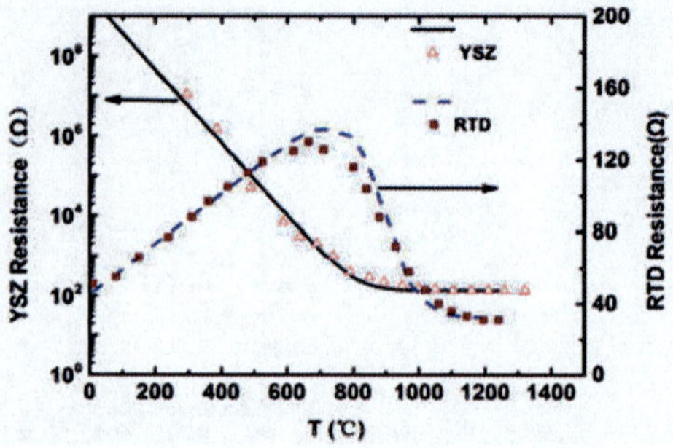

Fig. 4.16 Resistance–temperature curves of reduced conductivity under high temperature for the ceramic (empty triangle) and the sensor/ceramic/metal sandwich structure (solid square). Isolated points are experimental data. The solid and dashed lines are fitted curves for ceramic and sensor resistance

the electric current flowing from the high potential node toward the lower potential node. In this simulation, the metallic substrate (Ni–Cr alloy) is 2 mm thick, and the YSZ ceramic insulator film ranges from 10 to 500 μm. The Pt thin-film thickness is 500 nm, and the thread width and length are set to fit the measured resistance value (~ 50 Ω). Electrical conduction between electrodes (node 1 and node 2) may occur in three paths: via a Pt thread resistor, via a ceramic layer, and via the base metal substrate. At room temperature, the main conduction path is mainly via the top Pt resistor since the ceramic is a good electrical insulator. Assuming *S*, *L*, and ρ to be the cross-sectional area, length, and resistance coefficient of the Pt thermal resister, then the resistance of the Pt resistor can be described as:

$$R_1 = \rho L/S = \alpha T + \beta \tag{4.3}$$

In this formula, we also see the linear dependency of the Pt thread resistor with temperature according to our previous result in the ideal alumina substrate. The coefficients α and β are 0.155 and 32, respectively, where *R* is in Ω and *T* in °C in Eq. (4.3).

Additionally, the accurate *L* and *S* of the Pt thin-film (~ 500 nm) RTD fabricated on the top of the YSZ ceramic film could be difficult to estimate when the surface roughness is still too high when the sensor is built on the top of the YSZ surface (the surface roughness of the YSZ after polishing is ~ 1 μm even after polishing, compared to the ordinary silicon wafer surface roughness in the nm range). However, our close view (not shown here) of the surface topology of the Pt RTD ensures that the whole coverage and connectivity of the Pt thread on YSZ, even with this surface roughness, are also verified by our further electric characterizations.

Under high temperature, ceramic-induced electrical conductance may be caused by:

1. Reduced intrinsic resistivity of the ceramic.
2. Increased electrical conductance at grain/defect boundaries [53, 54].

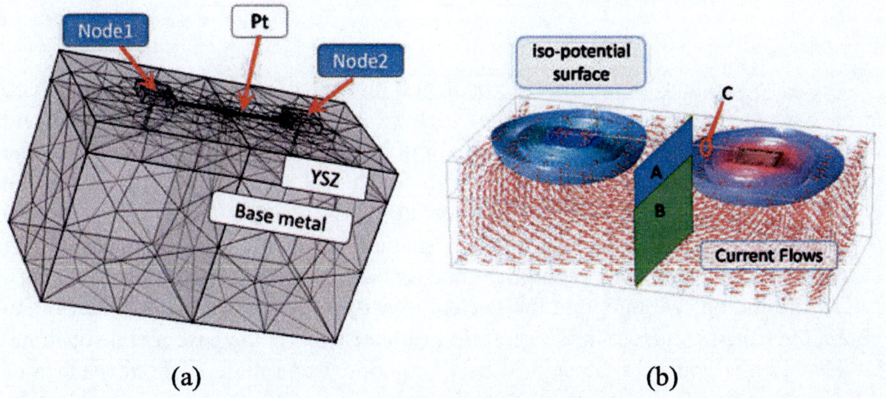

Fig. 4.17 **a** Mesh of the simulation structure of sensor/ceramic/metal composite structures for electrical-thermal resistance simulation. **b** Simulated potential profiles together with the current density distribution when applying a voltage between node 1 and node 2

3. Conduction path via the metallic substrate through the ceramic thin layer.

The reduced electrical resistivity adds the parallel conduction paths to the top Pt thread resistor, which will cause measurement error of the thermal sensor, and needs to be properly addressed. There are two main challenges to evaluate these parallel resistance errors:

1. Conduction paths are irregular in shape in these composite structures; theoretical evaluation using a general formula could be difficult.
2. Different conduction mechanisms have their own temperature dependence due to their different conduction mechanisms (see discussion below).

In our finite element analysis [55] simulation, we are able to analyze the resistance performance of the sensor/ceramic/metal composite structures by interactive electrical-thermal multi-physics models to consider both the electrical and thermal impacts on the electrical conductance at high temperatures.

The simulation was conducted by solving the basic Poisson equation as follows on each grid point illustrated in the mesh to obtain the potential and current density on each grid point inside the ceramic insulator. The mesh is denser near the boundary of different materials because abrupt changes in the potentials usually occur near these boundaries, and more grid points are needed to make the iteration convergent. The basic Poisson equation is:

$$\nabla^2 \emptyset = -\frac{\rho_{\mathrm{f}}}{\varepsilon} \tag{4.4}$$

where ϕ, ρ_{f}, and ε are the electric potential, fixed charges, and permittivity, respectively. For the boundary condition during the simulation, a DC voltage 1 V is applied between node 1 and node 2, and for the rest of the surface, the boundary condition is:

$$\vec{n} \cdot \vec{j} = 0 \tag{4.5}$$

The zero product of normal vector $\vec{n}$ and current density $\vec{j}$ defines that there is no current flowing out of the other surface. The electric potential and current density were then calculated (Fig. 4.17b), and integration was performed across the middle cross-section of A, B, and C to obtain the total conduction current in the ceramic insulator, substrate metal, and top Pt thread resistor (section C, too small to be seen), respectively. The arrows indicate both the flow direction and amplitude of the electric current from the high potential node toward the lower potential spots. In the simulation, we integrate the current density over cross-sections A and B to obtain the flowing current in the ceramic insulator and via the base metal substrate. Divided by the applied voltage 1 V between node 1 and node 2, the resistance of different layers with different conduction mechanisms with varying temperature can be obtained.

In this simulation, electrical-thermal multi-physics models are integrated to consider the interactive effect of electrical properties with increasing temperature. The temperature dependence of the three materials in the sensor/ceramic/substrate structure is integrated in the simulation. The temperature dependence coefficients of the Pt layer and the ceramic layer were obtained from our high-temperature experimental measurements (below). The thermal-electrical coefficient for the standard Ni–Cr-based alloy is from the standard database and incorporated for the base metal substrate simulation.

Detailed Study of Electrical Insulation Improvements of Ceramic Coating

The thermal insulation coating is made by a binary coating system consisting of two parts: a bonding layer and a ceramic layer. The bonding layer is a mixed alloy of MCrAlY (M is Ni or Co), which is usually formed by conventional PVD or vacuum plasma spray processes. The different material components in the alloy enable the matrix and the ceramic layer to have good stress matching and bonding properties under high-temperature conditions of more than 1000°. At the same time, the substrate is effectively protected against corrosion and oxidation. In addition, 6–8% Y_2O_3 mixed with partially stabilized ZrO_2 material, referred to as YSZ, is used to conduct TBC and plays a role in the thermal protection of the matrix body.

A thermal barrier coating was formed on the surface of the circular steel block by a plasma spraying process. The metal or ceramic powder was sent to a high-temperature plasma flame and heated to a molten or highly plastic state. Then, the material was propelled forward by an air current and finally impacted the surface of the substrate with a certain kinetic energy to form a coating and instantaneously solidify. A NiCrCoAlY transition material was deposited on the nickel–chromium-based aero-engine turbine blade material to increase the high-temperature thermal stress buffer between the blade metal substrate and the ceramic layer. The YSZ ceramic material was then sprayed on the transition material by plasma spraying. In the specific spraying process, the surface was first blasted with alumina powder, and then, ultrasonic cleaning was carried out before spraying was performed in ethanol.

The bottom layer was commercially available CoNiCrAlY powder (AMDRY 9951, USA). During the spraying process of YSZ, an atmospheric plasma spraying (APS) technology system was used for spray deposition and adjustment of the spraying parameters. This caused the coating material to be layered in a molten state on the surface of the sample to form a relatively dense coating. The particle size of the powder was 30–70 μm. The spray angle, speed, and other process conditions were adjusted multiple times to make the coating thickness as uniform as possible. The thickness of the buffer layer is approximately 50 μm, and the thickness of the YSZ surface coating is approximately 300 μm. The spray equipment was a Sulzer Metco UniCoat (spray gun F4-MB, Switzerland).

Three TBC formulations were selected as follows:

1. Spray powder formed by a 30% Al_2O_3 + 70% YSZ ratio.
2. Spray powder formed by a 50% Al_2O_3 + 50% YSZ ratio.
3. Spray powder formed by the ratio of 70% Al_2O_3 + 30% YSZ.

Figure 4.18 shows the high-temperature resistance characteristics of different Al_2O_3-doped YSZs. As we can see, as the doping ratio increases, the resistance of the TBC sample also increases. In addition, as the temperature increases, the resistance of all YSZ samples decreases exponentially. When the temperature reaches 800°, the resistance of the 30% sample is 4385 Ω, which cannot meet the requirements for MEMS sensors. At the same temperature, the 50% doping sample is 70,000 Ω, and the 70% sample is 10^7 Ω. According to the result, as the temperature value rises, the resistance value of the 30% doped sample drops sharply, which means that the insulation cannot be guaranteed. Therefore, the 30% doped TBC is not suitable for use as an insulating layer. The 50% doped and 70% doped samples maintain good insulation before 800°, so they can meet the requirements of MEMS sensors. However, in the actual test, the 70% doped sample showed a phenomenon of TBC shedding, which was due to the deterioration of the coating adhesion after the content of Al_2O_3 was increased, which is consistent with the literature report. For the above reasons, 50% doped TBCs are the best choice for MEMS sensors.

In summary, the electrical installation of the original TBC ceramic thin film at high temperatures can be greatly enhanced by adding Al_2O_3 to its coating recipe, which can meet the insulation requirement to build a thin-film sensor on a metallic substrate. High-temperature characterizations of various doped TBCs were carried out by fabricating platinum joint resistors and point thermocouples on the TBC surface. The results show that the 50% doped TBC coating can meet the insulation requirements to build MEMS sensors. The adherence robustness and thermal barrier property of doped TBCs were also verified by a series of vibrating/shocking tests, combustion infield tests, and abrupt temperature shocking tests.

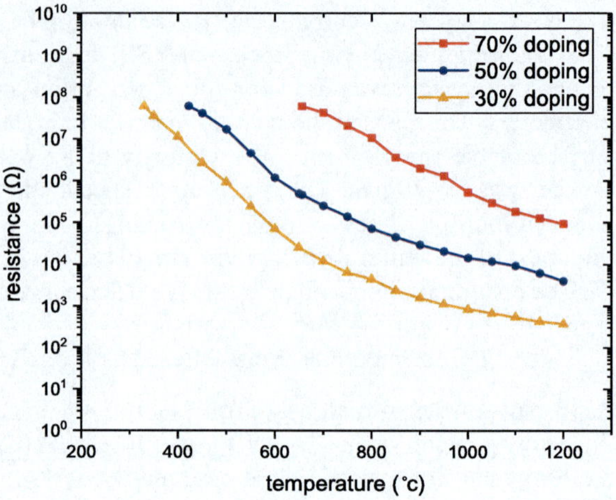

Fig. 4.18 Comparison of the high-temperature resistance characteristics of different Al_2O_3 doping ratios

4.3.2 Thermal Stress Mismatch

Thermal barrier coating (TBC) is a thermal protection technology [43, 56] that increases the service temperature of alloy materials by 100 ~ 150 °C [57]. It is also an electrical insulation layer, which can also be used to build a smart thin-film sensor on its surface to achieve accurate surface temperature measurements of key parts of aero-engines, such as turbine blades. The thin-film sensor and TBC form a multi-layer structures attached to the surface of the turbine blade. Due to the difference in the thermal expansion coefficient of different layers, thermal stress occurs at the interfaces between layers and this may cause the peeling off of the TBC and TFTC thin films, as illustrated in Fig. 4.19. This issue must be properly resolved before the TBC and TFTC sensors be implemented in the aero-system.

The simulated 3D thermal stress distribution at high temperature using a simplified model is illustrated in Fig. 4.20. The thermal stress is on the order of several hundred MPa, which is close to the yield limit of the material. Once this stress value exceeds the yield limit, it will cause deformation, even causing the film to fall off from the substrate or other failures. The thermal stress issue of this multi-layer structure under high-temperature stress must be properly addressed to avoid peeling off due to thermal stress mismatching, which causes the reliability problem of the system and the malfunction of the thin-film sensors. Therefore, how to reduce the thermal stress of smart high-temperature thin-film sensors on the surface of aero-engine parts to ensure effectiveness is a very important issue.

In the previous work, we noticed that thin-film sensors made on a rough TBC surface seem more robust than those made on a smooth surface. This inspired us new thought that the roughness may have a buffer effect on the thermal stress. In fact, further research on the surface roughness of the TBC indicates that the higher the surface roughness value is, the less serious the surface stress concentration and the lower fatigue damage resistance. Therefore, increasing the surface roughness will

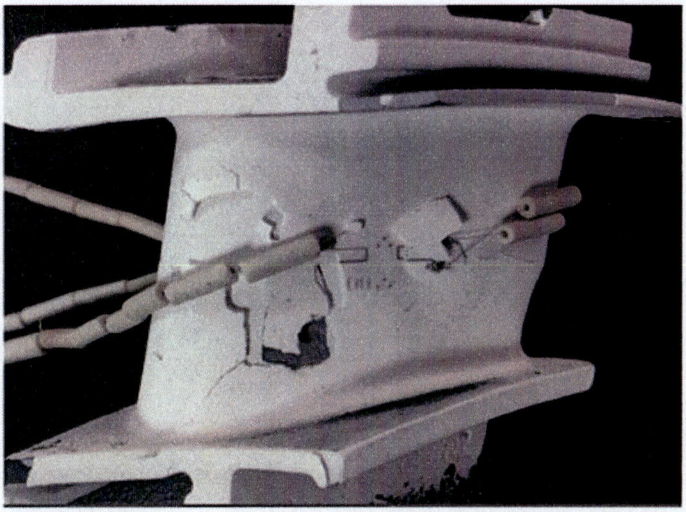

Fig. 4.19 Peeling off of the TBC due to the thermal mismatch stress

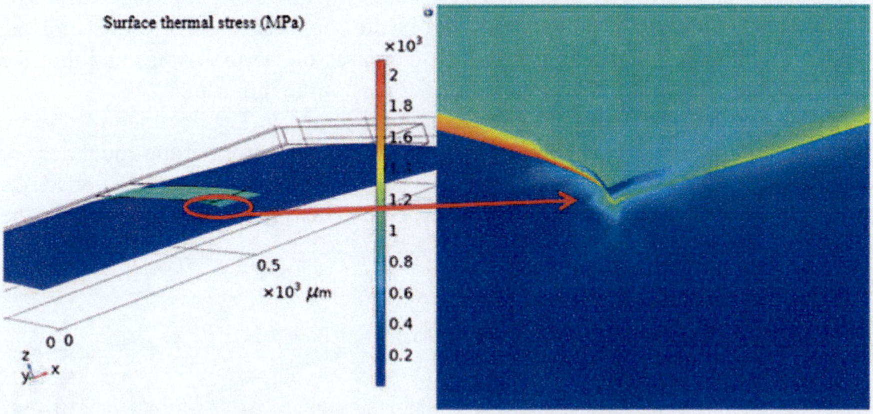

Fig. 4.20 Simulated structure and result of thermal stress of the metallic thin film on TBC at high temperature (1000 K)

reduce the fatigue strength of the parts [58], and the roughness has a buffer effect on the thermal stress of the thin film.

Simulation Proof of the Reduced Thermal Stress on Rougher Surface

The simulation experiment is performed by COMSOL Multi-physics software. The heat transfer module is selected to establish a simulation experiment environment [59]. A simplified structure representing the roughness is established in Fig. 4.21. The roughness is defined as the small spacing and unevenness of small peaks and

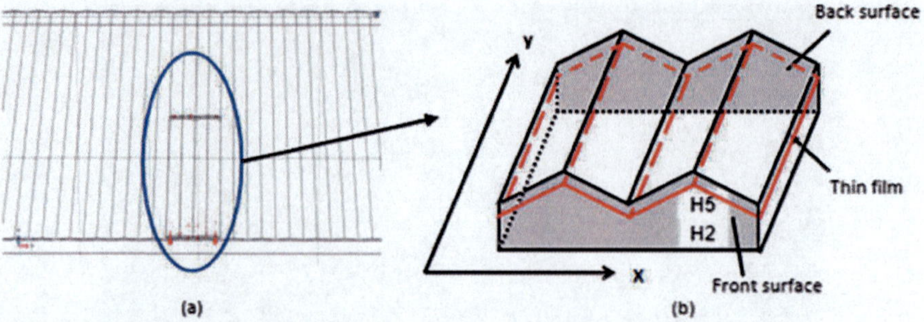

Fig. 4.21 **a** Surface roughness structure for simulation; **b** enlarged topology. H2 is the horizontal width and H5 is the height

valleys on the surface in a zigzag shape on surface. In this case, it is expected that the thermal stress will be reduced with more roughness.

We obtain a 3D colored map of the thermal stress of these two cases. The stress of smooth case H5 = 0 and the rough case H5 = 0.1 * H2 are compared in Fig. 4.22. From the figures, it can be clearly seen that in the smooth case, the maxima of the tangential component of thermal stress are larger than those in the rough case. The range is also significantly larger than that in the rough case. The results indicate that the increase in roughness can reduce the value of thermal stress and that the roughness has a buffer effect on the thermal stress of the thin film.

We also studied the thermal stress along the *x* direction. The obtained results are shown in Fig. 4.23. When H5 increases, it can be considered that the roughness of the TBC layer increases. It can be seen that when roughness increases, the von Mises stress values of the interface gradually decrease. This proves that the rougher the TBC layer is, the stronger the buffering effect on the thermal stress of the film.

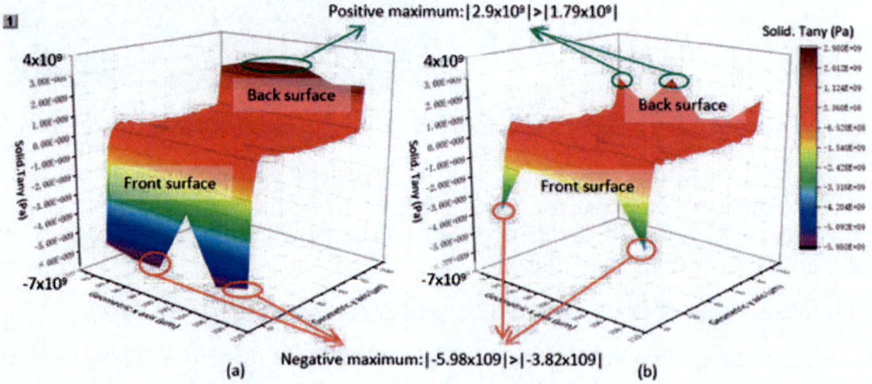

Fig. 4.22 Thermal stress of different roughnesses. **a** H5 = 0 (smooth); **b** H5 = 10% * H2 (rough)

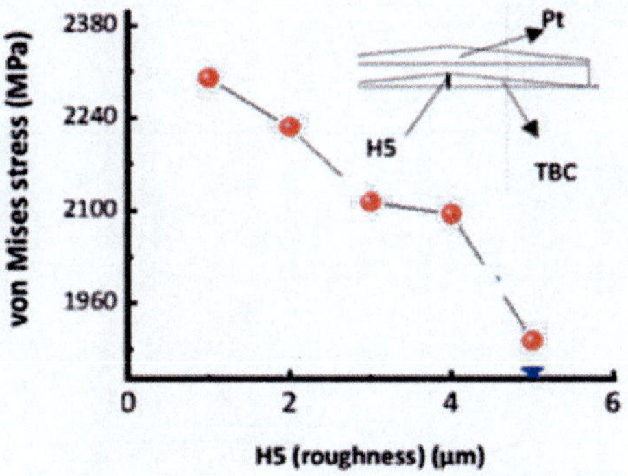

Fig. 4.23 Maximum and average stresses with the change in H5

Experimental Comparison

Thin films were coated on both rough and smooth surfaces. The rougher surface is made by a sandpaper rubbing half of a smooth quartz wafer and then deposited the Pt thin film on both sides as shown in Fig. 4.24. The surface roughness of the smooth quartz is 10 nm, while the roughness of the rough surface is ~ 3 μm, measured by a step meter as shown in Fig. 4.25. Also seen on the figure, Pt thin-film lines on both sides of the quartz wafer with a thickness of approximately 500 nm were created using an ordinary MEMS process, i.e., photolithography + magnetron sputtering + thin-film lift-off. After the oven heating, the platinum film on the smooth side was peeled off, while no peeling occurs on the rough substrate. The comparison proves that the roughness has a buffer effect on the thin-film stress—the film grown on the rough substrate is not prone to failure.

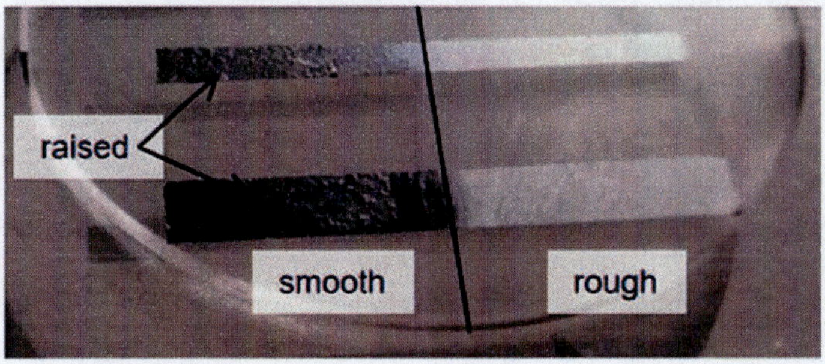

Fig. 4.24 The Pt thin-film structure deposited on both the smooth and rough sides of the plate

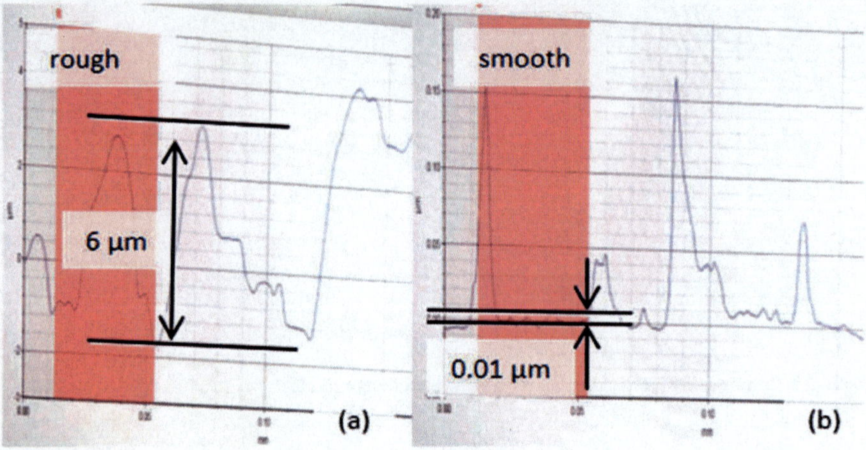

Fig. 4.25 Surface roughness of two sides

4.3.3 Minimizing the Test Error

Fabricating a thin-film sensor on a metal surface requires an insulating layer. However, there is an inevitable error due to the thicknesses of this layer (usually 300 µm)—the temperature measured from the TFTC is not the exact temperature on the metallic surface (at the interface of the TBC and NiCr alloy). This TBC insulating layer should be as thin as possible. However, in most cases, such a layer cannot be too thin since the electrical installation quality cannot be guaranteed.

On the other hand, PtAl is a promising turbine material for future gas engines and is expected to replace nickel–chromium alloys. Thermally grown Al_2O_3 is a much better electric insulator under high temperature compared to ordinary YSZ-based TBCs, and much thinner alumina layer is needed to build a TFTC sensor on its surface. Since this Al_2O_3 film is just 3 µm thick, the test error of the TFTC sensor is greatly reduced.

In the following, the Pt/PtRh thin-film thermocouple sensor was fabricated on the PtAl substrate with Al_2O_3 thin layer, and its overall performance was measured. For the first time, high-temperature insulation has been achieved with a very thin alumina film of only 3 µm. It is experimentally confirmed that such a thin layer of alumina can provide ideal high-temperature insulation and the sensor made on its surface has a satisfactory performance. Simulation studies show that the temperature error caused by 3 µm is just 3 °C. Using such a thin film as electrical insulation, the temperature on the surface of the metal can be measured using TFTC sensors fairly accurately, in real time and with good accuracy, thereby providing a comprehensive evaluation of the reliability of the work piece.

Experiments and Results

The raw material is CMSX-4 single crystal with a higher endurable temperature resistance up to 1150 °C. The thickness of the PtAl alloy is 50 μm and that of the alumina film is 3 μm, which is obtained after heat treatment for hundreds of hours. Thermocouples were fabricated using a modified MEMS process—the hand-drawing method is used to create the TFTC patterns as in Fig. 4.26.

After fabrication, the TFTF was characterized. The electrical insulation properties of the 3 μm alumina film are verified in the following chart. The thermal resistance of the alumina thin film versus temperature approximately follows a decreasing logarithmic pattern, and at high temperatures up to 1000 °C, the insulation resistance is still 1 kΩ, which is good enough to guarantee that the TFTC functions properly on the metallic surface, indicating that the thin alumina layer is able to provide insulation properties above the PtAl substrate (Fig. 4.27).

COMSOL software is used to explore the influence of the Al_2O_3 film thickness on the device temperature distribution. Results indicate that when the thickness of alumina is 3 μm, the temperature difference between the upper and lower surfaces is just 3 K, which is almost the same as the surface temperature of the metal. This is the verification that the temperature measured on the 3 μm alumina thin film can be treated as the surface temperature of the metal.

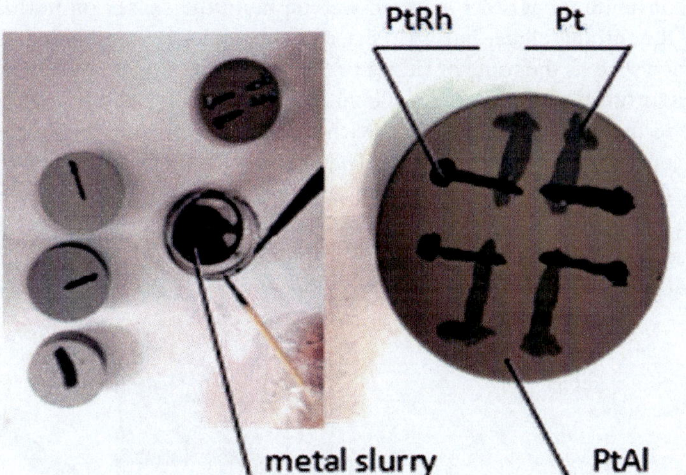

Fig. 4.26 TFTC samples fabricated by the hand-drawing method

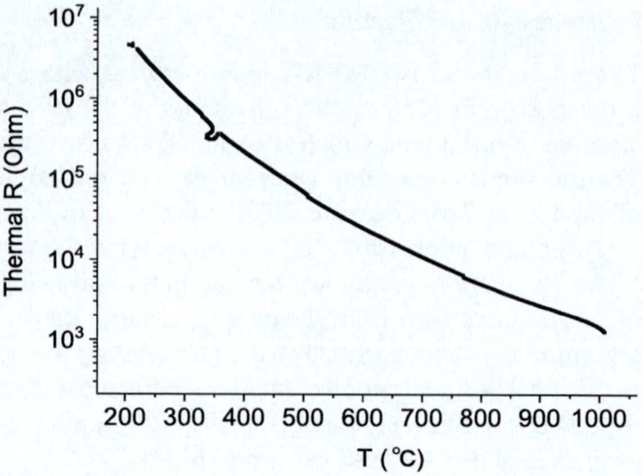

Fig. 4.27 Thermal resistance–temperature curve of the PtAl surface thermocouple

4.3.4 Sharp Corner Patterning

The conventional MEMS fabrication of TFTC is basically a planar process, i.e., to make a sensor on a flat and smooth surface, such as a silicon wafer. In reality, however, the AIAA scenarios require us to make the thin-film sensors on a curved surface. With the conventional MEMS method, we can build the senses on normal curved surface of the turbine blade, but it is very difficult to do the same things on a very sharp surface such as the roots or ridges of the blade. Special fabrication techniques are needed to fulfil such a task. As shown in Fig. 4.28, an array of thermocouples can be made by hand-drawing Pt and PtRh slurry on a spine or ridges at the sharp curvatures position of the turbine blade. The Pt and PtRh slurry consisting of metallic particulates mixed in polymer solvent is used to draw the pattern using two extremely fine brushes. Test results show that such TFTCs on these sharp surface in this way can accurately measure the temperature distribution on each point in the TFTC array (more details in Chap. 5).

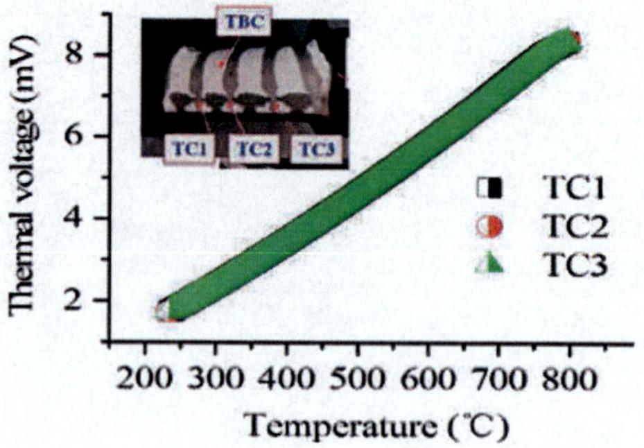

Fig. 4.28 Sensors built on the ridge of the blades

The Hand-Drawn TFTC Sensor Method

A new method is proposed and practiced to customize the Pt/PtRh TFTC pattern by hand-drawing without the comprehensive steps of photolithography and thin-film deposition. Points of interest can be easily located, and thermojoints of the thermocouples can be made accurately by delicate painting using soft pens. Connection of these thermocouples with the welded lines can be made elsewhere using the extended lines to the edge on the blades. In this way, the airflow and thermal transfer characteristics on the point of interest will not be affected by the embedded sensor (one of the drawbacks the wired TC used). This method is easy to operate and can accurately measure the surface temperature of the blade under high-temperature conditions. As shown in Fig. 4.29, three such hand-made thermocouple sensors are drawn as a smiling face. The three thin crosses on the middle line are three thermocouples.

We have successfully used this method to build a TFTC sensor array on a turbine blade surface as well, capable of obtaining a high temperature up to 1200 °C under various tough working conditions [60]. When the curvature is too complicated, this new method can successfully resolve the difficulty of patterning of TFTC using the

Fig. 4.29 Thermocouple "smiley face" fabricated by the pain brush using the Pt and PtRh slurry

conventional MEMS method and our results also show that these new TFTCs have comparable results to ordinary TFTCs made by conventional MEMS technology. These new devices are more suitable and more feasible for collecting real-time temperature values at various points of interest of aero-engine parts.

4.3.5 On-Site Fabrication

Many AIAA scenarios require us to make the thin-film sensors on site either because the part is too large or just difficult to carry, such as in the missile launch field. Instead of bringing the parts into the MEMS fabrication laboratory to build the sensors on its surface, the on-site fabrication method is needed to make the TFTC sensors.

The on-site fabrication method is different from the on-site sensors, which bring the purchased sensors and then install it on the machine part. The latter option has a drawback of poor thermal stress properties, which may bring the peeling off of the attached sensors by thermal paste. Therefore, proper derived on-site method is necessary based the traditional MEMS process. Our novel on-site methods include mask-free photolithography and the hand-draw painting method.

Mask-Free Photolithography Patterning Technique

Conventional photoresist (PR) is liquid form, and it is very hard to apply this PR evenly on the curved a complex surface like we did on a plenary silicon wafer. Dry film photoresist (DFPR) provides us a flexibility to build the sensors on site. Thin-film photoresist (DFPR) with excellent flexibility can be applied on a curved surface to ensure an evenly coated PR layer. However, traditional MEMS process uses the photolithography machine to perform the exposure process to the PR or DFPR under a flexible mask in the cleanroom. Such prerequisites limit us to use the photolithography technique on site. Here, we proposed a new DFPR method allowing us to make the sensors on site. The first new proposal is the mask. We can either used the pre-exposed DFPR to attach on the substrate or just use hand-draw pattern with an oily marker pen (Sharpie 32001). Second, the exposure is performed with a cheap UV portable lamp instead of using the expensive photolithography machine in the MEMS laboratory. The portable UV exposure and the process are shown in Fig. 4.30. The UV light intensity of the portable lamp is 2.2 mW/cm^2 and the exposure time is 10 s in our experiment which is comparable to the conventional photolithography machine. Of course, the accuracy and precision are not as good as the expensive photolithography machine. But for most of the AIAA requirements, the size of the sensor is much bigger than just 1 μm, and this new method is just doable to build the MEMS sensors on site. Either the pre-exposed by mask or hand-draw ink pattern will block the DFPR from exposing UV light, and the pattern will be created after development.

Figure 4.31a shows the pattern development and Fig. 4.31b shows the TFTC built after thin-film deposition and lift-off process. TFTC contains two layers of metals,

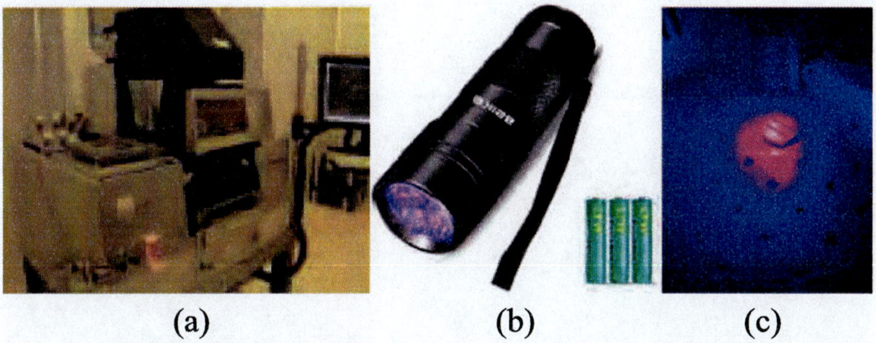

Fig. 4.30 **a** Ordinary photolithography machine, **b** portable UV light lamp, **c** on-site exposure using this portable UV light

Pt and PtRh; therefore, two steps are needed to form the thin-film thermocouple (TFTC), as shown in Fig. 4.31b.

Hand-Draw Sensor Directly on the Machine Part

In case of difficulties to do the thin-film deposition to build the TFTC onsite, another on-site fabrication technique is invented. We use the metal slurry to paint the TFTC on the machine part either covered with the DFRP mask or just draw it like a brush painting (Fig. 4.32). The first method is equivalent to pave the signs on the road, and the second way is like to draw a painting of van Gogh's "Sunflowers". For a TFTC pattern in millimeter size, this is achievable by either way. We use

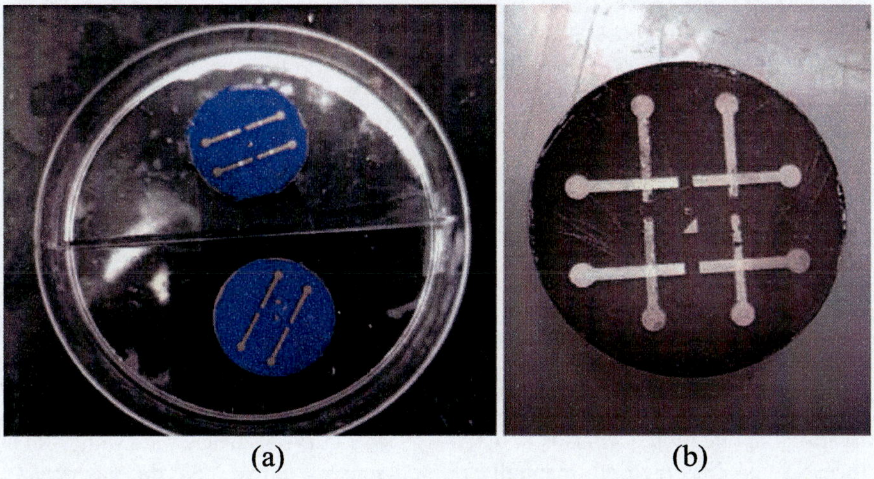

Fig. 4.31 **a** Patterns after the dry film photolithography, **b** TFTC structure made by maskless DFPR method

Fig. 4.32 Hand-drawn TFTC pattern using the Pt and PtRh slurries

the high-temperature flaming gun to perform the thin-film hardening/welding to probe to the extension wires after the sensor is built. Although this hand-painted patterns of the TFTC may not be as beautiful, it works fairly well. Good repeatability/linearity of the thermocouple performance and excellent robustness of high-temperature stress/thermal shocking were achieved, which fulfilled the engineering need for aero-engine propulsion smart sensors [61].

In short, these on-site modified MEMS techniques are very suitable for manually building any functional patterns on complex surfaces where ordinary photolithography and PVD sputtering machines are not available. The fabricated TFTC can monitor on-site high temperatures ~ 1000 °C with good repeatability and linearity and can withstand cycling high-temperature calibration and multiple thermal shock tests with a good binding force, which fulfills the engineering need for aero-engine propulsion smart sensors. The method is very useful to make sensors on site on the tough corner or the root bottom of engine components or large engine parts that we cannot bring them into the MEMS laboratory.

4.4 Technology and Process

Below, we describe in detail a few distinct processes to build the sensors to fit in various application scenarios in AIAA: PDMS curved surface patterning, RTD process, mask-less method, manually printed sensors, BTW YSZ thermistor, and SAW wireless sensing technology.

4.4.1 PDMS Curved Surface Patterning

The core of MEMS technology is the transfer of graphics. The PDMS process is an embossed graphics technique using a seal made in a silicone rubber material. Previous research shows that the technologies used to prepare nanoseals usually include electron beam, extreme ultraviolet light, focused ion beam, or reactive ion etching. The preparation of nanoseals requires high cost, so it is not suitable for mass production in industry. The transparent soft template made of the elastic material is more feasible and is used in our soft lithography patterning to replace the traditional mechanical contact nanoimprinting scheme, which can solve the above nanoimprinting limitations to the greatest extent.

PDMS is polydimethylsiloxane (also known as silicone rubber, English name is polydimethylsiloxane, abbreviated as PDMS). Because of its low cost, simple to use, good adhesion with silicon wafers, and good chemical inertia, PDMS has become a polymer material widely used in micro-fluidics and other fields. The basic principle of PDMS MEMS processing is to coat the surface of the original hard mold prepared by the traditional lithography process with the elastic material PDMS and remove the PDMS after drying. The pattern on the original hard grinding surface is transferred to PDMS to form a PDMS soft film, which is the key material for transferring the pattern to other substrates in subsequent soft lithography technology. Next, the PDMS soft template is used to transfer the soft lithography pattern on the surface of the engine blade to form the MEMS sensor pattern structure. In this process scheme, PDMS is used to transfer graphics on the curved surface to make a reliable micro-sensor on the metal curved surface.

There are basically two steps: preparation of the PDMS soft template and application of its graphics on the blade surface. The common method for preparing PDMS templates is to cast a mixture of PDMS prepolymer and the corresponding curing agent onto silicon wafer hard motherboard prepared in advance through photolithography and etching processes. After curing, the PDMS peels off the motherboard, and the pattern on the motherboard is copied to PDMS; that is, the preparation of the PDMS soft template is completed. In Fig. 4.33, we show several PDMS patterns using our school logos.

A KV-X250 Laser Microscope System from KEYENCE Company is used to observe the three-dimensional morphology and size. Compared with traditional microscopy, laser microscopy has the following advantages: it can obtain a structure with higher image resolution than conventional optical microscopy and can suppress the scattered light outside the focal plane, so it is possible to obtain a picture with higher contrast. Figure 4.34 shows the 3D morphology and measurement size of the silicon wafer hard motherboard, respectively. It is seen that the snake-shaped resistance wire pattern of the original silicon wafer motherboard is partially convex, that is, the non-graphic area is etched in the reactive ion deep silicon etching.

From the 3D measurement parameters, we can see that the width of the serpentine resistance wire pattern is approximately 52.6 μm, the spacing between adjacent resistance wire patterns is approximately 48.7 μm, and the height of the convex

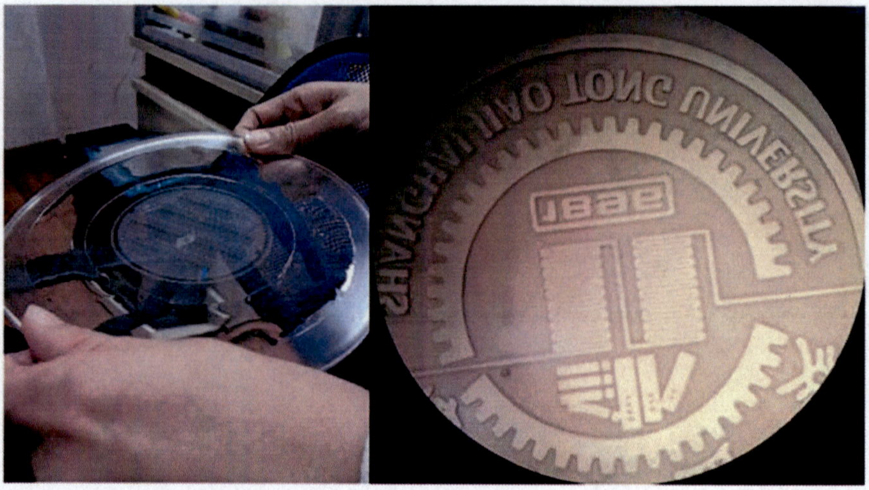

Fig. 4.33 PDMS soft mask molded using the silicon wafer

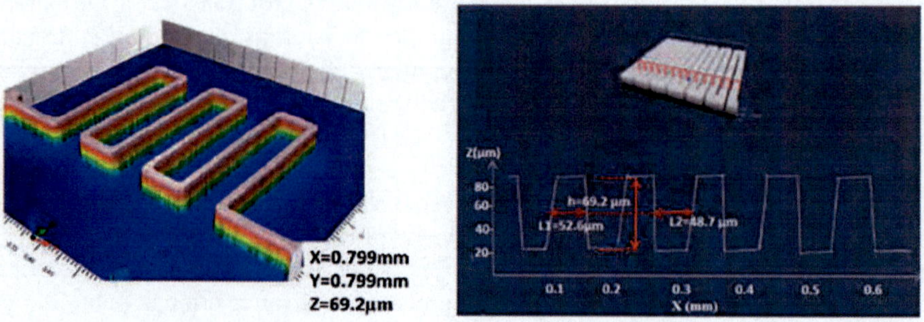

Fig. 4.34 (Left) Three-dimensional topography of the silicon mask structure, (right) 3D topography of the deep-etched silicon surface

pattern and the instant etching depth are approximately 69.2 μm. In the original design size, the width of the serpentine resistance wire pattern is 50 μm, the spacing between adjacent resistance wire patterns is 50 μm, and the etching depth is 70 μm. Compared with the pattern on silicon hard motherboard, it is found that the width of the resistance wire pattern on silicon wafer increases by 2.6 μm, while the sum of the resistance wire pattern and spacing basically does not change. The reason is the error caused by the diffraction effect during lithography. The etching depth of the silicon wafer is only 0.9 μm less than the designed 70 μm, indicating that the etching process parameters basically meet the requirements.

How to transfer the graphics on PDMS to the blade surface to realize the RTD sensor is also a technical challenge. To do this, we coat on PDMS a α-ethyl cyanoacrylate and then quickly imprint PDMS onto the surface to form a pattern. α-Ethyl cyanoacrylate is the main component of 401 glue, which is a high strength and fast adhesive α-ethyl cyanoacrylate; due to its viscosity and high infiltration, α-ethyl cyanoacrylate generally does not flow quickly when poured on the plane, while compared with ordinary liquids such as water, anhydrous ethanol, and even photoresist. In addition, α-ethyl cyanoacrylate is soluble in acetone [62] which is a very important quality in MEME process, that makes it possible to do lift-off technique after the thin-film deposition (the ordinary PR is solvable in acetone).

The specific process flow of PDMS soft template and soft lithography is as follows:

(1) Spin on the cleaned 3-inch silicon wafer surface with α-Ethyl cyanoacrylate while controlling the rotational speed of the glue slinger to 1000 r/min, the time to 30 s, so as to the thickness of ethyl cyanoacrylate is controlled at about 10 μm.
(2) Paste the prepared PDMS soft template graphics onto the silicon wafer completely for 20 s.
(3) Quickly remove the PDMS soft template and stamp it on the blade surface, and put it into the 60 °C oven for 30 s.
(4) Remove PDMS after cooling.
(5) On the basis of the previous step, 200-nm platinum (Pt) was sputtered on the blade surface, and then, a complete MEMS sensor pattern was obtained on the blade using the stripping process.

Figure 4.35 shows the RTD sensor patterned based on the school logo—the comb-like wire forms the RTD temperature sensor pattern made on the blade. The 3D morphology is obtained as the optical images from Zeiss Microscopy Company. It can be seen that the graphic copy effect is good. Sensor patterns were clearly observed by PDMS method. Using the Zoom5 optical microscope of Zeiss Company, the distortion caused by the graphic transfer from the plane to the curved blade surface is estimated. The sensor at this position on the curved surface has a height difference of about 1 mm within a scale of nearly 8 mm and a curvature of about 5%. The measurement result of edge line graph shows that the outermost layer width is 98.64 μm compared to its design dimension of 100 μm. The reason is that during the process of pasting PDMS on this position of the blade, the bending of PDMS on the curved surface will cause the extrusion of lines, reducing the line spacing of PDMS soft template at this curvature angle. However, the error is within the acceptable range, so it can still achieve good transfer results at the position of the minimum curvature radius of the blade. The PDMS method cannot guarantee a very narrow width patterning. The maximum acceptable width for PDMS is 50 μm which is good enough to fit the need for most of the AIAA scenarios.

PDMS soft lithography can be used to realize the graphics of turbine blades, and the results laid a good foundation for subsequent metal film deposition and lift-off processes. The PDMS soft template replaces the hard mask used in traditional photolithography and can create a sensor pattern on aero-engine blades.

Fig. 4.35 Three-dimensional topography microscopy of the school logo (an RTD sensor) patterned on the blade surface [63]

4.4.2 DFPR Photolithography

Dry film photoresist (DFPR) consists of three layers of films (Fig. 4.36) [64]. Among them, PE polyethylene film is the covering layer on the dry film surface to prevent the pollution of dust particles and oxygen to protect the PR film; PET is the support film for the PR to prevent oxygen from damaging the free radical inside the photoresist film and will be removed before development. Photoresist film is the key component of dry film for pattern transfer. When ultraviolet rays irradiate the photoresist film through the mask, resin cross-linking and curing reactions will occur in the photoresist in the area irradiated by ultraviolet rays to form a macromolecular structure, and the color of the dry film will change from light to dark. During development, the photoresist in this area will be retained, while the photoresist not irradiated will be dissolved by the developer to obtain the required pattern on the substrate surface.

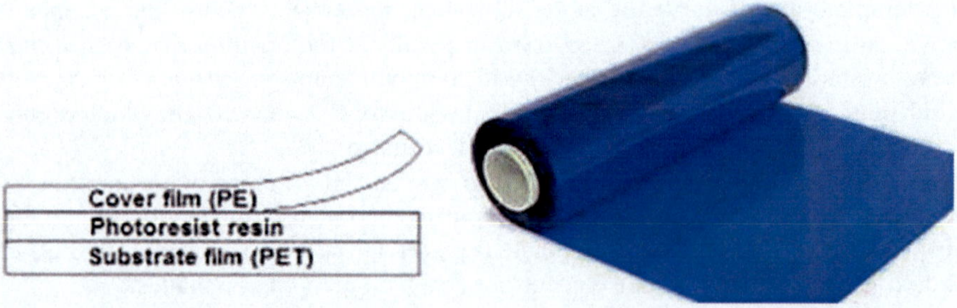

Fig. 4.36 Structure of the dry film photoresist (DFPR)

The whole dry film lithography patterning process for turbine blade TFTC is described below. Such process is applicable to make MEMS sensors on similar parts or surface in AIAA scenarios as well.

(1) Substrate clean. The substrate surface is polished with sandpaper to reduce the roughness and prevent the punctuation of the dry film from causing seepage during sputtering. Acetone and deionized water are used for 2 min to remove oil and dust.
(2) Prebaking. The cleaned substrate is heated to 70 °C in an oven for 20 min to remove the residual water.
(3) Filming and pressing. The protective film of the DFPR is removed first and then quickly attached evenly onto the substrate surfaces. Then, a rubber ball is used to smoothly press the dry film in one direction to drive off the residual air between the dry film and the substrate surface to ensure close bonding between the two layers.
(4) Post baking. The substrate is put in an oven again to further improve adhesion at 68 °C for 20 min.
(5) Mask and exposure (Fig. 4.37a). A mask is used to block UV light when doing the exposure for the DFPR. The UV light intensity of the photolithography machines is 2–20 mW/cm^2 and the exposure time is 10–60 s.
(6) Development (Fig. 4.37b). Then, the DFPR is developed in a negative developer (25%AZ400 + 75%water) for 40 s. After drying at 40 °C, the pattern could be transferred into the DFPR pattern.
(7) Sputtering and lift-off (Fig. 4.38). A layer of PtRh film with a thickness of 500 nm was sputtered using a magnetron sputtering machine (Z550 Sputtering System, Leybold Heraeus, Germany). 500 nm was sputtered three times, with each round of sputtering power of 150 W, chamber pressure of 2×10^{-4} Pa, and sputtering for 210 s. Then, the DFPR on the substrate surface was completely peeled off by lift-off using a 4% NaOH solution and a 50 °C water bath to obtain the PtRh pattern.

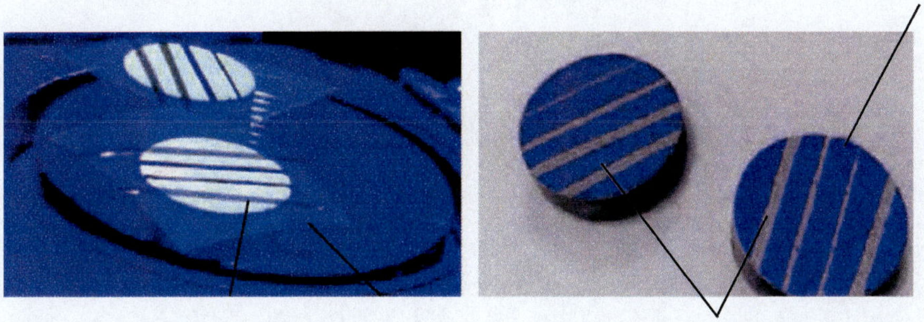

(a) (b)

Fig. 4.37 **a** Exposure of the DFPR; **b** DFPR after exposure and development

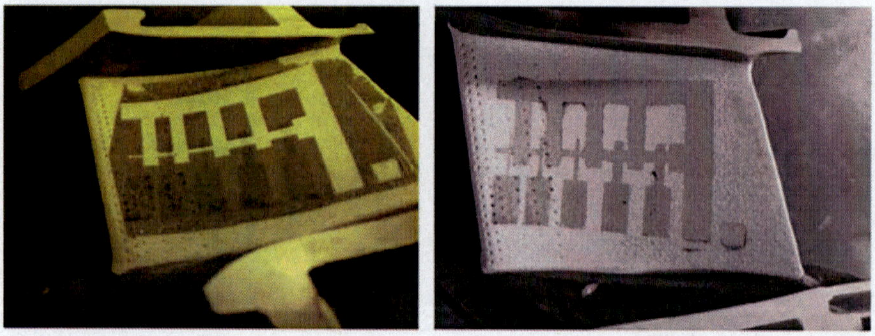

(a) (b)

Fig. 4.38 On the turbine blade surface: **a** TFTC pattern after the exposure and development of DFPR. **b** TFTC created after the DFPR process together with the thin-film deposition and lift-off

The following charts illustrates the key steps in the above process.

Figure 4.39 shows the resulting TFTC sensor array on the turbine blade surface and its measurement results. A very consistent thermal voltage versus temperature behavior is observed for all three TFTC sensors on the turbine blade. This result clearly indicates that the TFTC sensors built on this new method possess the same performance as the previous MEMS method.

One of the key applications of the DFPR photolithography is the mask-free approach described before. The mask-less photolithography process will eliminate the mask which is a key and tremendous difference from the previous traditional MEMS process. Mask-free MEMS allows more flexible micro-fabrication

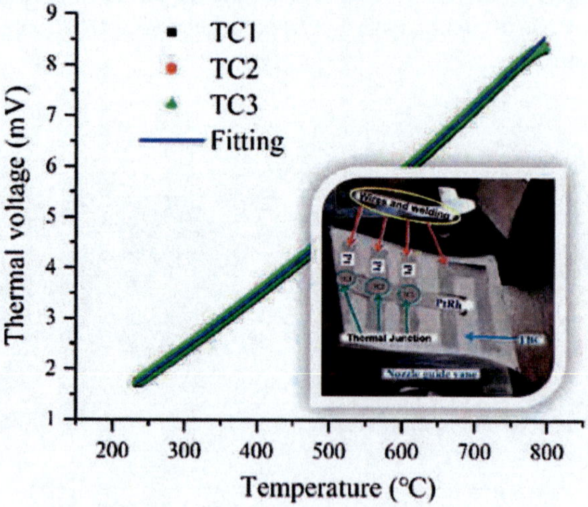

Fig. 4.39 DFPR made TFTC array and its measurement results

processing, such as on-site built TFTC sensors. In addition, the mask-free process is much simpler than the conventional MEMS process which requires expensive machines and clean rooms.

4.4.3 YSZ Thermistor

We can use the YSZ both as a thermal barrier insulator as well as a temperature monitor for turbine blade by taking advantage of its temperature sensitivity. Such a "by-the-way" sensor can be used as a long-term health monitor of the turbine working conditions such as the gas generators for a power plant.

The Fabrication of a YSZ Thermistor

There are only two simple steps to build an YSZ thermistor: YSZ coating and thermistor connection welding.

First, the turbine blade surface was sandblasted with alumina powder, followed by ultrasonic cleaning in ethanol before spraying. The bonding layer of CoNiCrAlY (AMDRY 9951 powder) was sprayed prior to the TBC layer by a Multi-Coat™ equipped with a DJ2600 gun spray (HVOF gun (HVOF)). The YSZ surface coating was performed by an SPS system from Mettech plasma thermal spray coating equipment. The SPS system uses the Axial III plasma torch and NanoFeed model 350 feeding device. The YSZ powder is a 7.5% yttria-stabilized zirconia (7.5YSZ) commercial submicron particle with a median particle size of 0.06 μm. In the spraying process, the spraying material is first heated to the melting or semimelting state. Then, the spraying powders were accelerated by high-speed gas flow to generate enough momentum toward the surface of the CoNiCrAlY bonding layer on the Ni–Cr-based superalloy turbine blade substrate and then condensed onto the surface after cooling. The thickness of the CoNiCrAlY bonding layer is approximately 150 μm, while the thickness of the YSZ surface coating is ~ 200 μm. Then, the YSZ-coated sample was treated by soaking the entire part in an acetone basin for 10 min of ultrasonic cleaning.

Pt paste was used to dot the Pt/YSZ sensors and other connections on the TBC surface. Pt soldering dots in an area of 1 mm^2 were carefully paved on the YSZ surface using Pt paste with a solid mass fraction of 82.6%. Solidification of platinum paste is required at 800 °C with high-temperature sintering for 10 min and cooling to room temperature naturally. The YSZ thermistor and thermocouple nodes were electrically connected to 0.1 mm-thick Pt/PtRh wires by platinum/platinum–rhodium paste and then hard-baked at 800 °C for 10 min.

A high-temperature extension cable is used as insulation extensions toward the outside digital multi-meters (DMMs) for thermal voltage and thermal resistor measurements. Multi-channel data were collected by the computer to obtain the readings during the whole temperature ramping process. Since the measurement of the resistivity covers from the Ω to MΩ range and due to the different configurations of the digital multi-meters, there were measurement glitches when the resistance range shifted from Ω, kΩ, to MΩ (Fig. 4.40). To circumvent this issue, a new strategy to measure the thermistor resistance ranging from Ω to MΩ is formulated as follows: to measure the electric current and convert it to the Ω value. A 9 V voltage was applied on the device under test (DUT, in this case "the YSZ thermistor") with a 10 kΩ resistance in the loop, and the current flowing through the DUT was monitored. The thermal resistance is converted using the measured current i in this formula:

$$R_{\text{thermistor}} = \frac{V_{9\,\text{V}}}{i} - 10\,\text{k}\Omega \tag{4.6}$$

where $V_{9\,\text{V}}$ and i are the battery voltage and measured electric current from the DMMs, respectively. It can be quickly estimated from this formula that when current i varies from 1 to 999 μA, the $R_{\text{thermistor}}$ is in the range of Ω to MΩ. The resulting thermistor characterization in these two methods is compared in Fig. 4.40 during temperature ramping from room temperature to 800 °C, corresponding to the thermal resistance from the Ω to MΩ range. It is clearly observed that the ohm measurement using the method of collecting the electric current data eliminates the measurement glitches and hence is more appropriate than the conventional ohm measurement using DMM for thermistor characterization.

YSZ thermistor devices need a one-time calibration to obtain a thermistor–temperature relationship prior to their first use infield, since the thermal resistance versus temperature dependence varies with the different YSZ compositions and thicknesses of the TBC layer. Once the sensor is being calibrated, its resistance versus the temperature dependence can be recorded as a temperature sensing database. Since the YSZ is a very stable material as temperature, its resistivity versus temperature relationship is very stable, i.e., YSZ resistor can be used as a reliable and stable long-term temperature monitor.

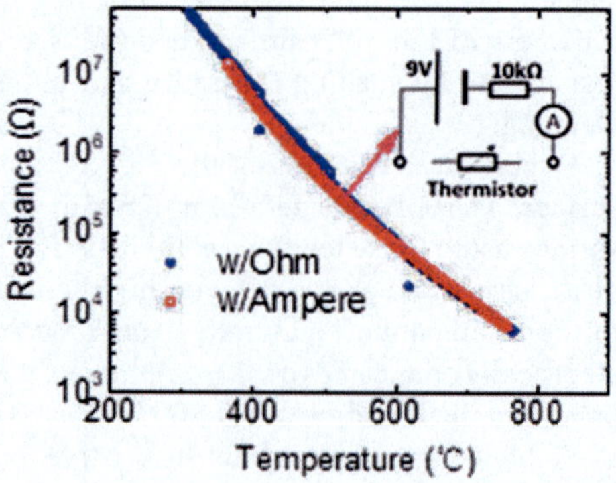

Fig. 4.40 YSZ resistance in temperature cycling test between 400 and 800 °C. The reading glitch appears when DMM changes the range from Ω, kΩ, and MΩ. A new method is formulated to avoid this measurement issue (inset)

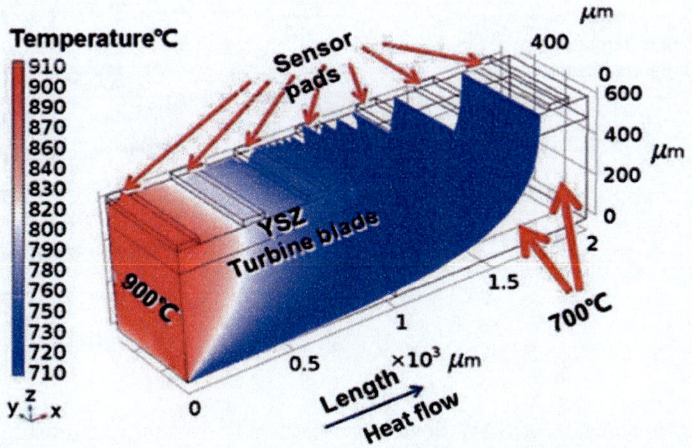

Fig. 4.41 Test array using the YSZ thermistor along the temperature field distribution on the surface

Test Array to Measure the Temperature Distribution Using YSZ Thermistors

By placing a series of YSZ thermistors, the temperature gradient distribution on the surface can be monitored and characterized. The test array is illustrated in the structure (Fig. 4.41) together with the simulated temperature distribution profile in YSZ. The thermal-electro interactive behavior of the YSZ thermistor on a metallic substrate was studied with the help of COMSOL Multi-physics simulation. The model considering the interactive effect of electrical properties with increasing temperature is incorporated in the simulator. The dependency of YSZ resistance on temperature is from the experimental data above and is integrated into the model. The dimensions and material properties (thermal conductivity, electrical conductivity of all layers) of the simulation structure follow the real case as much as possible for a better understanding of electrical conductance interacting with temperature. In detail, the metallic substrate (Ni–Cr alloy) is 2 mm thick, and the YSZ ceramic insulator film is 300 μm.

The YSZ temperature distribution is shown in Fig. 4.42. Most of the temperature variations occur near the high-temperature end, and the corresponding YSZ resistance in the YSZ thermistor follows this trend very well. This simulation result indicates that the YSZ thermistor array can be used to detect the surface temperature distribution on the turbine blade in an environment with varying heat flows.

4.4.4 SAW Wireless Sensing

Surface acoustic wave (SAW) devices are receiving a lot of attention since after White and Voltmer first invented the interdigital transducer (IDT) in 1965 [29]. The

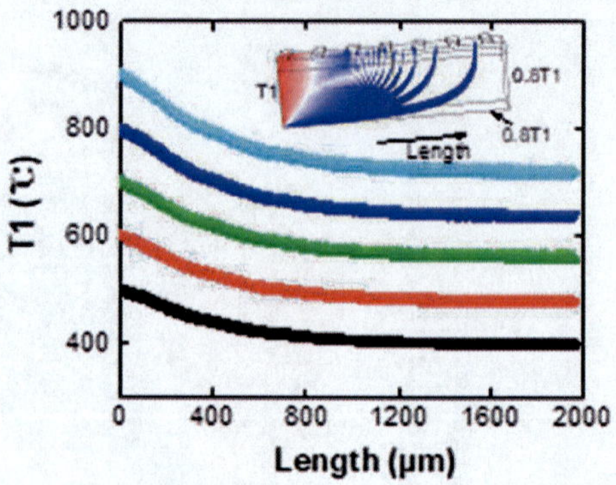

Fig. 4.42 Temperature distribution at different locations along the length direction

resonant frequency f_0 of SAW device possesses a consistent relationship with the ambient temperature, and by using such feature, the temperature can be measured by detecting the changes in f_0. The SAW signals can be transmitted as an RF signal to the outside network analyzers systems so as to achieve the wireless sensing capability. Manufacturing of the SAW device under and normal temperature (< 200 °C) is a matured process and is widely used in many fields such as monitoring the temperature of a rotating wheel. However, to use SAW sensor to measure the high temperature (~ 1000 °C), wireless is a big challenge, which is the target of our following efforts.

To achieve this goal, the manufacturing process consists of two parts: ordinary SAW device fabrication and SiO_2 passivation. Photolithography, magnetron sputtering, and lift-off techniques are used to fabricate the SAW device. Before that, research work on design and process optimization on SAW device is necessary by various mask options and fabrication technologies. Then, passivation technique is used to protect the SAW device from the high-temperature corrosions.

The SAW Device Design

Although there are quite a few options of the SAW device patterns, a single-port resonant SAW design is chosen as the interdigital transducer (IDT) due to its simplicity and good Wi-Fi signal transmitter ability. The central (resonant) frequency of the SAW device is:

$$f_0 = \frac{V}{\lambda} \tag{4.7}$$

where the wavelength λ is:

$$\lambda = 2(a + b) \tag{4.8}$$

a and *b* are the width and the space of the IDTs (Fig. 4.43). In most cases, the width and spacing are the same. Therefore, the central resonant frequency f_0 of the device is dependable on the width of the interdigital electrode and the surface acoustic wave velocity *V*.

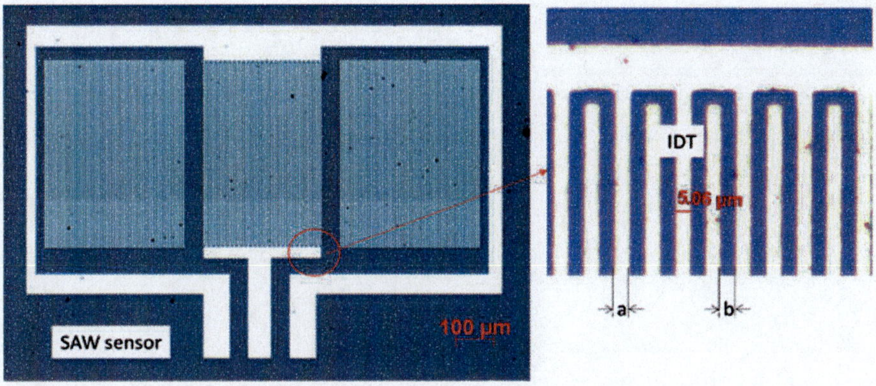

Fig. 4.43 Topology of a true SAW sensor and the interdigital transducer (IDT)

In a SAW design, if one wants to keep the center frequency within a 400 MHz when using the LN piezosubstrate with the $V \sim 3000$–4000 ms, then the λ is 4–6 μm. The choice of the LN substrate and the width of the IDT determine the exact frequency of f_0 the SAW device. For this purpose, the different masks of SAW topologies were formulated for the best SAW sensor structure. SAW devices were laid out at different angles/sizes to study the orientation dependence on S_{11}, bandwidth, and temperature frequency coefficient on different substrates. The mask design is illustrated in Fig. 4.44.

SAW Device Fabrication

There are two methods to create the IDT patterns. One is the etching technique; the other is lift-off. The lift-off technique is selected here to build the SAW devices, yet thicker photoresist is preferred to facilitate the lift-off process. A double photoresist technique was used to achieve a thick photoresist layer (~ 3.99 μm) to create a distinct minimum finger width (3 μm) in the IDT. The double photoresist technique is shown

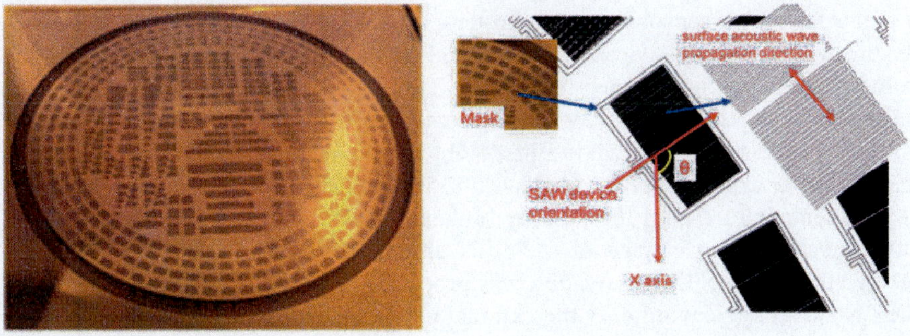

Fig. 4.44 Design mask of various SAW devices with different orientations

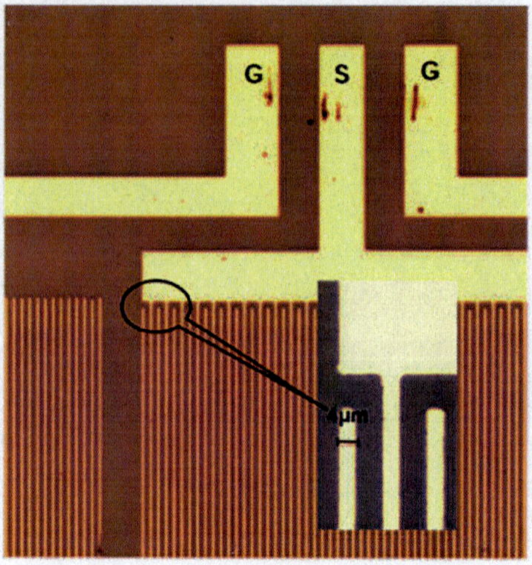

Fig. 4.45 Final metal IDT patterns after lift-off

in Fig. 4.45. Initially, a 1.3 μm PR layer is coated on the substrates and then exposed thoroughly without a mask. Then, another PR layer is coated onto this exposed PR to form a thick PR layer. Because of the mutual dissolution between the two PRs, the final thickness ranges from 3.8 to 4.2 μm. The exposure time was reduced to only half of the conventional time since the bottom PR layer was already exposed. Therefore, in the double photoresist technique, decreasing the exposure time can achieve a shaper trapezoid angle slope. With the help of the mutual dissolution of exposed resist and unexposed residue, the bottom resist can be developed easily.

Good SAW devices were successfully built with resonant frequencies in the range of 400–440 MHz. The line width of the IDT of such an SAW device is ~ 2.5 μm (the pitch is 5 μm, with a 2.5 μm line width and 2.5 μm space, and the thin film is ~ 30 nm).

SAW Sensor for High Temperature

We have studied the various design and process options above, and the next challenge is to make an SAW sensor for very high-temperature measurements (1000 °C). Since the line width of SAW device is ~ 2.5 μm and the thin film is just ~ 30 nm, it is highly vulnerable under high-temperature stress. This is the key reason for the previous failure to build LN-based SAW sensor exceeding the higher temperature limit of 450 °C [24]. In fact, as shown in Fig. 4.46b, we did observe tremendous melting of the IDTs after exposing the SAW above 500 °C; the IDT electrodes started shrinking at high temperature until they totally broken. XRD/EDS analysis also indicates a structure/component change of 41°Y-LN after exposure to high temperature. To resolve this issue, SiO_2 passivation was proposed to cover the SAW electrode and the LN substrate, leaving only the external electrode exposed for subsequent cable connection. Figure 4.46c indicates that the IDTs of the SAW sensor with 400 nm SiO_2 passivation look intact after 1100 °C, which ensures reliable SAW functioning

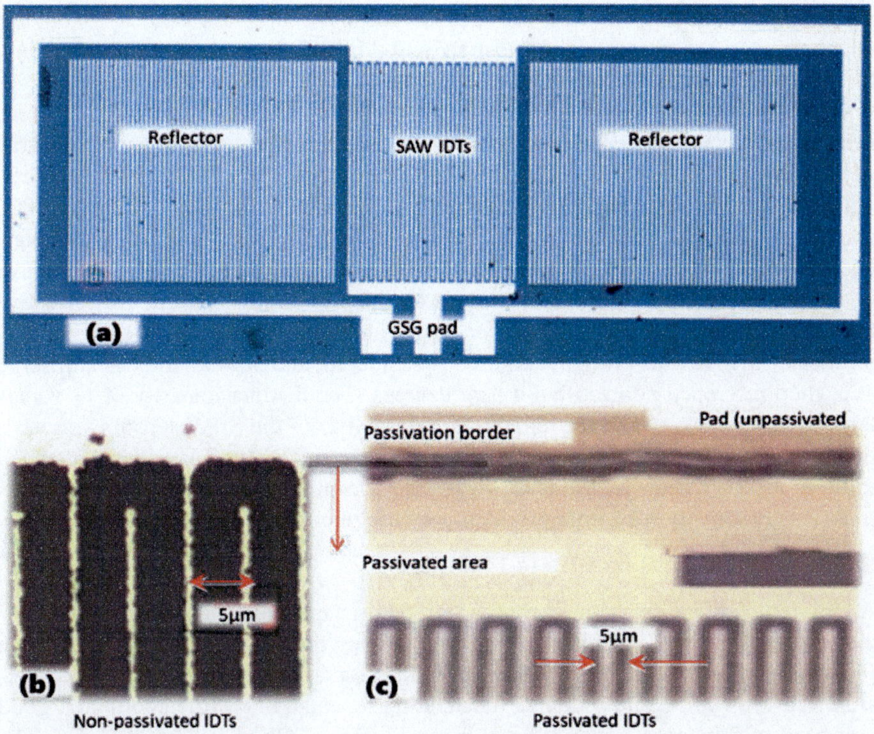

Fig. 4.46 Topography of the SAW device: **a** original SAW device; **b** SAW device without passivation after the 650 °C experiment; **c** SAW device with passivation after the 1100 °C experiment

as a sensor in a high-temperature environment. The exposed unpassivated Pt pad of the SAW device is soldered out to an extended Pt wire with Pt paste [65] and then connected to the external network analyzer for RF measurement.

Breakthrough—$LiNbO_3$-Based SAW Sensors Capable to Measure Up to 1100 °C High Temperature

Piezoelectric materials are the key building material used to form SAW devices. There are quite a few piezoelectric materials that can be used as SAW substrates for high-temperature measurement, such as lithium niobate (LN) [66–68], Langasite (LGS) [69, 70], and more advanced super high-temperature piezoelectric materials [27, 71]. Among them, $LiNbO_3$ is a mature commercialized material with good piezoelectric, ferroelectric, optoelectronic, and thermoelectric properties [72], while the others are expensive or still in the developing stage. LN's Curie temperature is as high as 1140 °C, which is suitable for most situations in aero-engine systems (~ 1000 °C). However, previous studies have shown that SAW sensors with $LiNbO_3$ substrates cannot operate at temperatures higher than 400 °C [30] due to dehumidification and recrystallization [73].

This bottleneck was broken with a SiO_2 passivation layer on the SAW device, which protects the piezoelectric substrate and the IDT electrodes. In the following, the design and manufacturing process of these high-temperature tolerance SAW sensors and the SiO_2 passivation technique, together with the high-temperature RF verification test using these SAW devices, are described. The SAW sensor is capable of measuring temperatures up to 1100 °C with good repeatability and durability, which is quite a great breakthrough. High-temperature sensing in the range of 800 and 1100 °C has a special significance for aero-engine R&D [9] since the tolerance of most of the turbine metallic alloy is not more than 950 °C. The active operation temperature of the turbine blade can be measured using our cost-effective $LiNbO_3$-based wireless passive sensor. For higher-temperature SAW sensors, new piezoelectric materials [74, 75] should be considered, and our method can still serve as a good reference point to build new devices. The distinct linearity of f_0 versus temperature together with the temperature durability of the SAW sensors was verified by conducting various high-temperature RF tests. Such an SAW sensor attached to an embedded near-field antenna was also formulated to enhance the wireless transmission quality for future high-temperature remote sensing systems.

SAW Sensor Performance Up to 1100 °C

A high-temperature experiment was carried out in a muffle furnace with the temperature ramping from 20 to 1100 °C and then cooling down. The results are shown in Fig. 4.47, indicating that the SAW sensor with passivation can measure high temperatures up to 1100 °C and retain an outstanding repeatability while cooling. A good temperature-frequency linearity was observed with a TCF of − 15.12 kHz/°C. The standard error of two measurements is less than 0.07%.

In summary, a $LiNbO_3$-based surface acoustic wave sensor was built using MEMS technology for the purpose of wireless high-temperature measurements in an aerospace system. Such a small-size SAW sensor is capable of measuring high temperatures up to 1100 °C. SiO_2 passivation is a key technique to ensure sensor function and robustness under very high-temperature stress. The SAW sensor has good repeatability and temperature endurance (~ 10 h).

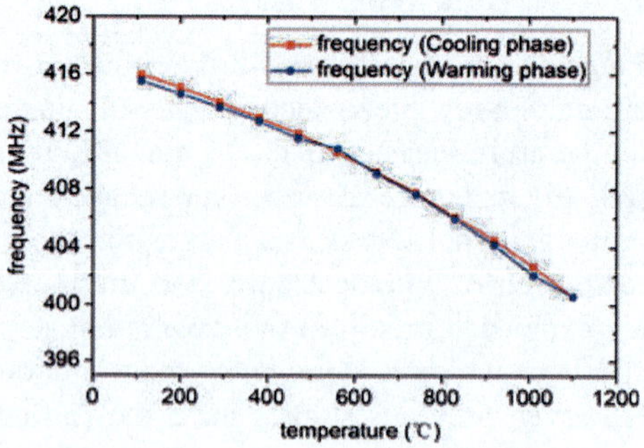

Fig. 4.47 Frequency responsiveness of the SAW sensor during the 1100 °C experiment

A near-field embedded antenna is needed by routing a Pt thin film on the TBC surface coated on a turbine blade substrate to amplify the SAW signals and ensure a proper Wi-Fi distance. Other issues, such as the impact of the Doppler Effect, the transparency of RF signal transmission crossing the metallic shell, and the in situ calibration of the SAW sensor, are also need to be addressed. More details will be elaborated in Chap. 5 (the test section). Together with the above efforts, the SAW-based wireless passive sensing system is expected to be able to measure inner temperature/stress inside the aero- and astro-system.

References

1. The Versatile Affordable Advanced Turbine Engines (VAATE) Initiative: An AIAA Position Paper, Program. AIAA Air Breathing Propulsion Technical Committee, Jan 2006. AIAA, Reston, VA (2006). https://www.aiaa.org/docs/default-source/uploadedfiles/issues-and-advocacy/policy-papers/position-papers/2006-vaate.pdf?sfvrsn=ff5746d3_0
2. AIAA Position Paper: The Integrated High Performance Turbine Engine Technology (IHPTET) Initiative, Aug 1991
3. Dan, M.: MEMS mechanical sensors. Sens. Rev. **24**(3), 319–320 (2004)
4. Adams, T.M., Layton, R.A.: Introductory MEMS (2010). ISBN: 978-0-387-09510-3. https://doi.org/10.1007/978-0-387-09511-0
5. Fernandez, L.J., Visser, E., Sese, J., et al.: Radio frequency power sensor based on MEMS technology. In: SENSORS, 2003 IEEE. IEEE (2004)
6. Zhang, Y.F., Duan, F.L.: IC Manufacturing Technology. Shanghai Jiao Tong University Press (2018). ISBN 9787313186515
7. Ballal, D.R., Zelina, J.: Progress in aeroengine technology (1939–2003). J. Aircr. **41**(1), 43–50 (2015)
8. Hunter, G.W., Neudeck, P.G., Okojie, R.S., Beheim, G.M., Powell, J.A., Chen, L.Y.: An overview of high-temperature electronics and sensor development at NASA Glenn Research Center. J. Turbomach. **125**(4), 658–664 (2003)
9. Hunter, G.W., Wrbanek, J.D., Okojie, R.S., Neudeck, P.G., Fralick, G.C., Chen, L., Xu, J., Beheim, G.M.: Development and application of high temperature sensors and electronics for propulsion applications. In: Proceedings of SPIE (International Society for Optical Engineering), Sensors for Propulsion Measurement Applications. International Society for Optics and Photonics (2006). https://doi.org/10.1117/12.668458
10. Zhang, S., Wang, B., Liang, H., et al.: Study on application of optical measurement system in high temperature environment. Transducer Microsyst. Technol. **27**(5), 73–75 (2008)
11. Zhang, X., Xue, X., Chen, B., et al.: Research and application of thermal indicating paint in aeroengine test. Meas. Control Technol. **27**(1), 21–23 (2008)
12. Xu, R., Xu, H., Shen, X., et al.: Application of infrared temperature measurement system in aero engine wheel test. China Meas. Test **038**(004), 32–35 (2012)
13. Li, Y., Yin, G.: Research on crystal temperature measurement technology for aeroengine turbine blade. Aeroengine **3**, 83–87 (2017)
14. Duan, L., Gao, J., Ding, G., et al.: A robust embedded MEMS high-temperature sensor for aero-engine turbine blade. Aeroengine **44**(192) (04), 60–64 (2018)
15. Seebeck: Magnetische Polarisation der Metalle und Erze durch Temperatur-Differenz (Magnetic Polarization of Metals and Ores by Temperature Differences), pp. 265–373. Abhandlungen der Königlichen Akademie der Wissenschaften zu Berlin (in German) (1822)
16. Parker, D.W.: Thermal barrier coatings for gas turbines, automotive engines and diesel equipment. Mater. Des. **13**(6), 345–351 (1992)

17. Wrbanek, J., Fralick, G., Blaha, C., Busfield, A., Thomas, V.: A multilayered thin film insulator for harsh environments. In: 38th AIAA/ASME/SAE/ASEE Joint Propulsion Conference & Exhibit, Indianapolis, IN, 7–10 July 2002. Nasa Sti/Recon Technical Report N, 2002, 3
18. Gao, J., Duan, F.L., Yu, C., Meng, W., Liu, L., Ding, G.: Electrical insulation of ceramic thin film on metallic aero-engine blade for high temperature sensor applications. Ceram. Int. **42**(16), 2 (2016)
19. Diehl, W.: Platinum thin film resistors as accurate and stable temperature sensors. In: Proceedings of Sensor'82—Transducer Technology and Temperature Measurement, pp. 89–101 (1982)
20. Lee, W.S., Park, J.S., Kim, N.H., et al.: Electrical and thermal properties of platinum thin films prepared by DC magnetron sputtering for microheater of microsensor applications after CMP process. Solid State Phenom. **124**, 267–270 (2007)
21. Ekkels, P., Rottenberg, X., Puers, R., et al.: Evaluation of platinum as a structural thin film material for RF-MEMS devices. J. Micromech. Microeng. **19** (2009)
22. Tiggelaar, R.M., Sanders, R.G.P., Groenland, A.W., et al.: Stability of thin platinum films implemented in high-temperature microdevices. Sens. Actuators A Phys. **152**(1), 39–47 (2009)
23. Maeder, T., Sagalowicz, L., Muralt, P.: Stabilized platinum electrodes for ferroelectric film deposition using Ti, Ta and Zr adhesion layers. Jpn. J. Appl. Phys. **37**(Part 1), 2007–2012 (1998)
24. Pereira da Cunha, M., et al.: Wireless acoustic wave sensors and systems for harsh environment applications. In: IEEE Topical Conference on Wireless Sensors & Sensor Networks (2011)
25. Pohl, A.: A review of wireless SAW sensors. IEEE Trans. Ultrason. Ferroelectr. Freq. Control **47**(2), 317–332 (2000)
26. Fachberger, R., Bruckner, G., Knoll, G., Hauser, R., Biniasch, J., Reindl, L.: Applicability of $LiNbO_3$, langasite and $GaPO_4$ in high temperature SAW sensors operating at radio frequencies. IEEE Trans. Ultrason. Ferroelectr. Freq. Control **51**(11), 1427–1431 (2004). https://doi.org/10.1109/TUFFC.2004.1367482
27. Zhang, S., Yu, F.: Piezoelectric materials for high temperature sensors. J. Am. Ceram. Soc. **94**(10), 3153–3170 (2011)
28. Yan, X., Zhang, K., Wu, X.: Opportunities and challenges of MEMS technology in the environment of intelligent aero engines. J. Aerosp. Power **27**(6), 1210–1217 (2012)
29. White, R.M., Voltmer, F.W.: Direct piezoelectric coupling to surface elastic waves. Appl. Phys. Lett. **7**(12), 314–316 (2004)
30. Frankel, D.J., Bernhardt, G.P., Sturtevant, B.T., Moonlight, T., Pereira da Cunha, M., Lad, R.J.: Stable electrodes and ultrathin passivation coatings for high temperature sensors in harsh environments. In: IEEE Sensors, Nov 2008, pp. 82–85 (2008). https://doi.org/10.1109/ICSENS.2008.4716388
31. Ryshkewitch, E., Richerson, D.W.: Oxide Ceramics, pp. 1–40. American Ceramic Society, Inc., Westerville, OK (1985)
32. Martin, L.C., Fralick, G.C., Taylor, K.F.: Advances in thin film thermocouple durability under high temperature and pressure testing conditions. NASA TM-1999-208812, Jan 1999
33. Wrbanek, J.D., Fralick, G.C.: Thin film physical sensor instrumentation research and development at NASA Glenn Research Center. In: 52nd International Instrumentation Symposium (2006)
34. Epstein, A.H.: Millimeter-scale, MEMS gas turbine engines. In: Proceedings of ASME Turbo Expo 2003 on Power for Land, Sea, and Air, Atlanta, 16–19 June 2003 (Paper No. 28)
35. Martin, L.C., Wrbanek, J.D., Fralick, G.C., Martin, L.C.: Thin film sensors for surface measurements, NASA TM-2001-21149. In: 19th International Congress on Instrumentation in Aerospace Simulation Facilities, Cleveland, OH, 27–30 Aug 2001
36. Lei, J.F., Will, H.A., Martin, L.C.: Thin film sensors for minimally intrusive measurements in harsh high temperature environment. AIAA-98-3610 (1998)
37. Grant, H.P., Przybyszewski, J.S., Claing, R.G.: Turbine blade temperature measurements using thin film temperature sensors. NASA-CR-165201, Pratt and Whitney Aircraft. East Hartford, CT (1981)

38. Shepherd, R.S.: Thin film sensor techniques for the instrumentation of ceramic/metal interfaces in next generation aero gas turbines. Ph.D. thesis, University of Salford, Salford, UK, July 1999
39. Gleeson, B.: Thermal barrier coatings for aeroengine applications. AIAA J. Propul. Power **22**(2), 375–383 (2012)
40. Duan, F.L., Gao, J., et al.: Simulation analysis of temperature distribution of thermal barrier coatings on aeroengine blades. J. Shanghai Jiaotong Univ. **51**(8), 915–920 (2017)
41. Martin, L.C., Holanda, R.: Applications of thin film thermocouples for surface temperature measurement. In: Proceedings of SPIE—The International Society for Optical Engineering (1994)
42. Sheng, C., Yu, Y., Yu, Y., et al.: Microstructure and thermal characterization of multilayer insulation materials based on silica aerogels. J. Inorg. Mater. **28**(7), 790–794 (2013)
43. Bose, S., Demasi-Marcin, J.: Thermal barrier coating experience in gas turbine engines at Pratt & Whitney. J. Therm. Spray Technol. **6**(1), 99–104 (1997)
44. Shen, Z., He, L., Xu, Z., et al.: LZC/YSZ DCL TBCs by EB-PVD: microstructure, low thermal conductivity and high thermal cycling life. J. Eur. Ceram. Soc. (2018)
45. Padture, N.P., Gell, M., Jordan, E.H.: Thermal barrier coatings for gas-turbine engine applications. Science **296**(5566), 280–284 (2002)
46. Jouanny, I., Palisaitis, J., Ngo, C., et al.: In situ transmission electron microscopy studies of the kinetics of Pt–Mo alloy diffusion in ZrB_2 thin films. Appl. Phys. Lett. **103**(12), 1–11 (2013)
47. Cao, X.: Application of rare earths in thermal barrier coating materials. Mater. Sci. Technol. (Engl. Ed.) **23**(1), 15–35 (2007)
48. Liu, C.B., Lin, F.N., Jiang, X.L.: Current state and future development of thermal barrier coating. Chin. J. Nonferr. Met. **17**(1), 1–13 (2007)
49. Shen, Z., Huang, G., He, L., et al.: Effect of surface treatments on lifetime and failure mechanism of thermal barrier coatings. Rare Met. Mater. Eng. (2016)
50. Cao, X.Q., Vassen, R., Stoever, D.: Ceramic materials for thermal barrier coatings. J. Eur. Ceram. Soc. **24**(1), 1–10 (2004)
51. Mishra, S.K., Jagdeesh, N., Pathak, L.C.: Fabrication of nanosized lanthanum zirconate powder and deposition of thermal barrier coating by plasma spray process. J. Mater. Eng. Perform. **25**(7), 2570–2575 (2016)
52. Clarke, D.R., Oechsner, M., Padture, N.P.: Thermal-barrier coatings for more efficient gas-turbine engines. MRS Bull. **37**(10), 891–898 (2012)
53. Coquil, T., Pilon, L., Nemanick, E.J., Tolbert, S.H., Reitz, C., Brezesinski, T.: Thermal conductivity of ordered meso porous titania films made from nano crystalline building blocks and sol-gel reagents. J. Phys. Chem. C **144** (29), 12451–12458 (2010)
54. Schlichting, K.W., Padture, N.P., Klemens, P.G.: Thermal conductivity of dense and porous yttria-stabilized zirconia. J. Mater. Sci. **36**(12), 3003–3010 (2001)
55. Reddy, J.N.: An Introduction to the Finite Element Method, 3rd edn., pp. 415–821. McGraw-Hill Science (2006). ISBN 9780071267618
56. Xu, H., Guo, H., Liu, F., et al.: Development of gradient thermal barrier coatings and their hot-fatigue behavior. Surf. Coat. Technol. **130**(1), 133–139 (2000)
57. Schulz, U., Leyensa, C., Fritscher, K.: Some recent trends in research and technology of advanced thermal barrier coatings. Aerosp. Sci. Technol. **7**, 73–80 (2003)
58. Zhang, D.C., Pei, X.M.: Effects of machining processes on surface roughness and fatigue life. China Mech. Eng. (2003)
59. Li, H., Sun, Y., Duan, L., et al.: Thermomechanical reliability of high temperature composite multilayer film structure sensors. Instrum. Technol. Sens. **11**, 7–10 (2015). https://doi.org/10.3969/j.issn.1002-1841.2015.11.003
60. Ji, Z., Duan, F.L., et al.: Temperature distribution measurements on turbine blade surface by the aid of simple dotted Pt/PtRh thermal couple test array. In: AIAA Propulsion and Energy 2019 Forum, Indianapolis, 19–22 Aug 2019
61. Xie, Z., Duan, F.L., Ji, Z., Qian, L., Bao, Z: A new method of making thin-film thermocouples at sharp surfaces of aero-engine hot component. In: AIAA Propulsion and Energy 2020 Forum, 24–28 Aug 2020, VIRTUAL EVENT. https://doi.org/10.2514/6.2020-3511

62. Hatzakis, M., Canavello, B.J., Shaw, J.M.: Single-step optical lift-off process. IBM J. Res. Dev. **24**(4), 452–460 (2010)
63. Gao, J., Duan, F.L., Wang, Y., et al.: Preparation of PDMS soft template and soft photolithography on turbine blade curved surface. Micro Nano Electron. Technol. **53**(5), 7 (2016). https://doi.org/10.13250/j.cnki.wndz.2016.05.010
64. Duan, F.L., Liu, S., Xie, Z., et al.: The various MEMS methods to build TFTC sensors for related aero-/astro-applications. In: AIAA Scitech 2021 Forum, p. 1399 (2021)
65. Han, J., Cheng, P., Wang, H., Zhang, C.C., Zhang, J.B., Wang, Y., Duan, L., Ding, G.F.: MEMS-based Pt film temperature sensor on alumina substrate. J. Mater. Lett. **125**, 224–226 (2014)
66. Smith, R.T., Welsh, F.S.: Temperature dependence of the elastic, piezoelectric, and dielectric constants of lithium tantalate and lithium niobate. J. Appl. Phys. **42**(6), 2219–2230 (1971)
67. Nakagawa, Y., Yamanouchi, K., Shibayama, K.: Third-order elastic constants of lithium niobate. J. Appl. Phys. **44**(9), 3969–3974 (1973)
68. IEEE standard on piezoelectricity. In: ANSI/IEEE Standard 176-1987. IEEE (2002)
69. Hornsteiner, J., Born, E., Fischerauer, G., et al.: Surface acoustic wave sensors for high-temperature applications. In: IEEE International Frequency Control Symposium (1998)
70. Naumenko, N., Solie, L.: Optimal cuts of langasite, $La_3Ga_5SiO_{14}$ for SAW devices. IEEE Trans. Ultrason. Ferroelectr. Freq. Control **48**(2), 530–537 (2001)
71. Yan, H.X., Ning, H.P., Kan, Y.M., et al.: Piezoelectric ceramics with super-high Curie points. J. Am. Ceram. Soc. **92**(10), 2270–2275 (2010)
72. Koos, G.L., Wolfe, J.P.: Phonon focusing in piezoelectric crystals: quartz and lithium niobate. Phys. Rev. B **30**(6), 3470–3481 (1984). https://doi.org/10.1103/PhysRevB.30.3470
73. Jiang, X., Kim, K., Zhang, S., et al.: High-temperature piezoelectric sensing. Sensors **14**(1), 144–169 (2013). https://doi.org/10.3390/s140100144
74. Chen, C.L., Lai, C.T., Lee, J.Y.: Three layer perovskite-like structured $Pr_3Ti_2TaO_{11}$ ferroelectrics with super-high Curie point. J. Am. Ceram. Soc. **97**(11), 3624–3630 (2014)
75. Li, B.W., Osada, M., Ozawa, T.C., et al.: $RbBiNb_2O_7$: a new lead-free high-Tc ferroelectric. Chem. Mater. **24**(16), 3111–3113 (2012)

Chapter 5
To Test the Smart Sensors for AIAA

In the previous chapter, the manufacturing and fabrication of various sensors were described in various smart methods. This chapter discusses the testing of these sensors in the context of the AIAA application scenarios. It has to be mentioned here that the testing of the smart sensors for the AIAA is equally challenging as manufacturing them since the sensor criterion must both meet the basic IEEE standard as well as to survive the tough AIAA working environment. The testing of these sensors is basically from two perspectives: sensor characterization mainly from the lab in IEEE's angle and infield-related testing from AIAA's angle. In Chap. 3, we have already discussed some infield applications. In this chapter, we focus on the details of the characterization and robustness verification tests of these sensors before sending them to the infield use. The outline of this chapter is as follows:

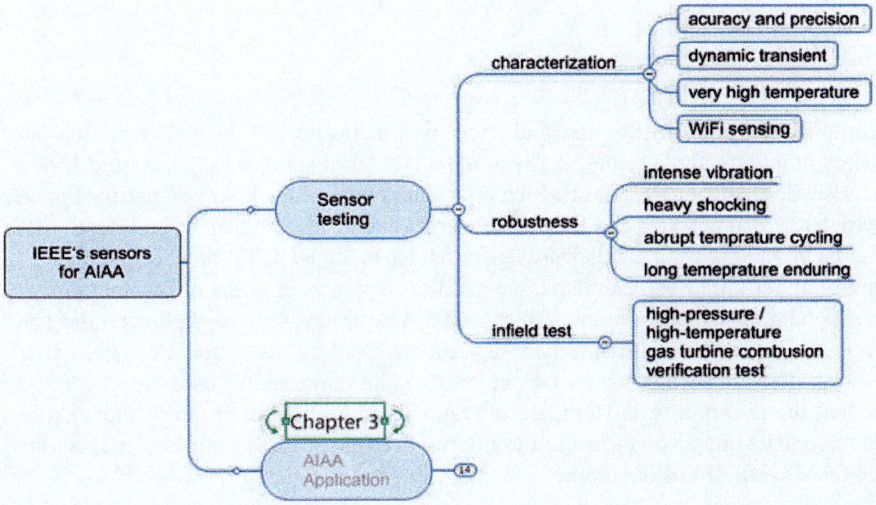

© The Author(s), under exclusive license to Springer Nature Singapore Pte Ltd. 2023
F. L. Duan, *When AIAA Meets IEEE*,
https://doi.org/10.1007/978-981-19-8394-8_5

AIAA's tests are usually more stringent than the general IEEE characterization. For example, it is not difficult to get a sensor curve in IEEE's lab of a TFTC sensor. However, this TFTC may not pass the abrupt temperature cycling test in AIAA (the TFTC thin film may peel off due to the thermal stress mismatch which frequently occurs in gas turbine machines). IEEE people care more about top performance records, while AIAA people care more about durability, reliability, and robustness. In addition to the basic parameters such as highest temperature limit, test accuracy, and test error, the IEEE people also need to address these issues: endurance and cycling reliability, transient and distributive behaviors, vibration and shocking tolerance, and some additional features such as real-time test instead of after-test reading (temperature paint, Table 4.1 in Chap. 4), undisturbed test from TFTC versus traditional wired TC used for cooling test (or embedded instead of armor insulated wired TC, Sect. 3.3). In this chapter, we first discuss the basic performance tests from IEEE's lab, and then we discuss a few typical AIAA-requested robustness tests. Finally, we show an infield test that is a comprehensive verification of our TFTC sensor performance combining electrical, thermal, and mechanical strength—the TFTC inside the gas turbine engine.

5.1 Performance Test—IEEE

Regarding the sensor's test performance, there are the following characteristics: the highest temperature limit, test accuracy and precision, and dynamic response. To compare with other temperature test methods, TFTC sensor exhibit the obvious advantages (check Table 4.1 in Chap. 4).

5.1.1 Test Accuracy

Temperature tests involve thermal transmission in the test system (machine part, wires, ambient), and thermal energy transfer is not as responsive as electronic devices. Unlike the electronic response which is prompt, temperature involves thermal transfer which has strong space and time effects and depends on the thermal conductivity of media of and between the thermo-sensor device and the DUTs (device under test). In high-temperature measurements, we need to overcome a series of technical bottlenecks related to measurement and calibration accuracy, such as the uncertainty and repeatability, and the measurement deviation caused by the temperature field distribution. The following efforts are addressing the temperature calibration—how to reduce the uncertainty and temperature field distribution of high-temperature experiments and its impact on the accurate calibration of thin-film thermocouples integrated on the surface of turbine blade.

The Problem Description

In our early work, we incurred many measurement accuracy issues when calibrating our thin film TFTC/RTD sensors. The errors are mainly from the temporal and spatial differences in the temperature field between the DUTs and oven temperature readers (thermal sensor of the muffin furnace). We did a simple temperature test to illustrate this test error. We placed a standard R-type thermocouple (TC) in the middle of the oven (heated chamber area is approximately 15 × 15 × 10 cm^3), the location where we usually put our test sample. We then raise the temperature of the oven from 50 to 1300 °C, and compare the difference of our thermocouple versus the oven reading. Table 5.1 shows the discrepancies between the oven reading and the thermal couple reading. The temperature difference between the heating element sensor and the standard R-type TC can be as high as 90 °C. If we use the oven reading as reference temperature, the test error occurs and the test results are not as repeatable each time we do the temperature test. For example, this difference depending on the oven temperature ramping rate—the faster the temperature ramping, the more this difference becomes. Usually, it takes approximately 3 h to ramp the temperature up to 1300 °C in a muffle oven, and the temperature drop rate is much slower. Increasing the time for ramping and holding time may reduce this spatial variation, but it is still not a perfect solution. The reason lies in that the temperature readout of the oven is usually from the embedded thermal couple placed close to the oven's heating element. There is a special difference from the temperature on the spot where the embedded sensor sits. Therefore, an accurate standardized thermal couple sensor is necessary for the correct characterization. We should place a dedicated thermocouple just on the site of our testing sample for the correct sensor calibration purpose instead of using the oven reading as our reference temperature.

Another spatial difference in temperature measurement is shown in Fig. 5.1. Two standard thermocouples (OMEGA R-type Pt/PtRh thermocouple, P13R-008-40) with 1 cm spacing are placed inside the oven (temperature ramping rate 100 °C/h) and are measured synchronously. It is seen that the two TCs have a very good linear dependency, yet the intersection is not absolutely zero, an indication that there is an intrinsic spatial thermal difference of ~ 30 °C between these two TCs. The two RTD sensors embedded on the same turbine blade surface with a spacing of 1 cm have similar spatial/delay behaviors as the two TCs, i.e., the intersection of the two thermal resistors does not coincide at zero as they should.

The uncertainties caused by this space/time issue must be properly addressed for accurate calibration of our TFTC sensors. Our initial effort is to place a dedicated standard R-type Pt/PtRh thermocouple at the exact location of our MEMS sensor to avoid temperature discrepancy caused by the oven. This did improve the situation a lot and yet still not as perfect. The main error originates from the space–time thermal conductance difference between the TFTC sensor and the standard thermocouple—there is an air gap between them. Later on, we have successfully circumvent this issue with a new Pt-Dot-TC methodology. The results show that the Pt-Dot-TC method can totally eliminate the calibration errors caused by the temporal and spatial differences in the temperature fields between the DUTs and the reference temperature

Table 5.1 Difference in the oven reading and the in situ thermocouple reading

Oven temperature readout (°C)	Thermocouple readout (°C)	The difference (°C)
80	50	−30
158	100	−58
279	200	−79
390	300	−90
484	400	−84
584	500	−84
679	600	−79
771	700	−71
860	800	−60
954	900	−54
1045	1000	−45
1140	1100	−40
1234	1200	−34
1323	1300	−23

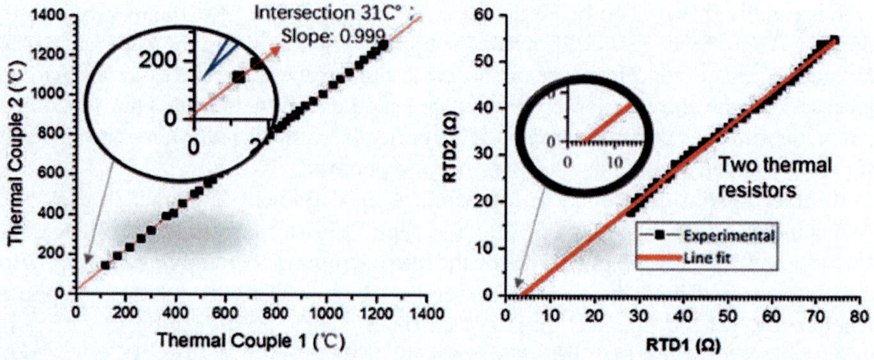

Fig. 5.1 Temperature reading difference caused by the spatial difference of two of TCs and two RTDs

readers, and we can effectively control the testing errors within 1% or ± 1 K in the 300–1500 °C range.

A New Temperature Calibration Methodology—Pt-Dot-TC

As shown in Fig. 5.2, by sticking the wired R-type Pt/PtRh sensing joint on the sample, an in situ temperature can be read out with much less spatial/thermal influence between the DUT and the reference temperature readout. Figure 5.3 shows the calibration curve by using this a Pt-Dot-TC for the TFTC built on CMC substrate, together with the ordinary reference readout using standard R-type TC. Over thousands of test points were collected over the whole temperature range and over a 10-h cycling duration. Statistical analysis is performed to evaluate the test error on each point using all the measured data. As seen from the result, the calibration by using the Pt-Dot-TC shows a much consistent thermo voltage versus temperature characteristics compared to using a separate wired TC as a temperature readout.

Figure 5.4 shows the statistical error analysis of these two calibration methods. As can be seen from this statistical analysis, the overall testing error from the ordinary wired separate TC temperature calibration method has much larger errors.

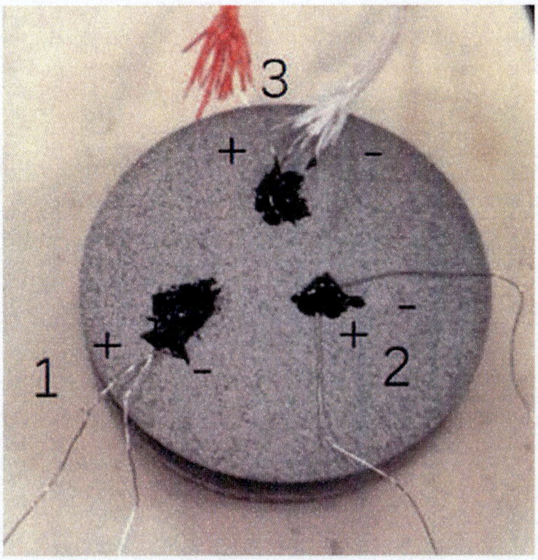

Fig. 5.2 Diagram of the Pt-Dot-TC for calibration thermocouple

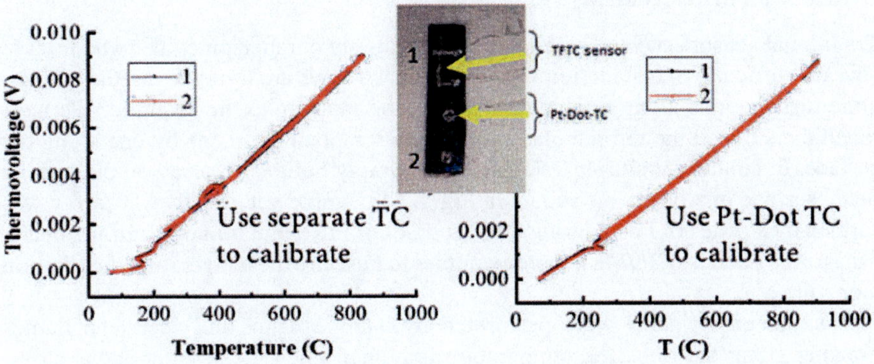

Fig. 5.3 Comparison of the temperature calibration tests by using two types of thermocouples

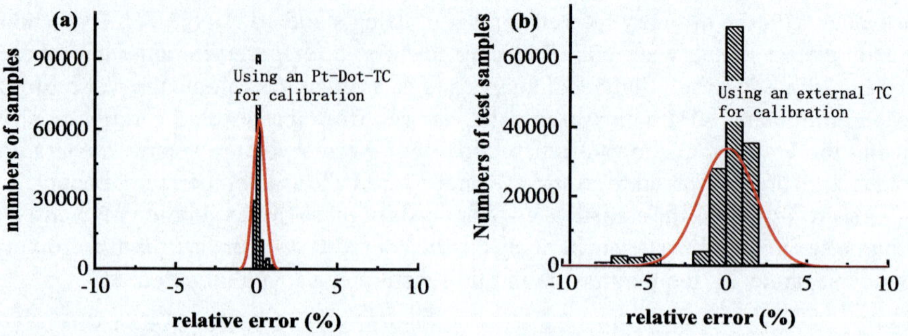

Fig. 5.4 The relative error of temperature calibration **a** using an in situ dotted Pt thermocouple. **b** Using the external standard R-type thermocouple

In summary, accurate and precise surface temperature measurements can be achieved by a new methodology of temperature calibration. Traditionally, the space–time difference between the DUT and the standard thermocouple causes the uncertainty and hysteretic effect of thermal voltage versus temperature in the ramping up-and-down temperature cycling process. Our newly invented Pt-Dot-TC can greatly reduce the time–space-delay-induced measurement errors. The TFTC built-in MEMS technology together with the new in situ calibration methodology provides a good candidate for future accurate and precise surface high-temperature test arrays for smarter aero- and astro-systems.

5.1.2 Dynamic Performance

Distribution Measurements

Traditional sensors such as wired thermocouples are cumbersome due to their large size and difficulty of installation. For example, when we are trying to test the temperature distributions along the blade, multiple thermocouples are required. We have to cut the slots on the turbine blade surface and implant them one by one along the surface. In contrary, multiple TFTCs can be simply built as an array on the turbine blade surface in a batch. As shown in Fig. 5.5, a simple yet effective TC array was formed by a series of Pt soldering dots on a slot of PtRh thin film on a turbine blade. This forms a series of Pt/PtRh thermocouples to measure the temperature distribution along the line.

Heating experiments were performed by using ethanol and acetylene flames blowing along the line. The temperature is higher when the sensing dot is closer to the beam. There are two ways to heat up the sensor array: blowing the beam from the right-hand side and from the center to the edge. The surface temperature will gradually decrease along the heat flow. This is verified by temperatures measured with this Pt/PtRh sensing array. The distribution of the measured thermal voltage

Fig. 5.5 Left: test array of 4 thermocouple sensors on the rim. Right: high-temperature beam flowing along the rim of the blade

from the sensor array agrees well with the temperature distribution caused by the ethanol heating flame. For the first case, the temperature on sensor #4 should be higher than the others, while for the second case, sensors #2 and 3 should have a higher temperature than sensors #1 and #4.

Figure 5.6 shows the distribution of the thermal voltage response along the rim when the heat flow is from right to left and from center to edge. The thermal voltage response of the sensors follows the same trend as the temperature distribution along the flame, i.e., the right sensor #4, which receives the highest temperature, shows the highest thermal voltage. When the heat flame flows from center to the edge, the temperature at the center is larger than that at the edge, i.e., the sensor#2 and #3 show higher temperature than that from sensor#1 and #4. Likewise, when the flame heats the sensors from #4 to #1, The sensing temperature follows the similar trend, i.e., the sensor#4 has a higher temperature than the others.

In summary, a thermocouple test array can be formulated and built onto the turbine blade surface to measure the temperature distribution very easily by MEMS method compared with conventional implanted armored wired TC methods. This methodology is very useful to monitor the dynamic temperature distributions on the turbine blade under the transient high-temperature ambient.

Transient Measurement

The TFTC sensor can quickly capture the transient temperature reading with responsiveness less than 10 μs. As shown in Fig. 5.7, the fast data collector (DEWE-43-A) can collect data at a 100k rate, corresponding to the 10 μs responsive time. If the data collector can capture every data point in a 100 K rate, then its dataset will reflect the responsive time less than 10 μs.

The transient test was performed using a flaming gun heating a series of TFTC sensors in the array. As shown in Fig. 5.8, when the flame heated the blade surface along the sensor array in the order of TC1, TC2, and TC3, TC1 receives the highest temperature, followed by TC2 and TC3.

This is indeed so as shown in Fig. 5.8. The actual detection temperatures of TC1, TC2, and TC3 in the TFTC array follows the heat flow as expected, i.e., the closer to the flame, the higher the surface temperature. This is a good indication that the TFTC

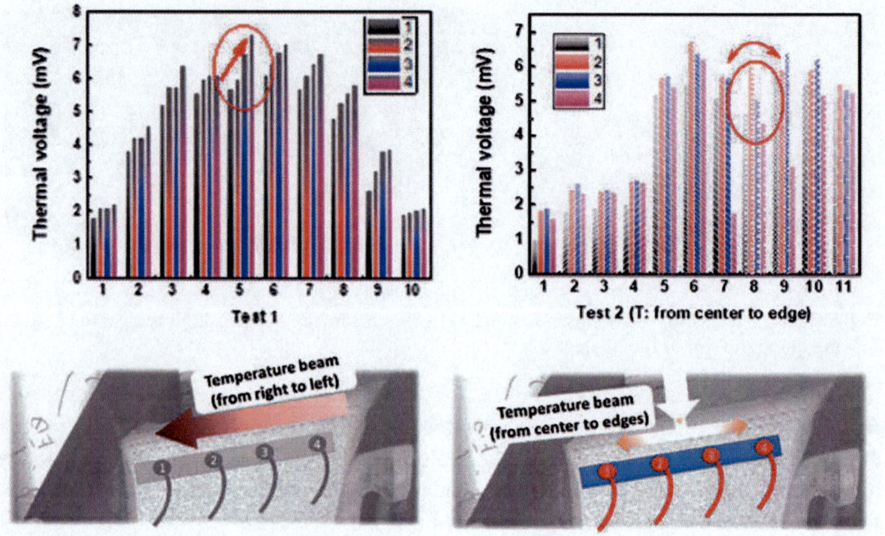

Fig. 5.6 Thermal distribution tests: the heat flow is from right to left (left), and the heat flow is from the center to the edge (right)

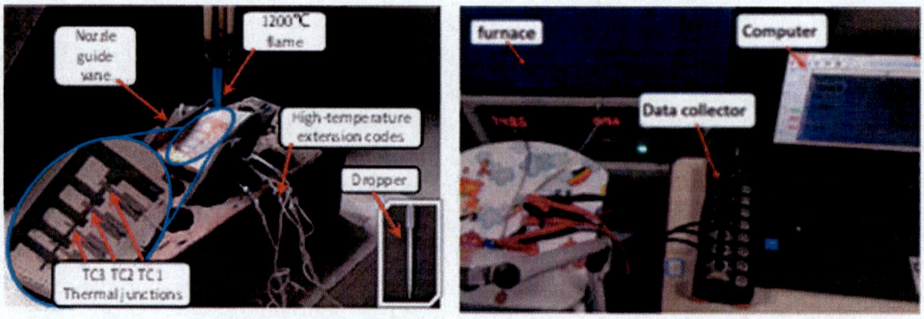

Fig. 5.7 Transient temperature test system

array is able to detect the temperature distribution on the blade surface. Figure 5.8 also shows a distinct discovery of how to measure the transient responsiveness. The temperature fluctuation trends of the three TFTCs are synchronized as shown in Fig. 5.8 right (same up, same down). If this fluctuation is a noise, it shouldn't follow the pattern systematically. This synchronized responsiveness is within 10 μs, which is the limit of our testing equipment (100,000 data per second). The above results suggest a new method to evaluate the response time of a sensor, which used to be a difficult measurement to quantify the time-responsiveness. Also, as seen from Fig. 5.8 (right), every 10 μs, the TFTC can collect one distinct reading of the temperature, which implies that the transient responsibility of the TFTC is no larger than 10 μs, due to the limitation of our data collector, the maximum sampling rate is 100 K (one

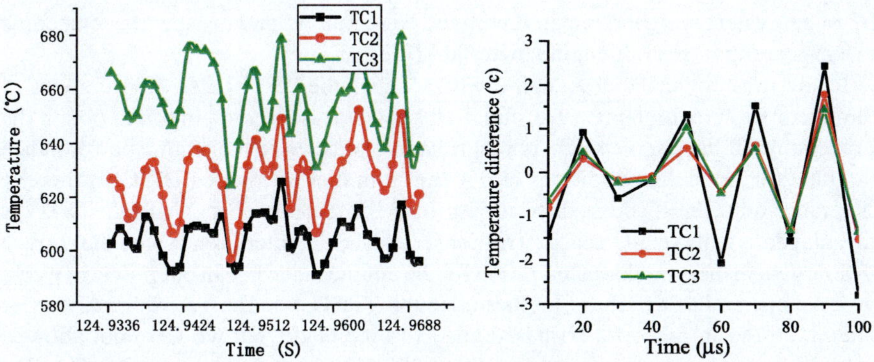

Fig. 5.8 Left: the TFTC sensors in the array are able to reflect the temperature distribution trend along the line. Right: the transient response speed is ≯ 10 μs

set of data per 10 μs). If the data collector can collect the data rate in a 1000 K rate and the data shows a similar symptom, then the response time is 1 μs. Therefore based upon the current data, the Pt/PtRh TFTC can capture the temperature at least as fast as 10 μs, which indicates a pretty fast responsiveness of the TFTC sensor. Compared with other methods to test the dynamic response ability such as a high-speed pulse laser, this methodology to characterize the responsive speed is more intuitive, much simpler, and more feasible.

In summary, the Pt/Rh/Pt thin film thermocouple sensors integrated on the blade surface are capable of measuring high temperatures with a nearly linear sensitivity of 12 μV/°C. Good stability and consistency are also achieved after 1 h of annealing at 800 °C prior to its first use in the field. The sensor array was capable of detecting transient temperature distributions with a fast responsiveness of less than 10 μs. This TFTC array can be in many AIAA occasions such as the calibration of other temperature measurement methods in infrared thermography, or can be used in cooling effect verification for optimized turbine blade design.

5.1.3 Ultra-high Temperature Measurements

The power-to-weight ratio of next-generation aeroengines has steadily been increasing, causing the inlet temperature to increase up to 1800 °C or higher [1]. The traditional nickel–chromium alloy metal-based blades can only bear temperatures up to 1000 °C, which must be cooled down by either the cooling gas and/or a thermal barrier coating layer, which increases the system complexity and eventually reduces the combustion efficiency of the engine. In short, traditional nickel–chromium alloy-based blade systems have difficulty meeting the needs of next-generation advanced propulsion systems. A ceramic matrix composite (CMC) material has become a new material candidate, which can both bear higher temperatures and be less weight (~

1/3 of equivalent weight compared with the Ni–Cr alloy) and is expected to become a next-generation aircraft engine material [2].

In the meantime, the aeroengine works under the "3H" harsh environment of "high temperature, high pressure, and high stress". It is very important to obtain the temperature of the engine blade in real time to characterize the engine performance and monitor its health conditions [3]. A thin film thermocouple (TFTC) is thereby integrated onto the blade surface in situ by MEMS technology. MEMS TFTC is advantageous in its small mass, fast response, and lack of interference with the surface airflow path in thermal characterization for the engine blade [4]. In our previous work, we developed a Pt/PtRh thin film thermocouple (TFTC) on the YSZ thermal barrier coating on the surface of a Ni-based alloy turbine blade, but we were not allowed to measure the higher temperature above 1000 °C simply because the Ni–Cr alloy substrate cannot bear it. In fact, the Pt/PtRh TFTC theoretically can measure the high temperature up to 1600 °C, but the high-temperature limit of Ni–Cr-based alloy of turbine blade limits us to quantify the TFTC performance up to 1000 °C. Higher temperatures will cause damage to the turbine blade. As shown in Fig. 5.9, the blade is substantially oxidized after a few above 1000 °C temperature cycles, and the TBC together with the thin film thermocouple surface was peeled off.

TFTC on CMC Substrate

Using CMC material as the substrate to make Pt/PtRh thin film sensors can effectively circumvent these two shortcomings. First, the CMC material has a higher temperature-bearing limit. Second, CMC is a ceramic-based material that already possesses good high-temperature electrical insulation quality, which ensures that TFTC functions as an electric sensor on its surface [5, 6].

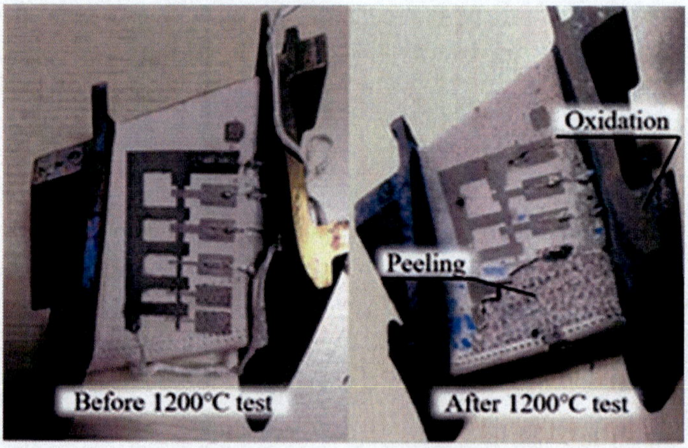

Fig. 5.9 The peeling off the TBC and TFTC on Ni–Cr alloy blade after the above 1000 °C cycles due to temperature stress mismatch

We used CMC as substrate to build Pt/PtRh thin film thermocouple using the MEMS process and perform high-temperature measurements in a muffle furnace, and obtained very good TFTC character up to 1500 °C as compared to just 1000 °C. We also perform thermal shock experiments using flame heating with immediate water cooling. The thin film thermocouple can work normally without peeling off and has good thermal shock resistance.

The following is the detailed process of the CMC sample:

(1) Preparation of raw material sheet. The modified phenolic resin, silicon carbide powder, and carbon powder were ball milled and dispersed with ethanol as the dispersing medium. Then, the slurry was evenly leveled and dried by a scraper.
(2) Lamination molding. Many layers of dried raw material sheet were put into the mold for hot pressing to make the preform.
(3) Carbonization. The preform was put into a high-temperature furnace for carbonization to obtain a porous body.
(4) Infiltration. The porous body and excess silicon powder were infiltrated in the graphite crucible, and the infiltration temperature was 1500 °C with a holding time of 30 min.

The following is the detailed process of TFTC fabrication on the CMC surface:

(1) Surface cleaning. Acetone and deionized water were used for ultrasound for 2 min to remove oil and dust on the CMC surface.
(2) Drawing the PtRh pattern. The diluted PtRh slurry was dipped with a soft pencil, evenly coated on the CMC surface according to the design shape, and baked in an oven at 100 °C for 30 min until the PtRh slurry on the surface was dry.
(3) Drawing the Pt pattern. As in step 2, make a dry Pt pattern on the CMC surface.
(4) Pin connection. Pt and PtRh slurries are used to connect the Pt wire and PtRh wire at the pin to the Pt end and PtRh end, respectively.
(5) An R-type wire thermocouple is also welded on the same CMC surface by the Pt slurry used to calibrate the TFTC sensor. In this way, the temperature difference between the thin film thermocouple and dotted R-type thermocouple can be greatly reduced, and the calibration accuracy can be greatly improved.
(6) Welding. The CMC material coated with slurry was placed into a high-temperature furnace, heated to 800 °C for 60 min, kept warm for 20 min, and then naturally cooled down to obtain the thin film thermocouple and dotted Pt thermocouple shown in Fig. 5.10. The film thickness of the thin film thermocouple is measured to be approximately 11 μm with a stylus profiler.

The Pt/PtRh TFTC is integrated on the CMC surface and is capable of measuring a high temperature of 1500 °C or higher with a measurement error within $\pm$ 0.2% and $< \pm 1$ °C. The TFTC has good binding with the CMC material, which can withstand multiple thermal shock tests. The Pt/PtRh TFTC and CMC materials constitute a perfect pair of high-temperature measurement systems for next-generation intelligent aeroengines (Fig. 5.11).

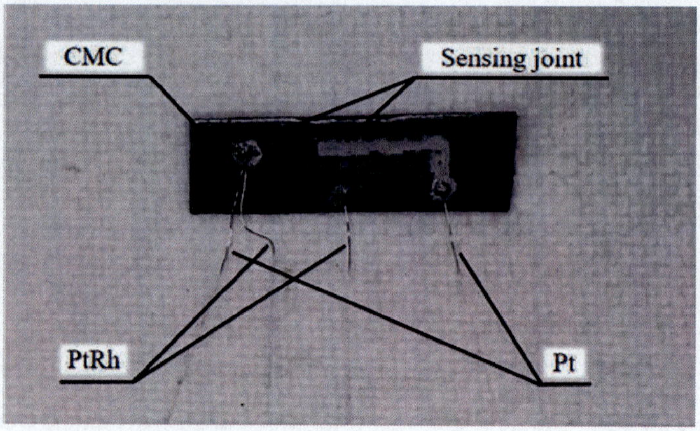

Fig. 5.10 The thin film thermocouple built on the CMC substrate together with dotted Pt thermocouple as temperature reference sensor

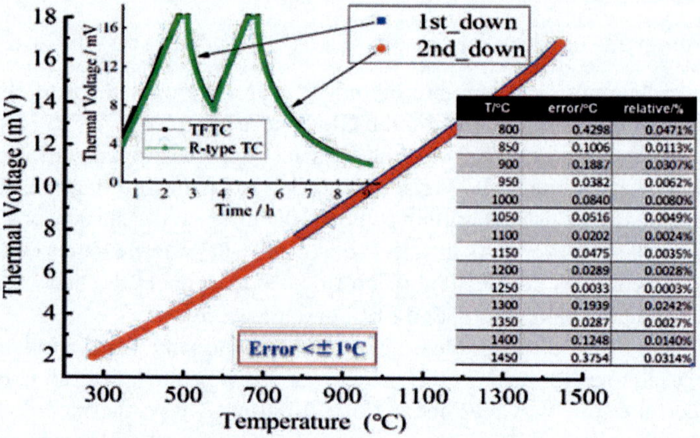

Fig. 5.11 TFTC performance on CMC substrate

5.1.4 *Wireless Temperature Sensing*

SAW-Based Wireless Sensing System

In Chap. 4, we successfully built a SAW sensor based on cheap lithium-niobate-based (LN) piezoelectric substrate and platinum-electrode IDT. Such SAW sensor is capable of measuring up to 1100 °C high temperature. However, using such a sensor to fulfill wireless sensing requires an external antenna to amplify the SAW signal. As shown in Fig. 5.12, an RF signal was emitted from an outside antenna, and this Wi-Fi signal was received by the near-field antenna. The SAW sensor received the

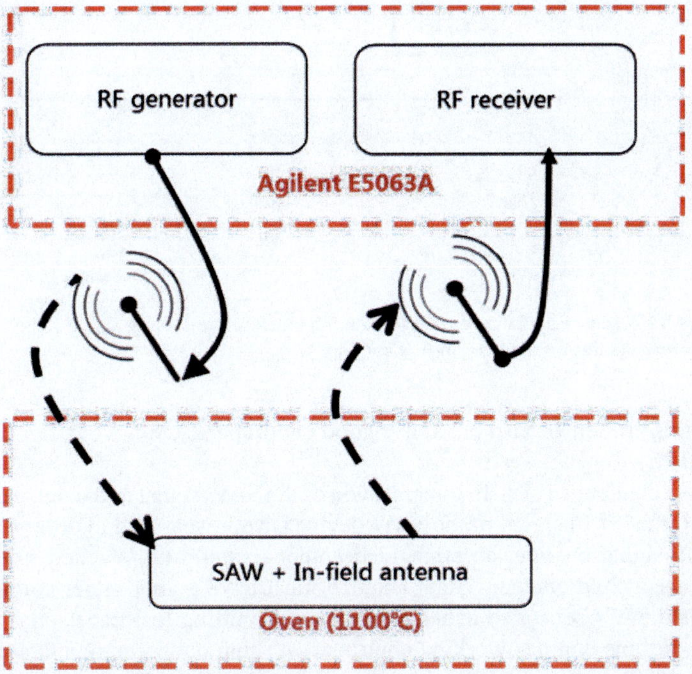

Fig. 5.12 Schematic plot of the wireless sensing systems using the SAW device and in-field antenna

Wi-Fi signal from the in-field antenna. After sensing, another RF signal from the SAW sensor is transmitted back to the near-field antenna. This near-field antenna emits the RF signal to the outside antenna for RF signal receiving and processing. Therefore, the whole sensing process is divided into two parts: a high-temperature SAW sensor and RF signal transmission.

We have already conquered the first challenge in the last chapter by a SiO_2 passivation layer covering the LN SAW device to avoid high-temperature damage. In the previous chapter, we have already seen in Fig. 4.11 that SAW sensor possesses a very consistent f_0 versus temperature behavior which is a good indication that our LN SAW sensor is capable to stand for up to 1150 °C high temperature. Here, we perform a more extensive test, i.e., the temperature cycling test that we repeated temperature from 800 to 1200 °C a few times in a 6 h span. As seen from Fig. 5.13, the SAW sensor's signal follows the temperature change quite well, and the repeatability is also very good. During the over 5 h temperature cycling shown in Fig. 5.13a: from room temperature (1100 °C) (phase 1), from 1100 to 800 °C (phase 2), then ramp up to 1100 °C again (phase 3) and eventually cool down naturally (phase 4), the sensor's performance as shown in Fig. 5.13b shows a good linearity and repetition of the resonant frequency versus temperature relationship from 800 to 1100 °C with a relative error less than 0.1%. The same SAW sensor was tested afterward at 800 °C for 3 h, while the SAW sensor's performance remained stable, indicating that the

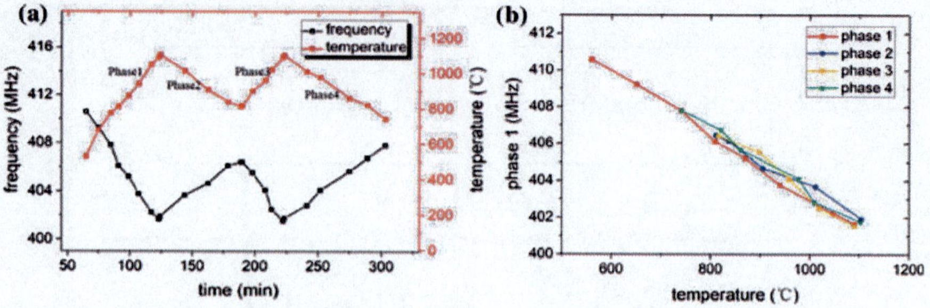

Fig. 5.13 **a** SAW sensor performance during the 6-h temperature cycling test. **b** Repetitive results of the SAW sensors during the temperature cycling

passivated SAW sensor also has fairly good endurance ($\not<$ 8 h when temperature $\in$ 800–1100 °C).

The next challenge is Wi-Fi transmission of the SAW signal to the outside readers. In general, the Wi-Fi signal of the SAW device is very weak and in order to transmit its sensing signal outside, an embedded antenna should be attached to it locally, we call it near-field antenna. Such antenna must have a good inner coupling with the attached SAW sensor to avoid the inner transmitting loss especially under the tough gas turbine engines working environment (high temperature and combustion ambient), which reduces the RF signal strength of the infield antenna as well as the SAW device. Figure 5.14 shows a preliminary work by connecting an SAW device with a copper rod antenna inside a high-temperature oven. Since the copper cannot withstand high temperatures with such a system, the Wi-Fi test temperature did not exceed 400 °C. More work is needed to develop high temperature near-field antenna in order to achieve the complete Wi-Fi communication of the SAW sensing.

We did some initial studies of the near-field antenna. At first, we need to understand what the suitable shape of the near-field antenna is and which direction has most transmission power? We have found that the ring-type antenna is most suitable to amplify the SAW signal. We did a simple experiment by building a ring antenna using our school logo on a LN substrate and measure its RF performance with a RF probing system (GSG prober and Cascades RF probe station). The probing of the school logo antenna and the result is shown in Fig. 5.15a and b, with a very good RF performance. We also studied its directional amplitude of the Wi-Fi signal by simulation and we found that the center vertical direction has the most powerful Wi-Fi signal as shown in Fig. 5.15c. This study helps us to build our future wireless SAW sensing system for the most efficient signal transmission between the SAW device and the near-field antenna. We had to keep in mind that this antenna must bear high temperatures (1000 °C) since the near future antenna is also in a high-temperature environment as the SAW sensor, which we plan to use the MEMS method to build an embedded ring antenna on the turbine blade surface (Fig. 5.17 (right)).

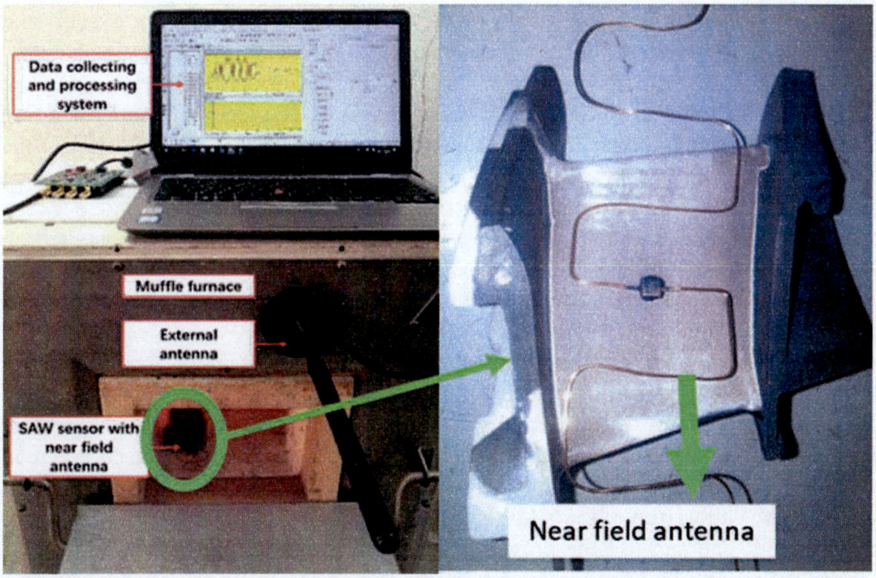

Fig. 5.14 Wireless SAW sensing system for high temperature testing

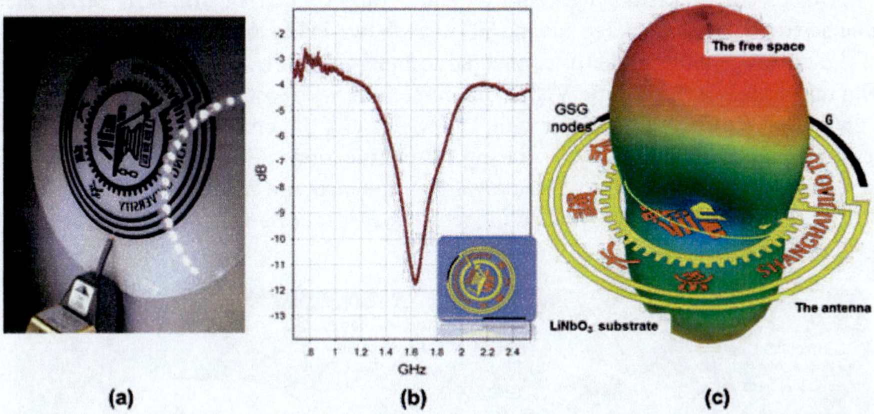

Fig. 5.15 The SAW's near-field antenna studies: **a** the probe for RF test, **b** the S11 performance; **c** the directional Wi-Fi transmission amplitude

Automatic Wi-Fi Temperature Sensing Using SAW Sensor

We usually use an expensive probe station and special RF pins to connect the SAW device to the network analyzer for the RF test (Fig. 5.15). Recently, we have developed an automatic Wi-Fi mobile system to measure the RF signal of an SAW sensor. The test system is much cheaper and it is a mobile device, which can replace the traditional way of RF testing using much comprehensive RF testing system and obtain the temperature onsite and in real time.

In this experiment, first we slice the SAW device on LN substrate using the DAD3650 Slicer into small dies and then use Pt paste to connect SAW nodes welding to a platinum wire and then connect to the pins on the package. We make use of the newly invented Pt-Dot-TC as the temperature calibration during the SAW device testing. The SAW with the embedded Pt-Dot-TC sensor is both on the same package plate to ensure the exact temperature reading. The SAW package was installed on the PCB board and then connected to external antenna.

Our goal is to use a small network analyzer (an economic NanoVNA purchase on Amazon or Taobao) to measure the RF signal of the SAW device quickly and automatically in real time. The script is written so that the user can obtain real-time f_0 and return loss versus time from the VNA. In the meantime, we use the embedded thermocouple to measure the thermal voltage versus time as the real-time temperature readout of the SAW sensor. We correlate the timing from these two datasets and then obtain the real-time return loss versus temperature curve as shown in Fig. 5.16. In this way, the calibration of the SAW is achieved by using the embedded Pt-Dot-TC, and later on, one can use the corresponding return loss reading as a temperature sensor. Automatic real-time temperature measurement can be achieved by using a very cheap portable Wi-Fi VNA as a temperature reader. Combined with the antenna, an economic real-time Wi-Fi temperature mobile reading device becomes possible.

Two TX433-JKD-20 wide-band antennas (approximately 400–470 MHz) are attached to the SAW package and the VNA, as shown in Fig. 5.17. The SMA interface of the antenna can be directly connected to the PCB interface of the SAW device. Figure 5.17 left is the mobile VNA communicating with saw package, and Fig. 5.17 right is the Wi-Fi communicate with The SAW with the embedded antenna on the turbine blade achieving the wifi sensing communication.

After the connection, the absolute value of the resonant frequency f_0 shifts from 400 MHz to close to 470 MHz, and the return loss is increased from – 15 dB to more

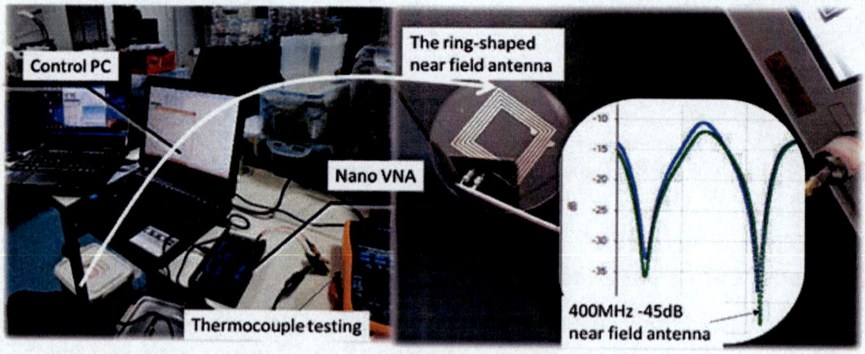

Fig. 5.16 (Left) Fitting curve of the SAW return loss value and temperature. (Right) Measurement system

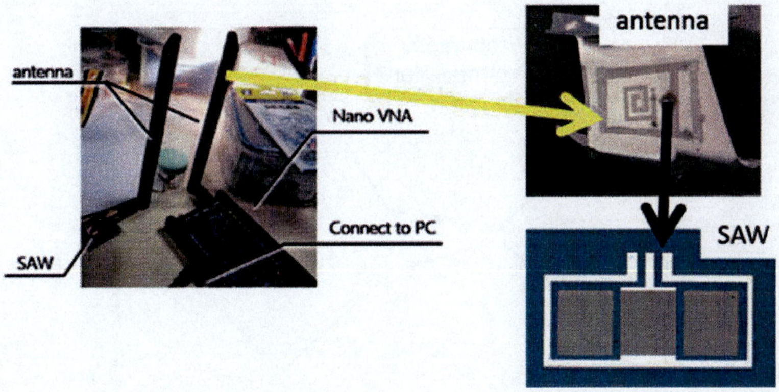

Fig. 5.17 Left: TX433 JKD—20 antenna. Right: wireless measuring system

than – 40 dB, as shown in Fig. 5.18. When the ambient temperature changes, the position of both f_0 and the return loss change accordingly. In this way, the temperature sensing is achieved.

Combining the above efforts, we are able to achieve automatic real-time Wi-Fi sensing from the SAW sensor that we built on an LN substrate, although the temperature is not as high yet. A heating plate is used to modify the temperatures of the SAW sensor together with the Pt-dot thermocouple (Pt-Dot-TC) from approximately 25 to 70 °C. The real-time temperature Wi-Fi sensing result is shown in Fig. 5.19.

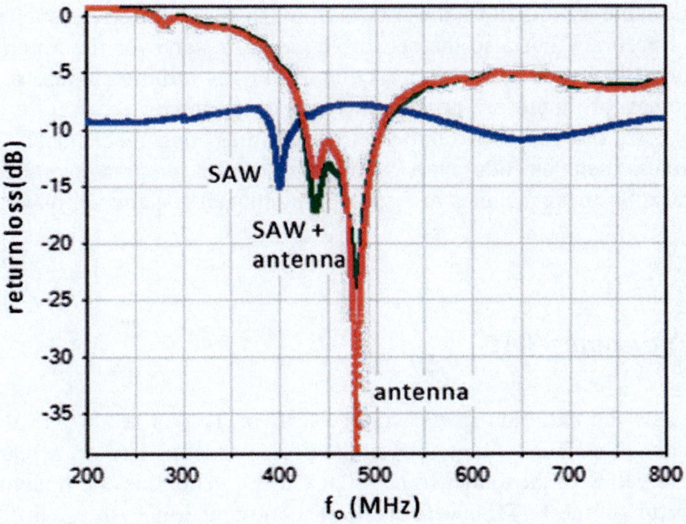

Fig. 5.18 The frequency curve of the SAW sensor (red) and antenna (green) and their connection (blue)

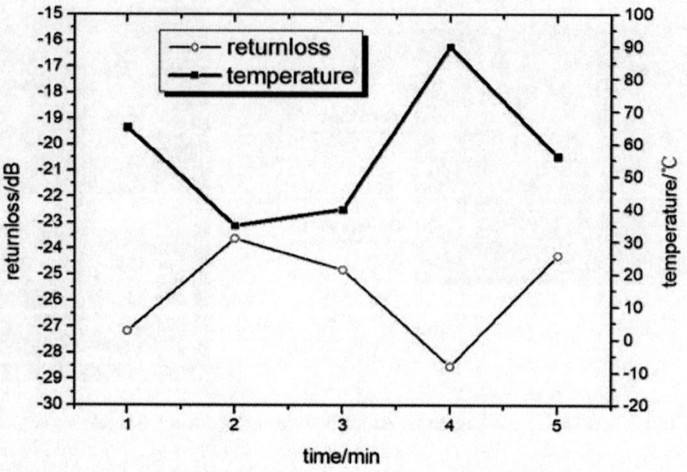

Fig. 5.19 SAW return loss and temperature curve trends for wireless connections

The return loss corresponds to the temperature very well, indicating that SAW Wi-Fi automatic measurements can be fulfilled.

5.2 Robustness Test—AIAA

Not only must the sensors fulfill the criteria from IEEE's perspectives, such as performance and accuracy but also the sensors be able to stand for the tough working ambient in AIAA, such as inside the combustion gas turbine engine. A series of robustness tests are therefore needed and have been conducted as following. The robustness tests include both temperature robustness and mechanical. Temperature robustness includes endurance, cycling, and abrupt temperature shocking test. Mechanical robustness includes heavy shocking impact test and intensive vibration test.

5.2.1 Endurance Test

Table 5.2 lists the variation in the voltage drift of TFTCs at 800 °C along with the endurance time. The voltage drift in the first hour is the largest, accounting for approximately 95% of the total drift of 130 μV in 3 h. After this one-hour annealing, the temperature of the TFTC is very stable. The initial thermal voltage drift is due to the oxidation of rhodium in PtRh films at 800 °C. This effect tends to be saturated after annealing, and since then, the Pt/PtRh TFTC has stabilized. This result suggests

Table 5.2 Thermal voltage drift of TFTCs at 800 °C

Thermal couple	Total voltage drift (μV^{-1})	Voltage drift (μV^{-1})		
		1st hour	2nd hour	3rd hour
TC1	176.78	163.89	10.72	2.17
TC2	138.45	133.12	6.28	0.95
TC3	134.62	130.48	2.37	3.47

that the newly fabricated TFTC should be annealed for at least 1 h to improve its working stability prior to its first use in the field. After that, the TFTC is highly endurable, with a drift rate of ~ 0.1% per hour.

5.2.2 *Temperature Shocking Test*

The turbine engine for commercial airplanes works under high-temperature stress conditions. At one hand, the higher temperature, the higher efficiency. However, the temperature tolerance of aero-engine materials is lower than the rated temperature [7]. For this reason, it is necessary to use cooling gas and TBC (thermal barrier coating) to cool the surface of the blade [8, 9], which increases the service temperature of alloy materials by 100–150 °C [10]. The MEMS-built sensors are multilayer structures of TFTC/TBC/Ni–Cr turbine blade alloy. Due to the difference in the thermal expansion coefficient of different layers, the films may peel off due to thermal stress caused by the mismatch of these expansions (Fig. 5.20). The thermal stress is on the order of several hundred MPa, which is close to the yield limit of the material if the layers are not properly coated, such as a lack of the MCrAlY bonding layer. Once this stress value exceeds the yield limit, it will cause deformation and the film will fall off from the substrate.

Thermal stress issue of this multilayer structure under high temperature must be properly addressed to avoid peeling off due to thermal coefficient mismatching, which causes the reliability problem of the system and the malfunction of the thin film sensors. Therefore, how to reduce the thermal stress becomes a very important challenge that we must conquer. Fortunately, we have successfully developed a new TBC coating recipe, together with the good TFTC sensor technology, which ensures the thin film is robust enough under the rough temperature shocking test conditions.

A thermal shock test was conducted as shown in Fig. 5.21. The sample was directly heated by a flaming gun, and immediate cooling was achieved by spraying or dipping cold water on the hot blade surface to achieve an abrupt temperature shocking effect. This is to verify the robustness of the thin film TFTC sensor and the TBC layer under such tough thermal stress. The test result shows that under such an abrupt temperature change, the TFTC sensor holds firmly on the blade without peeling off. The cooling temperature behaviors are shown in Fig. 5.22.

Fig. 5.20 The peeling off of the TBC and TFTC due to the thermal stress from thermal expansion mismatch

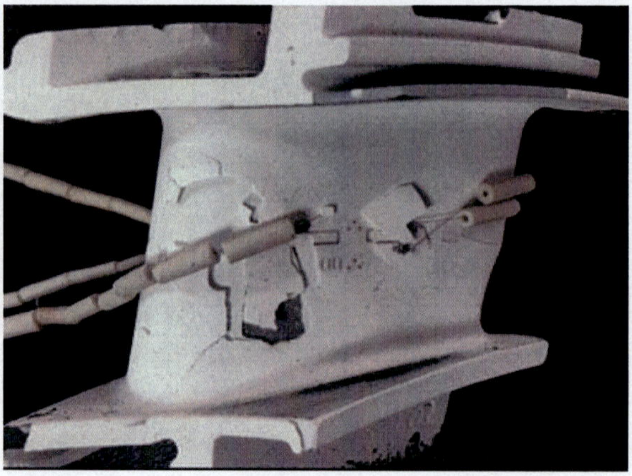

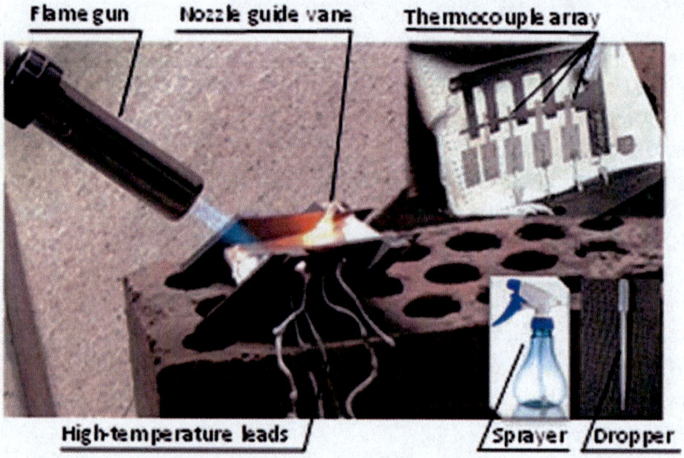

Fig. 5.21 On-site thermal shock test

Also shown in Fig. 5.22, the three thermocouples have similar temperature response curves when water is dropped at the hot junctions of TC3, TC2, and TC1. For example, when water droplets touch TC3, the temperature immediately drops from 700 to 200 °C. After the water drops evaporate completely, the hot substrate slowly heats up TC3 through TBC, and then the temperature of TC3 rises back slowly. The TFTC sensor arrays can capture the sharp and rapid temperature changes in this thermal conduction process. This test also validates the robustness of the TFTC in this abrupt temperature shocking test: no thin film peeling of the TFTC and TBC layers is observed. As seen, the TFTCs still work after this tough heating/cooling cycling test—a good indication that the thin film can be firmly bonded to the TBC surface.

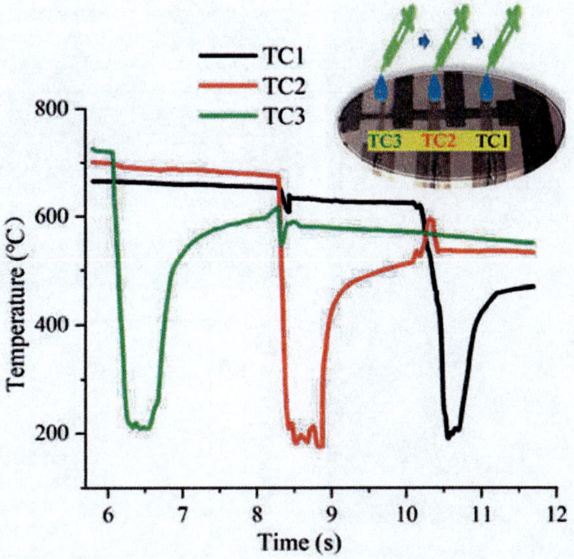

Fig. 5.22 Temperature response curve of TFTCs under water dripping cooling

5.2.3 Mechanical Shocking Test

Figure 5.23 shows the mechanical robustness test system. The maximum load of the vertical impact test rig is approximately 100 g, the duration of the impact pulse is 1.5–40 ms, and the peak acceleration is up to 600 g. The impact system can execute half sinusoidal waves to simulate the shock waves and the impact energy exerted on the turbine parts. These shocking and impact tests are intended to verify the survivability and integrity of the sensor systems, including the thin film sensor, soldering and wiring on the turbine blade. Visual and electrical connection tests are used to validate the sensors after the shocking.

Electrical connections and resistance before and after the impact test are verified by visual and multi-meters. If the connection is not broken and its resistance value is unchanged, then the robustness of the TFTC sensor system is valid. This sensing system embedded on the turbine blade includes: 1) the thin MEMS sensor, 2) the solder joints, and 3) the bonding wire. We have performed 5 times shocking tests and no broken connection is found after the tests.

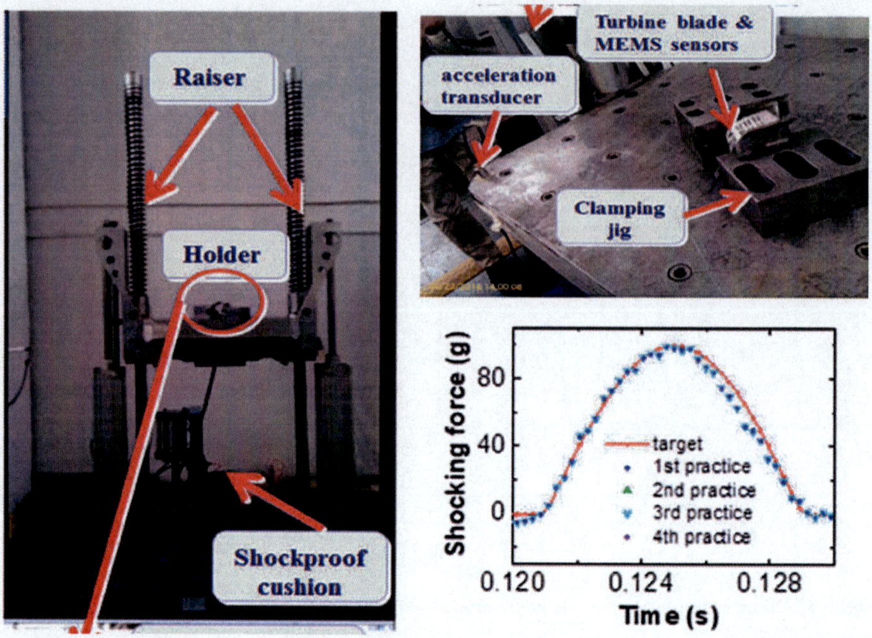

Fig. 5.23 MEMS sensors and wires together with the vertical shock testing systems. Below is the shocking test plans and practice. Shocking data agrees with the test plan very well

5.2.4 *High Vibration Test*

The embedded sensors must also survive harsh vibrating working conditions in gas turbine engine machines. Possible reasons for vibration could be the rotor imbalance, the gas flows through the channel, the gas engine excitation force, unbalanced force of the rotating parts, etc. The centrifugal load, the aerodynamic load, and the alternating load and compressor surge [11] may exacerbate the vibration state of the turbine system. These engine vibrations not only affect the working state but also make the structure of the engine suffer great vibration stress, which will have an adverse effect on the reliability of the engine. In the starting process, the maximum vibration value can reach 100 mm/s. Studying these documented experiments from aeroengine vibration provides a reference to set up our experimental scheme of vibration and shocking tests on MEMS-embedded sensors. We conducted a harsh vibration robustness verification test of our MEMS sensors built in Chap. 4, and the results show that the sensor system (including sensor and the wiring) is capable of standing for up to 40 g 10–2000 Hz vibration and 100 g shocking tests.

The experimental scheme of vibration intensity is shown in Fig. 5.24. A clamping jig was designed and fabricated according to the curvatures of the nozzle guide vane to mount the sensor system firmly on the vibration/shocking test platform. Figure 5.24 also shows the picture of the mounting parts and platform. The tenon slot of the fixture corresponds to the edge of the vane, and the two ends of the fixture can be trimmed according to the width of the vane. A separate acceleration sensor is installed on the vibration platform to monitor the frequency and intensity of vibration. The MEMS sensors and their wiring on the vane together with the monitoring sensor and vibration platform constitute a complete vibration test system.

The procedure of the robustness test of the MEMS sensors on the turbine blade is to measure and compare the electrical connections and resistance before and after the vibrations and impact test. This is an indication that the sensor system is vibration- and shocking-resistant if the connection is not broken and the thermal resistance value is unchanged. The sensor system on the turbine blade component includes the thin MEMS sensor, the solder joints, and bonding wire onto the blade surface—an electrical test would fail if any one of them is broken during the vibration and shocking test. Results show that the sensors as well as the connection are as solid as before after these harsh tests.

The thin films sensors were tested in an oven environment after running the vibration and engine-environmental test to verify its functionality under the 4-h 800 °C

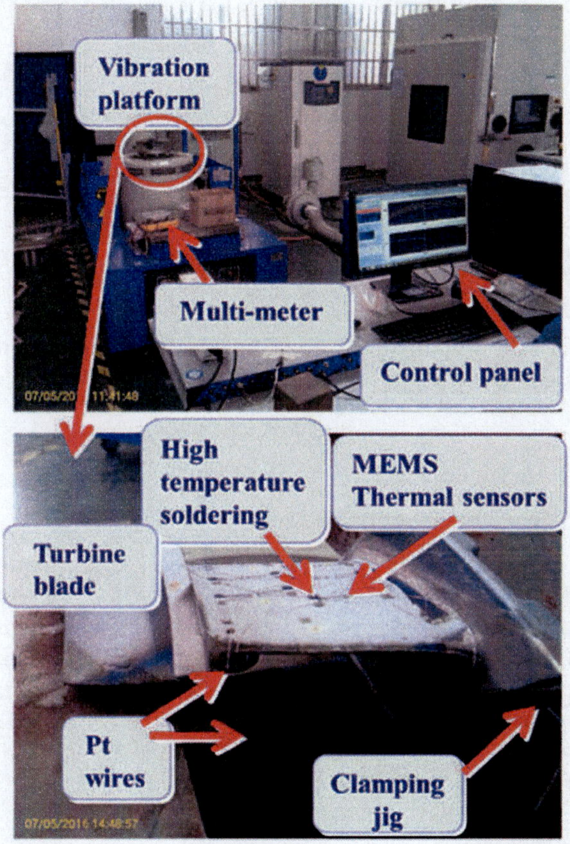

Fig. 5.24 The MEMS sensor, soldering, clamping jig, and experimental setup of the vibration platform with the controlling system

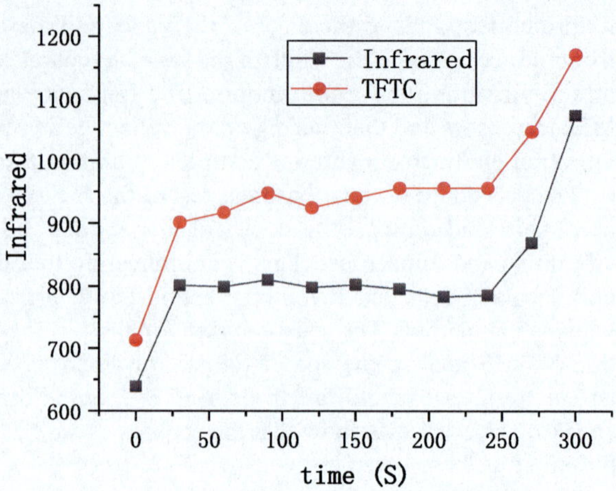

Fig. 5.25 The temperature responsive reading from the onsite built TFTC sensor and the infrared thermography under a very high-intensity kerosene heat flow

temperature cycling. Both the MEMS RTD and TFTC sensor exhibit a consistent and a stable thermal stability as the standard wired Pt/PtRh thermocouple device.

5.2.5 *Violent Heat Flame Rushing Test*

The test is performed by using a 400 M/s flow rate violent heat flow to flash the TFTC sensors built on the CMC surface. Thermal voltage is monitored with an external multimeter together with infrared thermography. The dynamic monitoring process of the TFTC together with infrared is shown in Fig. 5.25. The pretty consistent temperature difference between the two temperature test systems is observed, indicating that the thermography needs to be properly calibrated. The test process has a very high noise under such a violet heat flow rate, and an ear mask is needed for onsite professionals.

5.3 The Combustion Infield Test

Figure 5.26 shows the specific details of the infield test, including ① lead out of the thin-film sensor with a high-temperature connecting wire, ② fixtures to hold the blade on the high-temperature platform, and ③ high-temperature wires wrapped with qualified electrical insulation to the outside measurement equipment. Proper sealing of the connection wires is crucial in this test since the sensors are tested in the high-pressure/high-temperature ambient and no leaking is allowed during the experiment (the system will halt when leaking is detected).

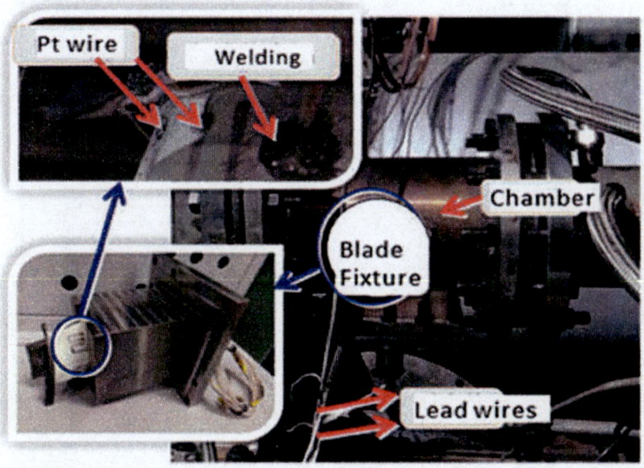

Combustion parameters		
air flow rate	kg/s	0.5
temperature	K	1476
gas/air ratio		0.024
pressure	Mpa	0.5
density	kg/m3	1.18
specific heat	kJ/kg.K	1.29
cross section	m2	0.008
rate	m/s	54

Fig. 5.26 The onsite test of TFTC inside the gas turbine combustion chamber [12]

Sensors survive after experiencing turbine engine hot and windy conditions and exhibit reliable temperature readings. The results show that the embedded MEMS sensor can detect surface temperatures up to 1200 °C under 0.5 MPa and 1.24 kJ/kg K heat flow rate environment. Sensors survive after two cycles of combustion while staying for 10 min at peak temperatures (~ 1200 °C). The outlook of the TFTC sensors before and after experiencing the high-pressure/high-temperature test looks the same as shown in Fig. 5.27, together with the mounting details of the testing system.

Figure 5.28 shows verification test after the sensor was taken out of the combustion test. It is seen that the RTD sensor is undamaged by the intensive heat flow in the combustion chamber experiment.

In summary, this chapter discusses the special tests requested by AIAA people in additional to the basic test in IEEE.

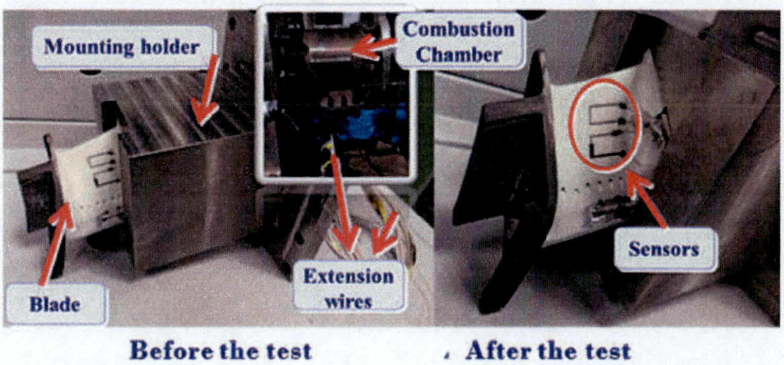

Fig. 5.27 Turbine blade with sensors mounted onto the hot engine combustion chamber

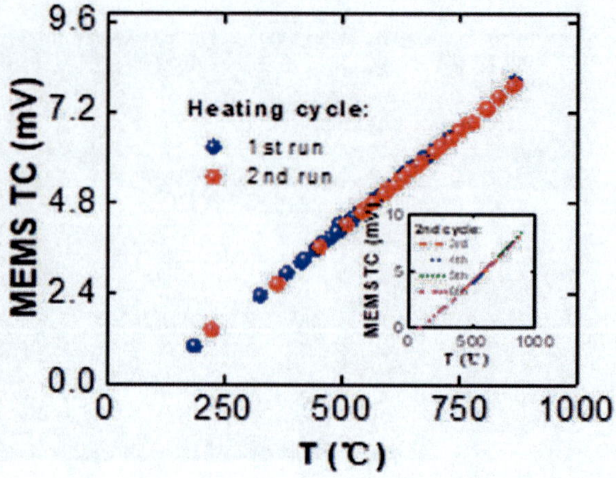

Fig. 5.28 Temperature sensing behaviors of MEMS sensors after the combustion test. The inset is the repeatability test

References

1. Zhang, G.Q.: Research and development of high temperature structural materials for aero-engine applications. Acta Metall. Sin. **18**(4), 443–452 (2005)
2. Verrilli, M., Calomino, A., Thomas, D.J., Robinson, R.C.: Characterization of ceramic matrix composite vane subelements subjected to rig testing in a gas turbine environment. In: Fifth International Conference on High Temperature Ceramic Matrix Composites, Seattle, WA, USA, Sept 2004 (2004)
3. Padture, N.P.: Advanced structural ceramics in aerospace propulsion. Nat. Mater. **15**(8), 804–809 (2016). https://doi.org/10.1038/nmat4687
4. Hunter, G.W., Neudeck, P.G., Okojie, R.S., Beheim, G.M., Powell, J.A., Chen, L.: An overview of high-temperature electronics and sensor development at NASA Glenn Research Center. J. Turbomach. **125**(4), 658–664 (2003). https://doi.org/10.1115/1.1579508
5. Franklin, L.D., Li, J., Gao, J., Ding, G., Cao, X.: Integrated fabrication of high-temperature microelectromechanical system sensor on aeroengine turbine blade. J. Thermophys. Heat Transfer **32**(3), 828–830 (2018). https://doi.org/10.2514/1.T5147
6. Gao, J., Franklin, L.D., Yu, C., Meng, W., Liu, L., Ding, G.: Electrical insulation of ceramic thin film on metallic aero-engine blade for high temperature sensor applications. Ceram. Int. **42**(16), 19269–19275 (2016). https://doi.org/10.1016/j.ceramint.2016.09.093
7. Hongbo, G., Shengkai, G., Huibin, X.U.: Progress in thermal barrier coatings. Mater. China (2009)
8. Bose, S., Demasi-Marcin, J., et al.: Thermal barrier coating experience in gas turbine engines at Pratt & Whitney. J. Therm. Spray Technol. **6**(1), 99–104 (1997)
9. Xu, H., Guo, H., Liu, F., et al.: Development of gradient thermal barrier coatings and their hot-fatigue behavior. Surf. Coat. Technol. **130**(1), 133–139 (2000)
10. Schulz, U., Leyensa, C., Fritscher, K.: Some recent trends in research and technology of advanced thermal barrier coatings. Aerosp. Sci. Technol. **7**, 73–80 (2003)
11. Gravdahl, J.T., Egeland, I.O.: Compressor Surge and Rotating Stall. Springer (1999)
12. Duan, L., Ji, Z., Weng, H., Li, J., Lin, Y., Cao, X.: MEMS high temperature measurement technology on the surface of turbine guide blades. Aviat. Manuf. Technol. **63**(5), 62 (2020). https://doi.org/10.16080/j.issn1671-833x.2020.05.062

Chapter 6
The Mindset Analysis of AIAA and IEEE

This chapter serves as an intermission like the long music drama "Sound of Music", intending to connect the previous session to the next session. This chapter intends to connect the AIAA and IEEE interactions from previous to future. From the corporative intimate interaction experience of AIAA and IEEE people on developing a specific project done—the smart IEEE sensor to test the surface temperature of the turbine blade in AIAA, we extrapolates this cooperation to the promising future of AIAA/IEEE in electrified aviation. In Part 2 (Chaps. 3–5), we use our six years' true experience of IEEE people with AIAA colleagues to explain the integrated process between the two groups of professionals, serving as a reference point for the future bigger cooperation—the electrified aviation in Part 3 (Chaps. 7–11). Electric propulsion and associated aviation electric power in Part 3 is much more challenging and more promising and may bring a more revolution to the whole aviation enterprise.

6.1 The Necessity of AIAA/IEEE Working Together

Initially, either AIAA or IEEE goes its own way of engineering and innovation because they belong to two independent branches in physics. AIAA involves mechanics, thermal dynamics, and material science/technology, and IEEE addresses electrical and electronic science/engineering. AIAA has two aspects, astronautics, and aeronautics, as does the IEEE, electricity, and electronics.

From 1960–1970, electronic control started to help the aviation industry with the advent of mature electric motors and transistors. The main help from IEEE to AIAA was mainly to provide electric power for the airplane and electric equipment from the electrical aspect of IEEE. From the perspective of electronics in IEEE, the control and automation from the devices and circuits made the aero-machines easier to operate and manipulate like the front control panels in front a pilot.

Since then, both AIAA and IEEE have made a great deal of achievements. Due to the advent of artificial intelligence (AI) and electric vehicle (EV), many useful

© The Author(s), under exclusive license to Springer Nature Singapore Pte Ltd. 2023
F. L. Duan, *When AIAA Meets IEEE*,
https://doi.org/10.1007/978-981-19-8394-8_6

technologies in IEEE were recognized by AIAA people to fit their need for advanced aviation purpose, for example, to use electric power shorten the takeoff distance of the big airliner. To achieve this goal, AIAA people need to work together with IEEE people, and this coworking not only involves technology expertise but also the man-to-man communications, namely, the IQ + EQ (Sect. 6.2.3) interlaced process. The motivation of this book is the integration of IEEE and AIAA, the two largest engineering organizations in the world, for more interactive and deeper cooperation involving the most recent advanced progress of both parties.

Recently, there have been strong interactive activities of these two teams regarding intelligent aero-engines and electrified propulsion [1]. Other than the activities from the academic and technical aspects, there exists the differentiation of the mindsets of AIAA and IEEE, together with their different technical habits and the multi-cultural differences since more and more future projects will be joint international for sharing costs/talents. Understanding these differences helps AIAA/IEEE people work together more efficiently. Part 2 (Chaps. 3–5) discussed the smart sensors built by the IEEE people to test the high temperature for various AIAA scenarios. This is the true experience from 2013–2022 of our IEEE team to work the AIAA people to develop smart sensors for jet engines. Not only this is still an active project between the AIAA/IEEE, here, we use this 10-year true working experience to illustrate the difference in mindsets, habits, preference and others. Our purpose is to use it as a learning curve for future deeper and more comprehensive topic—electrified aviation. Electric power and electric propulsion are two key technologies to enable electric aviation, and they all belong to the fields of IEEE. Both AIAA and IEEE people need to work together on drawing the blueprint of the next generation of aero-propulsion, the rim-driven Taichi fan jet as electric propulsion, and the 3D supercapacitor plus the lightweight gas generator as electric power that we will address in the following chapters. This new e-engine and its privilege is investigated in comparison to the old-fashioned propeller and the concurrent turbine gas engines. Rim-driven engines are small, light and compact, which are rotatable for flexible vertical/horizontal propulsion—a distinct difference from either the big jet engines or propeller drives.

From our previous learning curve of IEEE colleagues with AIAA fellows on developing the smart sensors embedded on turbine blades, we found that IEEE people need to adjust their focus to fit the need of AIAA, while the AIAA people need the eagerness to learn and adapt the new technologies in IEEE, especially in building intelligent aero-systems and powerful electric motors for electricity generation and propulsion. AIAA people need to learn how to define the problem for IEEE colleagues to solve and focus their efforts upon, i.e., to ask the right questions toward IEEE people. In many cases, being able to ask the right question is the rightful start of in innovation. For this, it is really worthwhile to read our previous book "The Introductory Engineering" published in Shanghai Jiao Tong University Press in 2019 on how to define a right question (the "IPO–Input–Process–Output" section in [3]). Since the AIAA/IEEE integration is not only a professional cooperation and involves a tremendous amount of integration efforts, a professional team may be necessary to aid the communications on both sides to coordinate their necessity and feasibility.

The members of this integration team shall be chosen to be as familiar with both knowledge of AIAA and IEEE and with both quality of IQ (knowledge and intelligence) and EQ (attitude and communication). For this, please check more interesting details in Sect. 6.3.

6.2 The Difference of AIAA/IEEE People

This difference originates from their different customers: AIAA used to be mainly serving the country or government, and IEEE is mainly serving people of the society. AIAA cares more about powerfulness and safety, and IEEE cares more about flexibility and innovation. This intrinsic feature determines their technical habits. In addition, IEEE is more multi-cultural and more international. The aspect of cultural habits should also be taken into account.

6.2.1 Difference in Technical Focus

These three pairs of keywords summarize the main features/focus of the IEEE and AIAA groups:

- The IEEE: economic, presentable, easy to use;
- The AIAA: powerful, robust, engineering.

These differences may misguide the initial IEEE's effort when developing the relevant technologies for AIAA applications. For example, in our early effort, we still follow the IEEE people's habit to make a small, tiny, and delicate sensor on a machine surface as shown in Fig. 6.1a and b. They look good and delicate and can show off our advancement MEMS technologies. However, in most AIAA scenarios, this delicacy is not necessary, and such tiny sensors are not as reliable at high temperatures as larger sensors. To make this fine lines on a rough turbine blade surface, the IEEE people spent much effort and time developing new techniques in soft photolithography processing. If they knew these beforehand, it could have saved them a lot of work by just making the wider line width sensor array as shown in Fig. 6.1c, which is much easier for them to make in the meantime make the colleges happy. Sometimes such learning can only be made by a few trial-and-errors. However, with the mature experience of cooperation, one could establish the 6th sense—a consciousness to know each other's needs to save work and time. To work more efficiently, both sides should learn to compromise the feasibility and necessity, the concept of the "Middle Way"—the idea of not seeking the "perfection or best" but to obtain the "optimization". The Middle Way is the ancient Chinese wisdom to balance the Yin

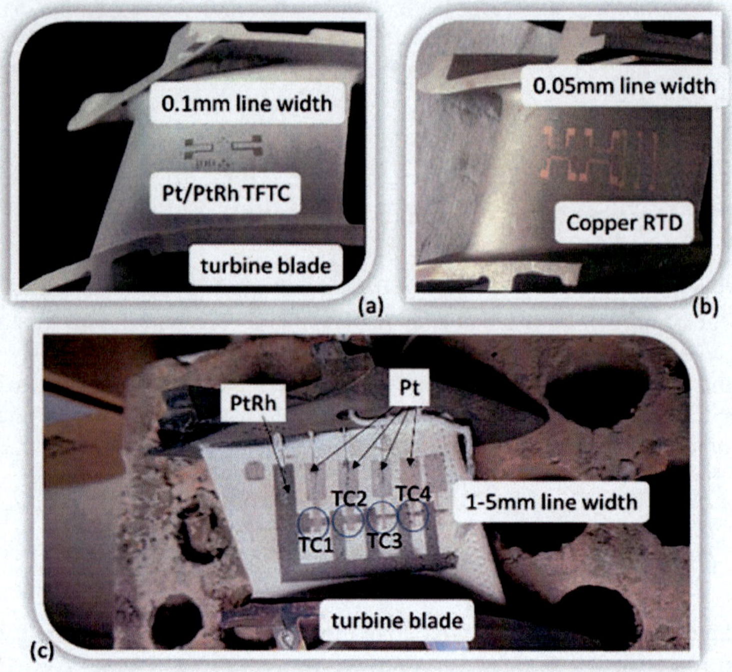

Fig. 6.1 **a**, **b** The IEEE people are good at making the fine lines (0.05–0.1 mm line width) of sensors. **c** The AIAA actually just need s wider (1–5 mm) sensor for robustness

and Yang,[1] which is illustrated in Sect. 1.3.2. In general, the IEEE people should focus more on "why" instead of on 'what" as they used to. They also need to communicate with AIAA people much more closely and frequently, such as weekly or monthly, and preferably face-to-face/onsite instead of online or by phone, to exchange their thoughts and results, to design/to perform the necessary infield verification test, and to adjust their efforts and progress based upon the feedbacks of the infields communications/experiments.

Very importantly, before a pattern of habit being established, one needs to reiterate it enough times in order to rectify the old one, for both AIAA and IEEE colleagues. Otherwise your old habits may appear frequently if without frequent reminders.

[1] The Chinese Middle Way (中国人的中庸之道). The Chinese believe in too much is as no good as too little. This is, literally, translated into the Middle Way, also translated as the Golden Mean, in Chinese language, Zhong1 yong1 zhi1 dao4,中庸之道. The middle way has shaped the Chinese life in all aspects, and it is rooted in the thinking of almost all the Chinese people. Throughout 3000 years' history till now, governors adopt this ideology to manage their country in China. The middle way is an unseparated part of people's life in China and it finds its deep roots in the Chinese culture and history. Overdoing one thing is as no good as underdoing one thing. Most Chinese believe in a medium amount of everything that goes to no extreme is the doctrine of the Middle Way or "Zhong yong" in Chinese.

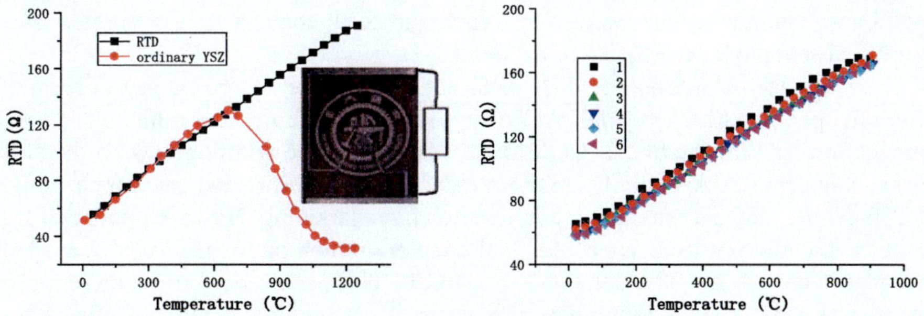

Fig. 6.2 **a** The IEEE's people job is to make sure that the RTD sensor works, **b** the AIAA people need to test the repeatability of the RTD in multiple thermal experiments

The other difference between AIAA and IEEE people is about the method of presenting the result since they have different focus points. We use the following example to explain this difference, which may happen very frequently in many AIAA/IEEE scenarios, especially at the beginning.

At the early stage, the IEEE people need to conquer a few technical barrier such that the YSZ insulation layer is losing its electrical insulation property at higher temperatures above 600 °C, which causes the deviation of the linearity of the RTD sensor, causing measurement error. This is purely IEEE problem and does not involve AIAA people. The IEEE people need to develop an improved YSZ to ensure its high-temperature insulation to guarantee a proper RTD thermal sensor performance. After this work, the IEEE's job seems done (as shown in Fig. 6.2 (left)) since their goal has been achieved, and the RTD sensor's temperature performance is good.

However, this is not enough for the AIAA people. They need the sensor's performance be highly repeatable, and they need it to be verified from multiple tests. Therefore, the IEEE people had to continue their efforts to perform multiple tests, as shown in Fig. 6.2 (right), to fulfill this need.

These multiple tests took them much more efforts than expected, and even this was not as perfect, since they still did not know the right way to calibrate the temperature reading of the muffin oven at that time. This causes a larger error deviation in the test results until they find the right way to deal with it later on—such error is greatly reduced by using the insitu Pt-dot sensors (more details in Chap. 5).

On the other hand, the AIAA people should learn to ask the right questions when imposing a request on IEEE people in the project. This is also a tough one since for most of them it is also the first time to face this question. For example, they propose a very stringent criteria on the vibration intensity tolerance of the TFTC sensor since they worry that the thin film attached on the turbine blade cannot stand for the heavy vibration in a gas turbine engine. The peeling-off of the sensor may cause safety issues inside the engine. Although the AIAA people qualitatively know that the temperature and vibration are high and intense inside the gas turbine engine, they do not know exactly the quantitative details, such as the vibration amplitude and the associated frequency spectrum. Therefore, their initial request may not be solid and legitimate. This needs a few back-and-forth conversation between AIAA and IEEE people (check Sects. 5.2.2 and 5.2.3).

In summary, AIAA people have different mindsets/points of interest compared to IEEE people. AIAA people like to say: reliable, reliable, and reliable..., more repeatable, and repeatable..., while IEEE people like the wording: newer, the first time, cheaper... AIAA people are more focused on reliability and endurance, while IEEE people may care more about innovation and marketing. However, both believe that the technology itself is not the final chapter—much more effort is still needed through team efforts. The delivered data from IEEE people have to be more engineering driven to fit the tough needs requested from our AIAA fellows—the sense that IEEE people should be aware of when starting to work on an AIAA project and dealing with AIAA colleagues. IEEE people need to adjust their focus and point of interest to fit the needs of AIAA, while AIAA people need to adjust their vision to adapt the advancements in IEEE, especially in building intelligent aero-systems and powerful batteries for electrified aero-engines. AIAA people should be able to ask the correct questions on their designated need in smartness, power, weight, and robustness when communicating with IEEE fellows; a professional integration team is suggested to aid the communications on both sides to coordinate their necessity and feasibility.

6.2.2 *Difference in Culture*

When AIAA is integrating the IEEE, it is just in the time of the twenty-first century. In this century, international cooperation among different nations/countries to cooperate on one project is becoming increasingly popular. The competing privilege of this cooperation is to take advantage of the intelligence, resources, and sharing costs. Therefore, the AIAA and IEEE peoples' cooperations should consider the multi-culture difference as well. In general, different cultures of the east and west are similar to apples and oranges. They are fruits and they taste sweet, but you may favor one and dislike the other. This is nothing to do with good or bad. One just needs to appreciate their difference and adapt; they should avoid unnecessary defensive thinking.

For example, the difference between a Chinese style and American style is as follows:

(1) The rule versus the flexibility

The rules/laws systems in the United States are perfect, and everyone abides by the spirit of contract. When everyone in the country does things according to the rules, the system can achieve the maximum efficiency as it designed. People just follow the rules and proceed. They don't use too much of their brains to think of the alternatives. They can just work hard "stupidly" and then take their vacation. Vacation and working are totally separated, and the distinction between them is very clear.

In China, the situation is different. People don't have too strict rules, and they are not get used to it. Most of Chinese follow their principles such as whether the things are righteous, duteous, or showing their dignities. They get used to think a 3rd way to handle the situations, and any laws/rules have exceptions/loopholes, and it is always to make the "impossible" into "i m possible" (more in Sect. 1.3.2). In history, the Chinese used to mix their business with pleasure. In old times, the cookery and small restaurants work every day. People work during the day and rest when the sun goes down. Working is your pleasure, and they are usually combined as one; business is always the pleasure, and there is no so-called weekdays and weekend. In fact, the concept of "seven days a week" of the Western culture was introduced to China for no more than 100 years. In old China, people enjoy their lives while working, for example, a chef enjoy cooking the delicacy dishes to his guest, while in the meantime, he gains his income in the restaurant business. Some Chinese people may even feel that the very long vacation is boring since an over-indulged life may be a boring life. This working pattern is just different from the Western style, and deep-rooted national habit that has nothing to do with good or bad.

Whether a product is made by a Chinese or an American company can be seen at almost a glance. The iPhone outlook is very simple and straightforward, just as their people. The GUI of a China smartphone has many touch icons to deal with the multiple cases. When you compare the Chinese airport versus United States' airport, you will also see this difference. You see a lot of multi-signs and you need to think and judge when you see the signs, while in US such signs are much simpler—you have only one choice. Different habits result in different product logics; difference in cultural attributes is the root of the differentiation in their products. The localization process of many foreign products in China must go through an integration process to adapt the situation in China. It is noted that this cultural attribute effect may not purely depend on the physical locations, be in China or in the US. For example, you are American company but situated in China, you must consider the Chinese attributes and try to adapt in your product from US to China because this cultural difference will be reflected in your product design concept, and will affect your future market and your strategy, otherwise your won't "survive" in China.

Looking at the future market place of AIAA and IEEE, international sharing, win–win, and multi-lateralism are the trends. Culture is for people to adapt but not to change. We just need to understand the cultural difference, and in fact, it is very hard to change them.

(2) Be spoken versus underspoken

The Westerns are outspoken people, but many Chinese people did not say words directly. They use imply and analogy a lot, even in the technical fields. Many Westerner do not quite get used to it. The American also imply but not as much as Chinese, which regard this as "politeness". There is a joke to vividly explain this cultural difference. When Marco Polo visited China in 1200, he liked Chinese pies very much, and he intended to copy this technique back to Italy. The Chinese make the pie by enclosing the stuffing inside it. However, when Marco returned to Italy, he forgot the recipe. He put the stuffing on the surface of the cake, and this becomes our pizza today.

It is very interesting to analyze a pizza versus a pie, which reflects the culture instincts: more or less, opened or closed, implied or outspoken:

i. More or less, pride or modesty, best or moderation. For a pizza, you can add a lot of coverings as you want; but inside the pie, you can just add the limited amount of stuffing.
ii. Openness versus closeness. The pizza is an open space that you can see all things without guessing. However, you can't see what the stuffing inside a pie is. Jokingly, in China, some indecent cook just inserts very little stuffing inside the pie to cheat customers while this is very hard to do for a pizza (stuffing is more expensive).
iii. The true reason of a pie becomes a pizza could be: (a) maybe Marco Polo intentionally wanted to do that because he just liked this way or; (b). maybe it was due to the culture instinct or; (c) he just forgot the recipe of making a pie.

Another modern story also illustrates this principle. There is a Chinese sales man selling paintings in gallery. A lady went to the art museum intending to buy an art painting. She hanged around for ~ 10 min, and then, sales man approached her and said, "Cannot make up your mind, right?", instead of saying "which painting you want to buy?" He assumed you want to buy the paintings but you are just hesitating. Then, I am going to remove that hesitation by sitting at your angle. When saying this, he should also stand beside the lady but not face straight to her or speak on her back, which will cause the confronting and protective feeling of the customer, and this does not mean physically but also mentally. If you stand in front of him, you have an equal relationship with him. If you stand behind him, he will immediately feel uneasy. If you stand on his side, he will be very comfortable. All these intimate details are crucial for the later success.

The second sentence is more powerful than the first: "What color is your wall?" instead of saying "which painting you want to buy?" It seems nothing to do with buying the art, but it assumes that you have already have it on the wall and you are looking at it. Let me help you think about which painting match your wall the best. The sales man gave the best advice based on the lady's answer, and the customer got her painting happily.

(3) Round versus straight

The Chinese use "TaiChi" to deal with relations and confrontations. Many times, Americans use power to win, a straight way to face confrontation. In contrast, the Chinese use a circle to circumvent you power, either away or direct back to yourself—this is the principle of Taichi marshal art. Chinese people usually do not use straight forces to solve problems. It is the characteristics of the nation to use the adaptation to resolve the problems instead of to solve it. The Taichi is round with circles while most Western colleagues are straightforward. Shown in Fig. 6.3, the Chinese use round circles to separate it into two parts, while the American uses a straight line to cut a circle into half. Most Chinese preferred a modest, implied, indirect way in their

Fig. 6.3 Vivid illustration of American and Chinese

expression. The Westerners are bold, open, and direct. To apply it to our fan blade design in our RDF jet (Chap. 9), the Taichi blade converts the confronting air to a smooth tangential flow and then jet as a thrust.

It is just not as fair to argue which is better but to say which is more appropriate or your preference—it is like the relationship between shoes and feet. You may be an American, but you may prefer a Chinese style or vice versa. This personal feature should be considered looking for your career. When a graduate student looks for a job, he needs to know which style is more suitable, either the Chinese culture or American spirit. Observe yourself and see which app you like to use: iPhone or Huawei, whether you are a rule-driven, like the simplicity, or you are a person who is good at finding opportunities between the lines. When you are in cooperation with international colleagues, you need to be aware of this cultural difference, which you may never experience before, and to try to adapt but do not expect to change—it is difficult and unnecessary. In America, it is impolite to ask which religion you believe because this may cause unnecessary disputes, and it is actually not arguable. The same principle applies to cultural differences. Culture is not arguable but must be addressed when different people mingle together. One should focus on the commonality that both want to achieve the same goals sincerely, and both have good motivations on it. It is just the way they do things are different. It is just as pragmatic as to learn, to appreciate, to respect each other, and be adaptable in order for a win–win result.

China's system is relatively flexible, including the laws/rules and ways to excise them. The Chinese are always good at finding the 3rd path without violating the rules. The 3rd way method is very useful for innovation, which will be further elaborated in Sect. 1.3.2.

6.2.3 2D Versus 3D Thinking

Stephen Covey proposed plenary quadrants regarding the time management of urgency versus importance [2]. This method is very effective in improving the efficiency of the task and responsibility of the team as well as the individual. In contrast, we propose 3D task management by incorporating the Chinese strategy of not only considering the objective side of the task but also taking into account the subjective aspect.

Stephen Covey prioritizes tasks by placing them into four quadrants in a 2D plane. He drew an *X*-axis representing the urgency and a *Y*-axis that measures the significance (Fig. 6.4). Stephen Covey's 1, 2, 3, and 4 quadrants are explained as follows:

Urgency refers to tasks or responsibilities requiring immediate action or attention, and importance refers to those with high significance or value to the final goals. The objective of Steven Covey's four-quadrant systems is to focus on improving both personal growth and professional accomplishment. Q1 is the top priority.

- Quadrant 1: Urgent and important
- Quadrant 2: Not urgent but important
- Quadrant 3: Urgent but not important
- Quadrant 4: Not urgent and not important.

Q1 involves responsibilities or tasks related to critical results and requires urgent attention. Q1 jobs have the following qualities: impending deadlines, direct relation to time-sensitive goals, and/or involve alleviating immediate risk.

Q2 involves the related capabilities to achieve the overall goals and to enhance your work force. Q2 quality requires delicate planning and long-term perseverance.

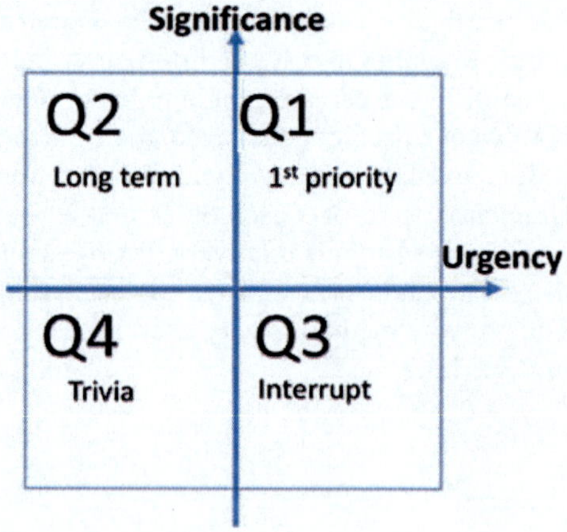

Fig. 6.4 Redrawn from Covey Leadership Center, Inc.

Q3 is urgent and assume some form of importance in the moment, which are likely reduced or removed from your workflow such as: some poor planning of items in Q1 and Q2, interrupting productivity or distractions.

Q4 is more likely able to be removed completely or reduced. It seems meaningless to list Q4 here because who is going to deal with the things which is neither important nor urgent? But, in reality, many of us very often dwell inside the quadrant 4 unconsciously or habitually. It is important to identify which items belong to Q4 consciously, especially when you are deeply engaged in some project. The conscious awareness is personal capability based on one's accumulated work experience. Q4 items are not directly related to neither overall nor time-sensitive goals, and our time shall not be wasted in that quadrant.

The benefits of using Covey's time management matrix include the following:

- Better productivity: help to organize and prioritize your task and to complete more and the most vital tasks in the same amount of time.
- Clear habits: help you to develop good habits of focusing only on Q1 and Q2 items.
- Work-life balance: help you to find time both for improving your working efficiency and to build up the workforce in the long run.
- Improved planning skills: effective people are proactive and understand the value of investing their time and energy into Q2 activities. Effective people respond to opportunities to do what's important, instead of reacting to urgent needs. When things come up, it is easier to determine what is important and what isn't once you have a clear goal and personal mission statement.

Stephen's four directions of emergency and essentiality may also have it shortcoming. From the teamwork point of view, one must consider both IQ and EQ. Team work is IQ + EQ, where both IQ and EQ have extended meanings here [3]. IQ not only represents intellectual quality but also includes the learning and adapting abilities, the way to organize your knowledge trees, be capable to find the 3rd path, etc. EQ not only refers to emotional quality but also the ability to be positive and active, passionate and compassionate, integrity and perseverance, etc.

In fact, there exists a third dimension: the timing must be right, such that "I am just in a good mood". Shown in Fig. 6.5, the best is the first quadrant in a 3D XYZ matrix. When being asked to do a favor in a wrong time, such as in a bad mood, fully occupied by something, or in a worry some state, thing will not get done either, and the bad trace may leave a bad memory affecting later-on communication. A good deed shall happen at a specific time and a specific place with specific people, i.e., the target is important, urgent and best timing to do it. If such timing is not yet ready, you may either wait or manage to create that opportunity. Trying to do it forcibly may just get things even worse. So, be patient!

To get the something done, one needs to control the proper balance of the three factors: urgency, significance, and the good timing. The Chinese call it (a tempted translation: chance, privilege, feasibility).

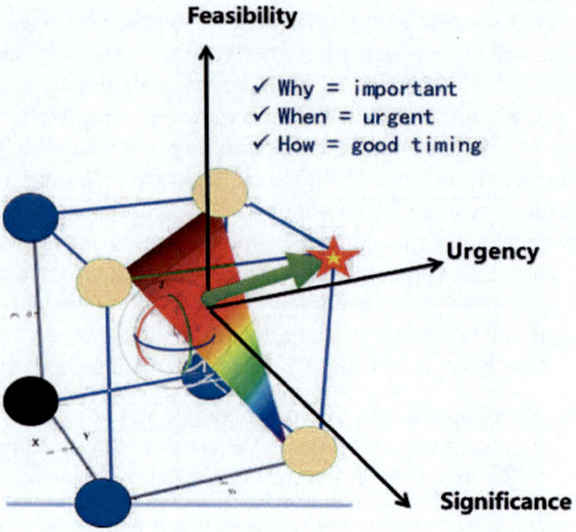

Fig. 6.5 Three-dimensional quadrants in the task management of a project

> *天时地利与人和*

it is more or less similar to the Western's saying (Louis Pasteur, a nineteenth century French chemist)

> *Chance favors the prepared mind*

Chinese people know that it is crucial to have the "3": the right time, the right place, and the right people in order to get one thing done. In old times, when a Chinese team was trying to win a war, they used the "天时地利与人和" theory; the resonance of the "3" is crucial to win a war. The physical situation, the necessity of fighting, and the readiness of the military workforce, corresponding but not limited to: war field conditions such as mountains and rivers, the motivation of fighting, and troops and weapons. They believe that just the physical or objective readiness is not enough. Currently, most human innovative efforts are not just AI work. Team work involves many emotions, such as passion, motivation, and willingness (EQs), together with their intelligence capabilities (IQs).

During the AIAA and IEEE cooperation work, although the technical details are very important and are favorites by both groups of people, it is relatively their weakness of not being good at "putting yourself in someone's shoes", i.e., to make an effort to imagine how they feel or act if you were in the same shoes. For example, imagine you are a sales person, you may have two options to improve yourself:

- What is the drawback of my product? What are the drawbacks of my survey? How can I improve them?
- What does a customer want? What are their needs? How can their needs be related to what we have?

To be honest, most of the engineers/graduates in the previous century favor asking the first types of questions, which is more suitable for performing dedicated work. However, in the twenty-first century, how customers feel and experience of your service becomes the 1st priority. To a certain extent, AIAA is the customer, and IEEE is the vendor. The cooperative work between them requires more intimacy. Purely objective-oriented Steven's four quadrants is not as advanced as the 3D quadrants by considering both the IQ and EQ. One must consider the best "mood" to obtain the best results, especially when dealing with international colleagues.

6.3 The Connection of AIAA/IEEE People

The main challenge of the AIAA and IEEE cooperation is originated from our traditional education pattern in college. Traditionally, our education system is scientific based but not the project based. For example, we often say "I am a student in the Department of Physics", but we seldom say that "I am a student who majors on electric propulsion". The physics is a science; the e-propulsion is a project, and there is no such department in the college. It is also very hard to arrange and manage the classes and textbooks for this. Not many universities like to challenge this like Shanghai Jiao Tong University in China, who dares to open the AI, 5G, and Micro/Nano as a Department or Institute. It is a very challenging job to find proper teachers who must be good at the hot and technological spots like IoT, UAS. What Jiaoda did is to seek for the graduate advisors who are good at these projects to deliver the lectures/practice in this class. Our new book "The Engineering Man" serves as a reference text book for them and to the up-to-date, this is the only book of its kind in the worldwide college bookshelf [3].

Both AIAA and IEEE people were trained professionally based upon the traditional college training programs. Traditional education system for college and for graduate students is the scientific-based education but not the project-based training. The scientific-based education is focusing on one specific field and then be good at it. It focuses on the depth instead of the width. Students graduated in this traditional training program are good at one distinct scientific field. But, in order to combine electronics with mechanics to fulfill an interdiscipline project, we are short of the professionals who are good at both fields. For a project-driven topic which combines a few scientific fields, such as electronics, mechanics, and material engineering, we need talents who are able to combine them organically in order to achieve a new product. Take the iPhone as example, the iPhone is a combination of many engineering achievements, such as electric circuits, RF electronics, computers, Internet, sensors, etc., even if all these are still within IEEE categories. Electric aero-propulsion involves more diversified physics fields covering both electrics and mechanics, which is almost impossible to find one professional who master both knowledge.

Therefore, the most crucial issue of the AIAA and IEEE cooperation resolves the conflicting targets between the traditional scientific based education and project oriented engineering education. Electrical science and aeronautic science belong to totally two different categories, and nobody is talented enough to master both. Besides one professional on one specific field does not want to take the risk to get laughed at in the new fields, especially when they become a pronounced professor in their comfort zones. Therefore, the integration of the multi-field project between the AIAA and IEEE becomes extremely difficult. There must be a certain methodology of strategy to handle this new issue, such that we have tried in Shanghai Jiao Tong University by establish a broader platform to accommodate the four major engineering schools: mechanical, electrical, material, and nautical science, considering in future such a multi-discipline phenomenon will become quite popular in many projects, such as the artificial intelligence (AI), the 5G (which we don't exactly know how to correctly define it), involving the professionals belonging to different physics fields, the chemistry and the biology. A new text book is tempted with many newer ideas on future college education with the original book name "To Be an Engineering Man" The word "Engineering" here has much broadly meanings and covers almost all the projects/activities which involved both the IQ and EQ, the only exceptions of which are the Albert Einstein and Leonardo da Vinci, etc., who just shut off themselves in their room to maneuver their own mind without cooperating with the outside world.

Another key factor for the AIAA and IEEE cooperation is the communication skill. Both sides should try to use the plain and oral language as possible to explain the technical terms. Avoid using the technical dragons or complicated technical terms as possible during the communication. Besides, try to use the analogy, common sense, and the first principles to explain the things.

As for a distinct project such as developing a dedicated RDF e-engine, it is highly suggested that there must be "a dream keeper" which is also a project coordinator and a project investigator (PI). He or she is a leader of the project such as the Steve Jobs for the iPhone. This PI should process both the engineering talents and management talents. Historically, the program manager is the professional MBA people. The shortcoming is that they have never get the engineering training neither be professional on any specific scientific field. Their mindset is management or economy but not the science and engineering. To better coordinate both AIAA and IEEE people, the preferred choice that he or she used to be a professional in some specific engineering, but he or she is also good at management and social communication. This is much better than just hiring a professional CEO as a pure program manager. Many professional MBA program manager in Silicon Valley are not graduated from the engineering schools but from the economics or MBA schools. There certainly a few unsuccessful cases that the company becomes a money machine instead of an enchanting innovation workforce. They are good at management, but they are short of the engineering habits/passion which build the basic instincts for the engineering project.

6.3.1 Customer Versus Vendor

Basically, to achieve a smarter AIAA by the technology from IEEE, IEEE functions as a vendor who serves AIAA as a customer. For IEEE, it is the technology waiting for the proper application (from bottom to top); for AIAA, it is the application looking for the suitable technology (from top to bottom). Regarding this customer/vendor relationship, let us use the example of Tesla and BYD, the two modern popular EV companies, to explain the difference between the two technological approaches: from top to bottom and from bottom to top. The battery and engine technologies from both companies are also highly related to our contents in Chaps. 7–11, electrified aviation.

Tesla and BYD are both developing and manufacturing electric cars but follow two different paths. Tesla starts from the design, i.e., an EV architecture. It is a "top-to-bottom" approach in which the design goes first and then to start finding the relevant and available technologies, and then combining them together to form a new product. It is new idea driven and then collects the necessary resources. BYD is different. It follows the "technology seeking for the product" approach. Even the brand name "BYD" was coined randomly at the very beginning of trademark registration in China, for the easiness of passing the trademark agency in China; now, the word "BYD" becomes "Build Your Dream" when the company gets prosperous nowadays. BYD was NOT a car manufacturer and almost had no experience nor the experts for automobiles. BYD was initially a battery company that had a solid background in developing a lithium iron phosphate battery ($LiFePO_4$). When the battery technology got matured on the market, the CEO of BYD decided to join the EV business to make the EV cars. To be honest, Li-ion batteries are the key for EVs, but a car is still too different from a battery itself since a car involves automobile architecture, mechanics, and electric engines. The early days of the CEO, Mr. Wang Chuanfu faced truly a big challenge, but he made it. In contrast, Tesla uses the mature ternary lithium battery system from Japan, which only packs the existing 18,650 standard cells together to form a battery package. It is much easier for Tesla to enter the EV market by borrowing the existing battery technology from Japanese vendors and e-engines from others However, from the battery's perspective, Tesla's technical background is not as strong and solid, and this weakness starts to reveal when both of them are competing the battery bottleneck right now. Tesla was not a professional battery company, and they just tried to modify the 18,650 (18,650 = ϕ18 mm in diameter and 65 mm long) standard battery into 4680 (ϕ46 mm × L80 mm), but this did not resolve the volume/weight efficiency from the root as BYD does. BYD is a professional battery company who is able to reconstruct the battery system from the root—layer-by-layer architecture instead of just packing the 18,650 one-by-one, saving a lot of space and avoiding many safety issues. Tesla just took a shortcut to circumvent the existing shortcomings, while BYD was able to enhance the system's ability from the root.

Both Tesla and BYD are successful in the EV market now, and both have their pros and cons. Likewise, either from AIAA's looking down to IEEE for technology, or from IEEE's going up to AIAA for playground, both approaches can make a success. Overall, it is the electronics who make AIAA more intelligent, and it is the electric motor and power who make the electrified aviation become true. However, a lot of integration journey must be gone through like we did in Chaps. 3–5, and we will continue this legacy in Chaps. 7–11.

6.3.2 The 3rd Way

When dealing with the conjunction joints between two professions, such as AIAA and IEEE, many innovations are involved, and the ability to find the third path is highly needed. Many Chinese people are good at finding the 3rd way in a dilemma. We use three stories to illustrate "the 3rd way" to obtain some insight into the third path.

- Story 1. In ancient China, there were many droughts. The irrigation of crops mainly depended on rain from the sky or manual irrigation either from the afar river or from the well. Other than the first kind which is totally depending on the heaven and fate, both methods were hard to fulfill, and there was not sufficient water to irrigate all the farmland. The ancient Chinese farmer found the third way to deal with this. The farmer dug out a small amount of soil at the bottom of the seedlings, poured a little bit water into it, and then buried it with dry soils. From basic physics or from common sense, we can easily understand why he does this way. Water buried under the soil provides nutrients to the plant and is not easily dried out by the sun. This delicate method of irrigation by leaking water into the soil to moisten its roots saves much water compared to sprinkling water on the leaves of crops, the conventional method of watering flowers. This is a "have-to-be" approach to achieve the optimal effect of irrigation; when you have the rich water you don't have to do this way. This is a 3rd way to deal with a seemingly impossible situation.
- Story 2. Another story to find the third path is from a Chinese TV series called "Top Secretes". The students are not allowed to take their notes out of the camp after listening to the Soviet military instructor, so how can they review their class work and pass through the exams? How can a person be so smart to remember every technical detail in a two-hour class? These guys did find a third way. The 30 students invent a strategy that each remembers only 1/30 of the class contents taught by the Soviet teacher, and after they came back, they reiterate the class contents and write them down into notes and share with each other. In this way, they incredibly passed the exam successfully. Since they did this secretly, the top officers and teachers felt it hard to believe how they can get such a good grade!

Even for teachers themselves, the used-to-be top students, could not do this. How could these soldiers who had not even graduated from junior high school have such a good talents? They certainly doubt that they had cheated the exam, but this is a military camp, and there is no chance of cheating in the whole process. However, it is quite plausible when this secretion is revealed in the end. There always exists a third path called "incredible"!

- Story 3 is the "third answering". When your boss asks you, "are you free now"? What would be the right answer? If you answer "yes", the boss may have the impression that you have nothing to do all day. If you answer "no", the boss will think you are arrogant and even ignore his request. Then, what you should say except yes or no? The 3rd way! You neither say "Yes" nor "No". You can say, "I'll come over right away". It is my business whether I'm free or not and I don't have to tell you that. What I need to do is do what you need, i.e., to come over to help. This answer circumvents the dilemma that may cause unnecessary misunderstandings.

Now, did you appreciate the beauty of the third word?

3rd way is important for innovation. The third way is a synonym of "a new idea". There are two popular ways to find a third way, "finding a gap and expanding it" and "finding a thin thread and thickening it".

(1) The microscope is the example of the 1st kind. You cannot see anything virtually on piece of smooth glass. Is there any gap on the glass? With a microscope and enough magnification, you can see the seams on the smooth glass surface. Likewise, in a very well-studied science topic, there always exists a gap to dig into and see more interesting things inside. This is how we enlarge the gap and open a new ground.

(2) "Find a thread and thicken it" is the method used to connect two or more interdisciplines to form a new science topic to study and expand it to a new playground. Micro/nanoscience involves technology in the micro-world, and the geometric scale is between 10^{-9} and 10^{-6} m. Aerospace is a technology in the macro-world, with a scale ranging above 10^{+4} m. Both are high-tech cutting-edge technologies, with a scale difference of more than 15 orders of magnitude. It is a wonderful thing to connect the two careers and build a bridge in between. Smart sensors for aero-engines built in the MEMS method in IEEE and then used for aero-engine optimization and health monitoring for AIAA are the linking bridges that connect two nodes together. It is a nearly zero-to-one (0-to-1) achievement after 6 years of endeavor since we first discovered the weak link (the thread) between the two. Then, we keep on working this project and expand it into a success story that we have successfully developed smart thin-film thermocouple sensors to measure the high temperature inside the gas turbine system. This is the example of "finding a thread and thickening it" to bridge the AIAA and IEEE.

"Be good at finding the 3rd way" examples above are from Chinese but not limited to Chinese. It is a habit of thinking that may not depend on where you are and who you are. However, one has to say there are more such happenings in China because this land cultivates this habit: to adapt and to survive in a stringent situation. The PRC and the USA have similar amounts of land but different ratios of resources per capita. The population is 15/4, and there is less arable land area in China than in the United States. Historically, there were fewer wars involved in the US than in other countries in the world. The situations help the Chinese be more alert and be good at finding a way among the "impossible" (in fact Impossible = I m possible).

6.3.3 A Dream Keeper

A dream keeper is a coordinator who should have enough authority to govern the resources on both the AIAA and IEEE sides to push things forward. The reason is that during the cooperation of AIAA and IEEE project, there are certain gray areas that do not have a distinct responsibility. Such a situation happens much more frequently compared to either AIAA or IEEE dealing with their own expertise. Let us use one example, the cooling effect test for gas turbine machines, to illustrate how this happens and why a perseverant PI (principal investigator) is necessary.

The target is to verify and use the TFTCs to replace the cumbersome wire TCs for the turbine blade cooling effect test. The IEEE is responsible for building the TFTCs on the blade and connecting them into wires. AIAA people are responsible for running the engine and reading the data. Initially, there is a gap—who and how to connect the wires and extend them to the outside data collector for thermovoltage reading and process the data into the valid temperature reading. This area does not belong to either side of AIAA and IEEE, and it is not the expertise of either side. AIAA people used to delegate this job to a third-party vendor, but now such a vendor does not exist. A dedicated PI is needed to look for a 3rd way to resolve this issue and fill in this gap. It takes much more effort, time, and patience as expected, and he or she must coordinates all the process. He/she must have the authority to get involved in the detailed engagement as well. Without enough passion, endurance, and perseverance of the dream keeper and without entitled authorities, it is hard to get things done!

A healthy ecosystem is also needed and must be cultivated. We should not always punish the failure and never encourage trial-on-error. For example, when you are doing things, you have a chance to make mistakes. Whenever you got mistakes, you got punishment, and your success is ignored. On the other hand, if I do nothing, I commit no errors, and I don't have any punishment. Which one you may choose? The logical answer is very clear. A good dream keeper should be aware of this basic logic and try to build a healthy ecosystem to activate the team members with more passionate to pursue the project.

References

1. AIAA Propulsion and Energy Forum since 2018, AIAA Propulsion and Energy 2020 Forum. eISBN: 978-1-62410-602-6. AIAA Propulsion and Energy 2019 Forum. eISBN: 978-1-62410-590-6; 2018 AIAA/IEEE Electric Aircraft Technologies Symposium, 2018. eISBN: 978-1-62410-572-2. https://www.aiaa.org/propulsionenergy
2. Covey, S.R.: The Seven Habits of Highly Effective People. Covey Leadership Center, Provo, UT (1991)
3. Duan, F.L.: The Introductory Engineering. Shanghai Jiao Tong University Press (2019). ISBN: 978-7-313-20777-7. https://lib.ecust.edu.cn/zh-hans/book/903714. https://baike.baidu.com/item/工程导学/59869213?fr=aladdin

Part III
The Smarter Aviation—Electrified Aero-engine

Electrified aero-engine brings about the smarter aviation:

- Smartness means VTOL, more efficient, more clever, less waste,
- Smarter means AAM, more versatile, more mobile, more flexible,
- Smartness implies cleaner and greener aviation.

Chapter 7
Why Electric Aviation—Versatile, Smarter, and Green

The outline of this chapter is:

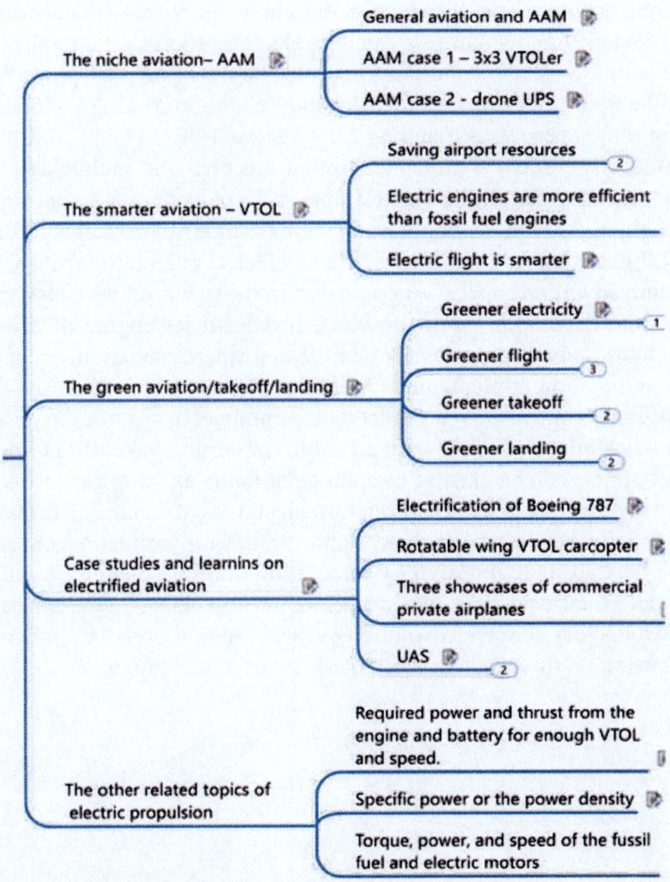

© The Author(s), under exclusive license to Springer Nature Singapore Pte Ltd. 2023

F. L. Duan, *When AIAA Meets IEEE*,
https://doi.org/10.1007/978-981-19-8394-8_7

In the Part I of this book, we use the IEEE's people's experience in developing smart sensors for AIAA fellows on high-temperature measurements of gas turbine engines, intending to explain the difference in their mindset and serves as a reference point for the very promising future of the AIAA + IEEE coproject, i.e., the electrified aviation. Obviously, IEEE's involvement is highly necessary since the e-aviation is highly electrified, and AIAA needs the expertise in electric motors and electric power from IEEE. From AIAA's perspective, electric propulsion has two distinct playgrounds: one is the eVTOL for small-scale AAM aviation, and the other is the e-STOL for large airliners. Both require an effective electric propulsion engine and high-quality electric power. From IEEE's perspective, there are two key technologies involved: one is the engine, and the other is the power. In this book, a new electric aero-engine—the RDF jet, and two powerful electric power suppliers—3D HK SC and LTG are proposed to fit in the AIAA's needs for e-aviation. These are the main topics in the following chapters in Part II.

The last 20th AD was a revolutionary century of human civilization containing many disruptive inventions; one is the airplane starting from the Wright Brothers in 1903. The main focus of the whole twentieth century is the capability for aviation, the big continental airliners, and the advance jet fighters. In twenty-first century, the key words are efficiency and smartness, such as electric propulsion and airplanes. The value of developing electrified aircraft does not lie in chasing the powerfulness but seeking for the smartness, i.e., the versatile mobile aviation (AAM [1]) for small-scale aviation and the clever takeoff/landing for large aircraft (STOL [2]). Pure electric aircraft is of indispensable significance for various civil and mobile aviation needs and for low-carbon green flight. For example, a 20-passenger minivan-type aircraft of 1000 m height, 400 km 3-h flight has a larger market share than a 400-passenger continental flight. Shorter takeoff aided by an electric engine to leverage gravity is a more optimized aircraft operation compared to the powerful yet noisy gas turbine propulsion. The developing work on electrified RDF jet engine in this article is intended to fit these two scenarios for flexible and smarter aviation.

Electrification of an airplane mainly relies on the engine, namely, the RDF jet, a small and lightweight rim-driven fan jet that is compact in size for easy and flexible orientation adjustment, which is indispensable for vertical takeoff/landing (VTOL) operation. RDF jets can be used as a supplement to the existing gas turbine or as an independent electric engine for UASs and small/mid-sized commune/drone aircrafts. Its smallness and flexibility offer many opportunities for various AAM scenarios as well for the best airplane redesign by integrating multiple e-engines with airplane (the distributed electric propulsion concept) for the best aero-dynamics performance.

In this introductory chapter, we discuss electric aviation from the application point of view, focusing on three key words: AAM, smarter, and green.

7.1 The Niche Aviation—AAM

In order to appreciate the value of term "AAM", let's at first compare its difference with "General Aviation", a well-accepted term in the aviation industry. The concept of general aviation from Wikimedia is as follows:

> *General aviation refers to aviation activities other than military, police, customs and public air transport flights, including operational flights in industry, agriculture, forestry, fishery, mining and construction, and flight activities in medical and health, rescue and disaster relief, meteorological exploration, marine monitoring, scientific experiments, remote sensing mapping, education and training, culture and sports, tourism and sightseeing. The aircraft used for these activities are collectively referred to as general-purpose aircraft. There are approximately 300000 general-purpose aircrafts in the world, and the number of general-purpose aircraft accounts for more than 90% of all civil aircraft.*

The focus of this definition is on *application scenarios*. One similar wording in general aviation is urban air mobility (UAM). UAM is recognized more by the public, but it also focuses on application and less on its technical aspects. Moore [3] once presented the concept of the on-demand air mobility, where the user can specify the departure place, destination, and departure time to best fit the private flights of general aircraft, charter flight, air taxi, and other operations. Comparing to the UAM, the term AAM has different focus.

Different from the above application-oriented definitions, the main focus of the term advanced air mobility (AAM) is "*technology*". NASA/NBAA defines the AAM term as follows:

> *AAM is a new* concept of air transportation ***using*** electric vertical takeoff and landing (VTOL) *aircraft to move people and cargo between places – local, regional, intraregional, urban – not currently or easily served by surface transportation or existing aviation modes. AAM* ***uses*** revolutionary new aircraft that are only just now becoming possible. (from: https://www.nasa.gov/aam/)

According to the above description, the word AAM focuses on its *technical features*. AAM technology implies the new concept of VTOL. Vertical takeoff and landing is a very advanced and ideal flight concept and is increasingly recognized by "aviation people" as the keyword "eVTOL", since the vertical takeoff/landing is most likely achieved by electric aero-engines driven by electric power. AAM technology fits in various "niche" occasions of the "air mobilities". AAM aims at highly specialized needs, such as a 3 $\times$ 3 VTOLer (3-men 3-m triangular e-aircraft) as a rescue plane for emergency or a catering drone for Amazon/eBay. Electrified aircrafting and aero-engines enable these more versatile, more mobile, more flexible scenarios. Let's review the general aviation and AAM in their technical details as follows.

7.1.1 General Aviation Versus AAM

The counterpart of general aviation is the mainstream aviation of military and commercial airplanes. Here, we use the method in the IC industry as a comparison that is, dividing the IC market into one main category and others as another category. In the integrated circuits marketplace, the main products are CPU and DRAM—the two key components for each computer with the largest market shares. The other category contains all sorts other ICs, such as interface circuits, buses, drivers, videos, audios… Similarly, we classify military and public flights into one aviation category. The other category is general aviation. Such a classification method may not be necessarily as scientific but is more practical to integrate the electric participation in the aviation technology.

General Aviation

General aviation covers a wide range of fields. The three main categories of electrically driven general aviation include purpose, scale, and technology, as shown in Fig. 7.1.

(1) The first category is from the perspective of usage, for private use or for public service. Private aviation refers to the air traffic and air vehicle used by individuals mainly as entertainment (now) and commune vehicle (future). Here, it should be mentioned that there are many synonyms of flying vehicles. One tends to define the flying object as an airplane, such as VTOLer, carcopter, and car-type aircraft, and the other defines it as automobiles, such as skycars and flying cars. Whether aircraft or car depends on the use and range. In general, the flight range of aircraft is relatively long, which generally refers to cross-continental and provincial transportation. The mileage of cars is relatively small, which may generally refer to indoor and urban transportation. At the current stage, the private aircraft is a luxury personal entertainment toy for high-income people. In the future, private aircraft will replace current cars as a popular personal transportation vehicle. The most prominent advantage is that it can avoid ground traffic. The VTOLer is most suitable for less than 500 kg weight and 300 km range. Its goal is to replace the current car as an ordinary means of transportation from home to the most nearby metro railway station and then to take the metro to downtown office. This is a very promising new marketplace, which has already appeared in science fiction novels. Imagination is an early arrival of the future reality. This car-type VTOLer is purely electric with mileage just enough to fly a person from home to shopping mall, maybe just a hundred kilometers away.

At the current stage, most of them are still using propellers driven by Li batteries. The price ranges from 300 K to 1 million dollars. The propeller drive is either

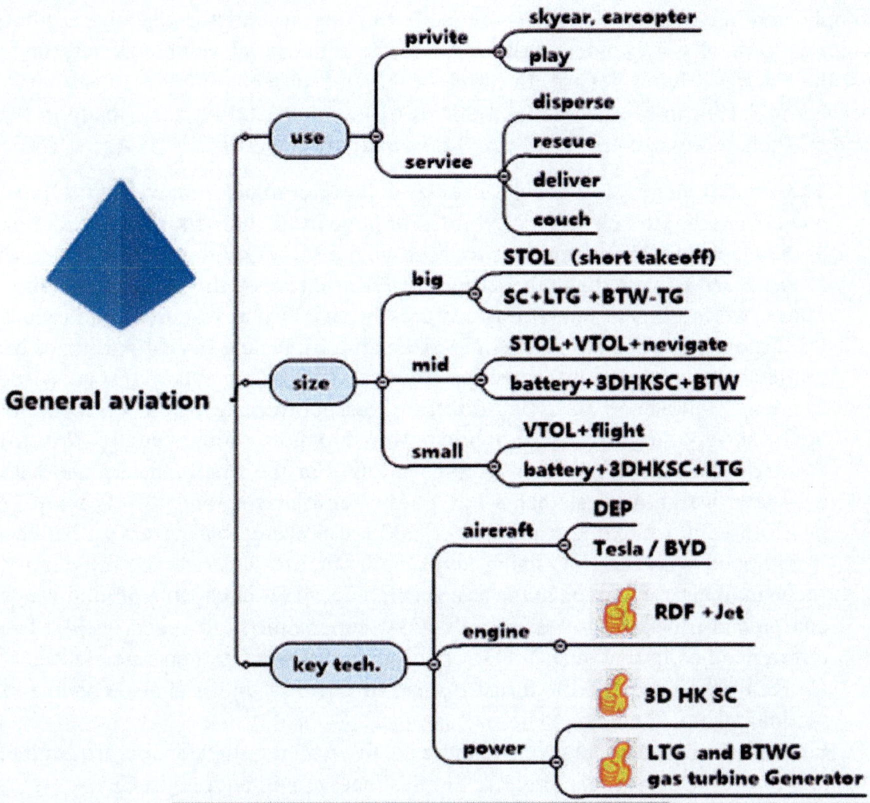

Fig. 7.1 The architecture of the general aviation industry

designed as helicopter type to hover in the air or by turboprop type for horizontal flying. The first kind of propeller is designed for maximum lift to conquer gravity, and the second kind is designed for high horizontal speeding. Both technologies are very mature but are not exchangeable. In comparison, turbofan technology makes it easier to adopt the rotatable wing technique to adjust its horizontal/vertical propulsion for navigation and vertical takeoff, yet further development work is needed.

As for the public service aircraft, its application fields are much broader and more extensive. There are four major fields. The first is dispersers, such as spraying pesticides/seed planting and mountain forest fire fighting. The second is an ambulance aircraft, which can replace the current helicopter. For example, the three-person ambulance aircraft (which we will discuss) is a 3-m VTOLer compared with the traditional 10-m helicopter, fitting the niche landing for rescue. The flight range is within 100 km to send the wounded to the nearest public hospital. The third type is cargo aircraft, which have a large market share and can be used as EMS/DHL and Amazon/eBay services. The delivery drone saves human payload and has less safety concerns, which has a huge future market places esp. in China with the very complicated geographies. The fourth type is training aircraft for coaching a pilot, which is generally a two-seater aircraft used to train aerial vehicle drivers or to test their aerial driving license. In terms of the current technology maturity level, all electric-driven unmanned aerial vehicles of less than 100 kg are already in the market, such as sky camera (DJI Mini 3 Pro) and farming drones (DJI Agras T40).

(2) The second category is the scale. The size of the aircraft determines which type of power is used, pure electric or hybrid? For large trunk jetliners, the contribution of the electric RDF jet engine is to leverage the MGTOW to reduce the takeoff distance and to slow down the jetliner by electric reverse thrust during landing. These two intelligent operations can greatly save airport resources and reduce the airport noise/pollution. The mid-sized aircraft has the largest portion of air traffic shares such as local traveling from city to city/from state to state. More frequent 20-passenge van-type aircraft are most appropriate to fit in this scenario. Such van-type aircraft must be hybrid powered for both enough range (~300 km) and the capability of vertical takeoff/landing. For the small car-airplane with a 2-seater weighting less than 1 ton, pure electric driving and VTOL are quite possible with a maximum journey of 100 km powered by current Li battery. If equipped with the LTG using kerosene as fuel to generate aero-electricity, such small aircraft can be made as a supersonic VTOLer capable of long-range continental private flight as Boom's XB-1 supersonic high-speed jet [4]. The difference lies in that such VTOL jet uses the kerosene to generate electricity for RDF jet to provide the thrust instead of turbojet, and it is also capable of vertical takeoff/landing.

(3) From the technology perspective, there are three major aspects: aircraft, engine, and power, which are the three key topics that we will address in Chaps. 9–11. One of the greatest features of electrified aviation is that the e-aircraft design is different from traditional airplane with dual or four engines. Electric engines are small and can be made multiple (more than 4), allowing more flexible and optimized engine/aircraft integration, namely, the DEP concept (distributed electric propulsion). Multiple electric engines can be synchronized by electricity and evenly distributed along the airplanes in different locations for the best aerodynamic performance. For example, the 3D orientation of each RDF jet offers the flexibility to maneuver the airplane moment by different rotation of the jet stream. Obviously, the RDF jet engine is the key for DEP, which is our second key technology elaborated in Chap. 9. The third key technology is the electricity to power this RDF jet engine. In Chap. 10, we review the various electric power options, such as Li and fuel cell batteries and ordinary supercapacitors, together with various fuel generators. Although Li batteries and FCs have made great contributions to the new energy automobile industry, such as Tesla and BYD, they are not as powerful enough when used for airplanes, esp. for VTOL purpose. We have to develop new energy sources, such as the 3D HK SC together with the lightweight hybrid generator as the ultimate aviation electric power source.

AAM

Figure 7.2 illustrates the AAM market places, the applications, and the technologies. The key AAM technology is VTOL. Vertical takeoff and landing offer tremendous advantages, and air mobilities and VTOL itself is a disruptive technology, which involves revolution e-engine and e-power. This certainly takes time to happen and the best fit-to-market option currently is the Li battery-powered propeller-driven 2-seat small airplane such as in RR's SOI and Eviation's Alice and DJI's camera shooting and agriculture farming UAS.

- The unmanned aerial system uses multiple propellers to hover in the air as well as a navigation powered with a Li battery. Lightweight (~ 30 kg) is its great advantage, which shall not exceed 100 kg.
- Propeller driving is not sufficient for the VTOL operation of heavier aircraft. Most flying or sky cars use Li battery-driven propellers to achieve horizontal takeoff and navigation. Generally, the marketplace is a 1–2-seat aircraft with a weight less than 1 ton, a height of 1000 m, and a speed of 500 mph. Most flying cars just

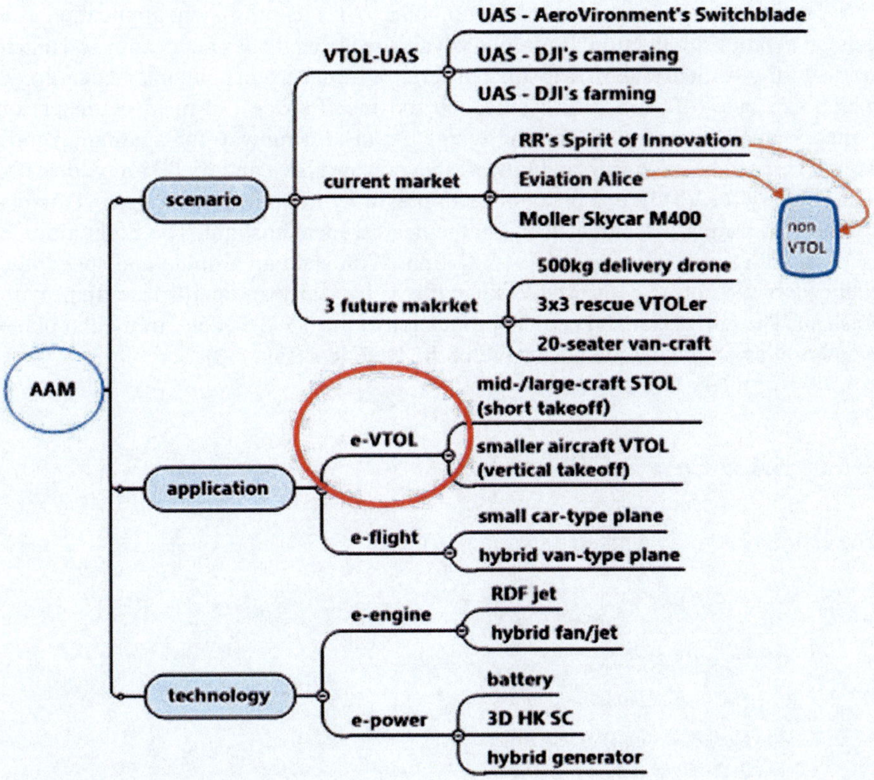

Fig. 7.2 The map of the advanced air mobility (AAM)

have a prototype to demonstrate/show off as an ad. The price range is between 300 K and 1 million dollars.

- Most of the private one seat electric airplanes carry less than 500 kg Li battery for power lasting 2 h for an ~ 300 km journey. XB-1 and other supersonic private jets are kerosene powered for long-lasting continental flight—from London to Washington DC in just 3 h. In the long-run kerosene-powered hybrid generator, LTG together with high-power output 3D HK SC offers a future alternative VTOL plus long-journey private hybrid electric aircraft.

In the following, we give two example cases of AAM: a 3 × 3 VTOLer rescue aircraft intending to replace the helicopter and drone delivery aircraft for Internet shopping—the IoTs.

7.1.2 AAM Case 1—3 × 3 VTOLer

It is a 3-m, 3-men, 3-engine aircraft as an emergency rescue vertical takeoff/landing "car-copter". The three-person VTOL ambulance has its prominent application as a general aviation application. Its unique advantage is that it does not occupy too much space during vertical takeoff, a distinct difference from current helicopter technology, which can take off vertically but occupy too much space (30 m in diameter) or ordinary small aircraft with folding wings but need a runway for landing/takeoff. This VTOL ambulance can land/takeoff in a narrow space and fly 80 km with a 50-kg Li battery for VTOL and 60 kg of kerosene for hybrid power to navigate. Patients can be sent from the accident scene to the nearest local hospital. The flight altitude is 1 km with a cruising speed of ~ 500 kmh. With such an altitude and speed, the flight journey is quite comfortable since this is basically an open-space flight with fresh air. The horizontal and vertical propulsion of the 3 × 3 3-seater triangular plane mainly relies on the unique advantage of the RDF jets (Fig. 7.3).

7.1.3 AAM Case 2—Drone UPS

The express cargo fly activates a nation's vitality:

> *The mobile cargo flyers*
>
> - *To activate the national mobility*
> - *To boost the national economy*
> - *To circulate the national currency.*

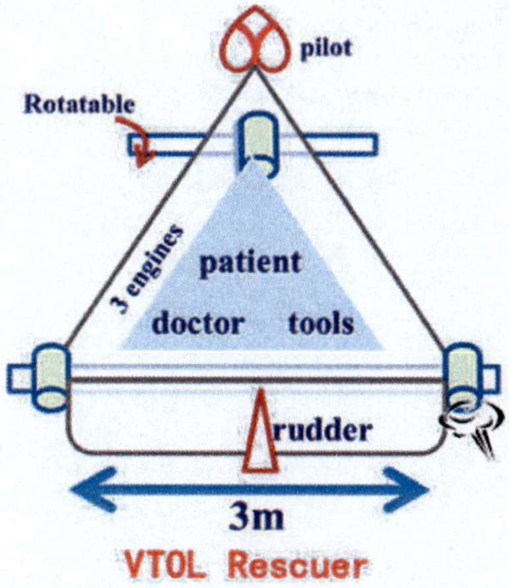

Fig. 7.3 3-men, 3-engine, 3-m 3 × 3 VTOLer rescue plane

Drone delivery makes the IoT become true. The IoT is a popular word after the "Internet"—to share the physical things in addition to the sharing the "mind things". The "Internet of Things" is communication in physical form based on visual/audio information from the Internet. For example, one can see and hear that a cake is tasty on eBay, yet the information conveyed on the "Internet" is just the message, not the taste. Through purchase and delivery, one can taste the real cake with their tongues—this is the "Thing". This is the original meaning of the IoT—delivering the wholesome experience with human's five senses: seeing, hearing, tasting, smelling, and feeling, with five organs: eye, ear, tongue, nose, and skin, instead of just sharing visual and audio information. A mid-sized drone can be used to serve the routine localized package delivery of UPS, EMS, DHL for Amazon, eBay in the US and Taobao and jd.com in China, which basically covers every corner of the land in the country. The business covers shipping, freight, logistics, and the supply chain. The drone delivery helps to activate the national economy and national IoT communication. Compared to conventional delivery services, drone delivery is fast and prompt, point to point. Compared to helicopter delivery services, it is much more economical and convenient (Fig. 7.4).

This triangular unmanned VTOLer is a handy drone delivery tool. The VTOL can be achieved by adjusting the RDF jet vertically and pushing the jet to the ground. After vertical takeoff, the electric aero-engines gradually turn to 45° to make the aircraft navigate while maintaining the aircraft hover in the air by vertical thrust and lift/drag ratio. The takeoff is directly driven by the battery and supercapacitor. It is feasible to carry 500-kg aircraft (lightweight airplane 140 kg, payload 200 kg, three electric engines 3 × 20 = 60 kg, battery and fuel 100 kg) for nearly a hundred miles, or a few tons weight larger delivery drone with hybrid electric power aided by lightweight gas turbine generator. The delivery distance can also be extended to 600 km.

The RDF jet also helps e-STOL for large commercial aircraft to achieve a smarter and greener aviation as follows.

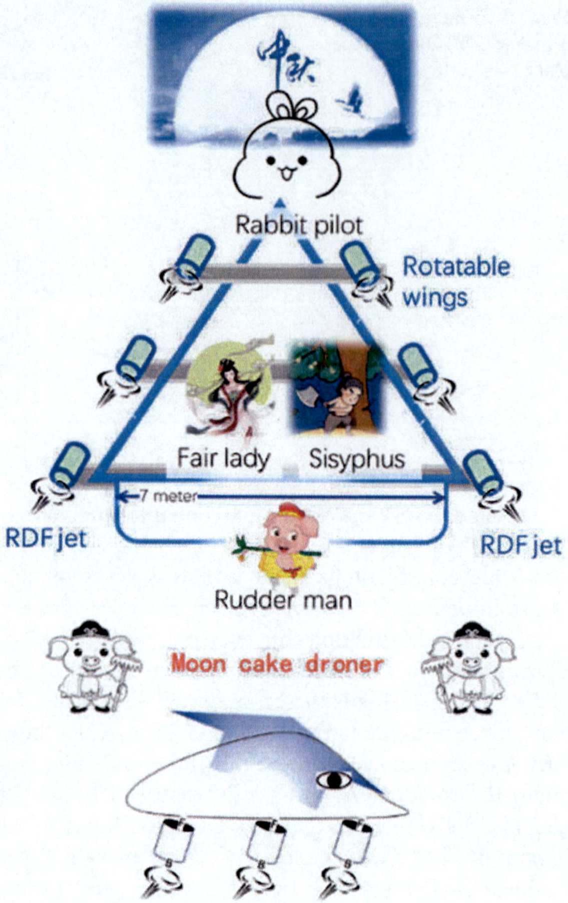

Fig. 7.4 Figurative 7-m moon-cake[1] delivery drone powered by 9 RDF jets (2 + 3 + 4 on each row)

7.2 The Smarter Aviation—STOL

VTOL stands for vertical takeoff and landing and has two synonyms electric vertical takeoff and landing (eVTOL) and short takeoff and landing (STOL) for large aircraft where it is not practical to conquer the gravity of 300 tons weight. To lift over

[1] This aircraft follows the Chinese fair tales that there are a few figures on moon, the beautiful goddess Change, a Chinese Sisyphus called Wugang, a rabbit of Change and a piggy as a steersman. Wugang is cutting a tree which automatically heals itself just like Sisyphus who is the loading a stone to a mountain top but rolling down again and again, as a punishment labor. In addition, the mid-moon festival is Chinese traditional holiday and people eat mooncakes on Aug 15 night (lunar calendar) to share the same bright moon afar (in Ancient Chinese far away friends/relatives cannot see each other so they use the moon as a remote mirror). This figure in this chart is just a parable that these four figures bought the mooncakes from the earth and flying back to the moon have a home festival party.

100 ton airliner requires more than 100 MW power, and the maximal power of one GEnX is just 45 MW. Two GEnX aero engines cannot lift this airplane vertically but can easily take off the airplane by horizontal acceleration. The VTOL also implies rotatable wing architecture that allows electric engines to rotate at various angles for flexible horizontal and vertical propulsion. In short, the VTOL is a comprehensive technology that does not rely on propulsion alone as the conventional fixed wing engine.

Vertical takeoff is a more advanced concept than the conventional Bernoulli's takeoff. Horizontal takeoff relies on the acceleration to use the wing to lift the airplane, needing a runway which brings about the friction (the heavier the aircraft, the more obvious) energy loss. Taking a 100 t aircraft as an example, the takeoff speed is 270 km/h (75 m/s); the required power is ~ 15 MW, and the friction power (to overcome) is 7.5 MW. Therefore, the power required for takeoff is approximately 23 MW. It takes 32.5 s to accelerate to takeoff; the takeoff distance is ~ 1000 m; the takeoff power consumption is ~ 200 MJ (= 56 kWh), and the friction power consumption is 100 MJ. So, the total takeoff power consumption is 300 MJ. For vertical takeoff, the power is $mgV_{\perp}$. Suppose the rising speed $V_{\perp}$ is 1 m/s; then, for a 100-ton aircraft, the power required is 1 MW, and the work done to 10-m height is $W = FS = 11$ MJ, which is much less than the energy consumed by horizontal takeoff and requires much less takeoff space.

However, to achieve a VTOL for this big aircraft, one needs to conquer the weight of 100 ton, which requires a 1000 kN thrust. This is not applicable even with the current gas turbine propulsion technology. One GEnX engine delivers 300 kN thrust, and four of them are needed to lift this airplane vertically. But, there is an intrinsic drawback of fixed wing GEnX which cannot rotate 90 degrees from vertical to horizontal. The rotatable wing technology does not fit to rotate a big and heavy gas turbine engine but quite applicable to shift the small, light multiple electric engines from vertical propulsion to horizontal.

Besides, the working principle of vertical lift up takeoff is different from horizontal acceleration takeoff. VTOL requires thrust, and horizontal takeoff requires speed. To compare the power needs of horizontal versus vertical force, we use the current automobile and aircraft as a comparison:

- The aircraft taking off needs to overcome gravity, so the vertical acceleration force must be above 9.8 m/s^2.
- For an automobile, to accelerate from 0 to 100 kmh in one minute, the acceleration rate is just 1.67 m/s^2; for an airplane, the takeoff speed is 75 m/s (270 km/h). To accelerate from 0 to75 m/s in 30 s, the acceleration is only 2.5 m/s^2, far less than 9.8 m/s^2.

Therefore, comparing the horizontal versus vertical takeoff for bigger airplanes, horizontal acceleration requires far less power than the vertical force needed for VTOL. It is pragmatic to rely on the wing to achieve this indirect takeoff at the current being for heavy aircraft. It is not as practical to VTOL an aircraft of more than 10t MGTOW. However, STOL is quite feasible with e-propulsion, which is discussed further in a later section.

7.2.1 Saving Airport Resources

The airport resources refer to space/time, noise, and pollution, i.e., how much space and how long an airplane stays before taking off or moves to inventory, the noise level during takeoff, and the harmful emissions during the takeoff/landing process. The *active time/space* of an aircraft in the airport during landing and takeoff is the determining factor of the overall noise and pollution level as well as the airport occupancy. VTOL and STOL help to save time and space during takeoff/landing, to reduce the emission and noise level by shortening this duration in takeoff and to reduce the emission intensity during landing. Let us first analyze the energy consumption, emissions and air pollution in the whole flight journey.

Energy Consumption

Aircraft engines use aviation kerosene as fuel to provide power. Various kinds of air pollutants are emitted during combustion, impacting human health and the environment. The emission level of NOx of A320 (150 seats) during one 90-min flight (1350 km) is approximately equal to the emission level of a compact car traveling 86,000 km. The emission level of cars and airplanes is comparable per person/per km ($150 * 1350 = 172{,}000$ persons km for airplane versus $86{,}000 * 2 = 202{,}500$ persons km for car). However, the airplane uses more fuels and emission intensity during the takeoff, and landing process is higher (Fig. 7.5). First, the amount of engine fuel and emissions is proportional to the thrust, which is full during takeoff and nearly 1/5 during navigation. Therefore, although the takeoff time is short, the emission brought to the airport due to the full thrust is at a very high level. The aircraft uses the maximum thrust to taxis and accelerates on the runway. At this time, the fuel flow rapidly rises to the maximum value, reaching 262.8 kg min^{-1}, compared to the average of 40.7 kg min^{-1} during flight. The takeoff and ramping span is 15.1 min, and the fuel consumption is 615.7 kg. During the landing phase, the engine thrust increases briefly, and the instantaneous maximum fuel flow reaches 74.2 kg min^{-1}. At this time, the engine uses the reverse thrust to decelerate to shorten the landing distance. The total flight time during landing is 9.0 min, and the fuel consumption is 215.4 kg [5].

Table 7.1 lists the fuel consumption at each stage in the whole journey of the A320's 90-min flight. The total flight time is 2 h 32 min; the fuel consumption is 6332.1 kg, and the flight distance is 1647 km.

The Noise Level

The noise level of the airport refers to the intensity and duration of the decibel levels versus distance, and the accumulated noise of air traffic of takeoff and landing. In general, the take-off time lasts approximately 0.8 min, and touch down takes 1 min. The engine is in full thrust during takeoff, with the highest noise level ~ 140 dB. When you are within 50 m of the takeoff plane, the noise level is ~ 110 dB, which is extremely harmful to health. Table 7.2 shows a few typical noise levels at various scenes.

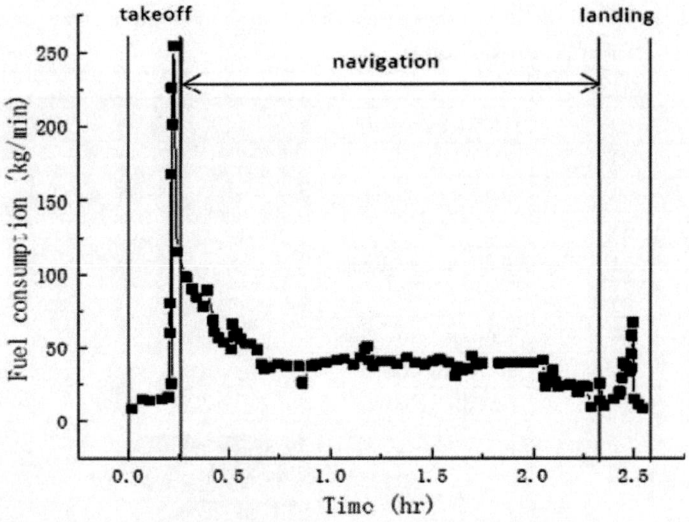

Fig. 7.5 Fuel flow during flight

Table 7.1 Flight time and fuel consumption analysis

Aviation stage		Duration (min)	Fuel consumption (kg)
Takeoff	Taxiing	12.4	182.2
	Takeoff	0.8	151.5
	Ramping to 1000 m	1.8	282.1
Flight	Ramping from 1000 to 10,000 m	21.1	1565.4
	Navigation	86.6	3519.9
	Descending from 10,000 m	20	415.6
Landing	From 1000 to 0 m	5.9	157.4
	Touch-down	1	29.4
	Taxiing	2.2	28.6
Total		151.8	6332.1

It is usually difficult to reduce the noise intensity during landing and takeoff since a large airliner must use the full thrust for takeoff and the reverse thrust in landing. However, reducing the takeoff duration is quite possible by making use of electric-driven jets that we proposed in Chap. 9 to leverage the takeoff weight for shorter takeoff distances. For landing, reverse thrust from the gas turbine machine can be replaced by the electrified jet thrust.

Table 7.2 Noise levels at different distance from the takeoff aircraft with the corresponding scenes

Distance (m)	Noise (dB)	Level (scene)	
0	143		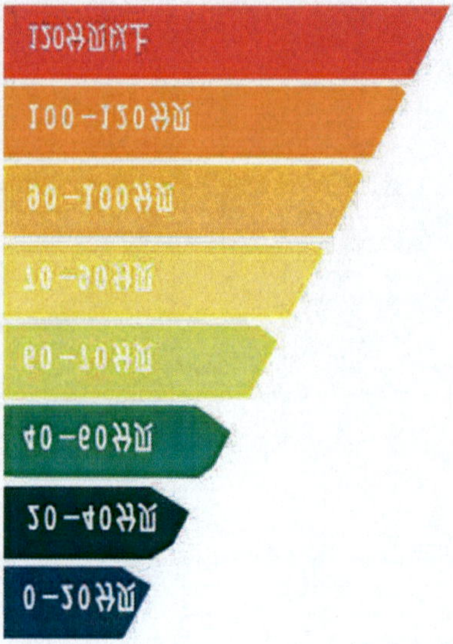
50	109	110 dB, damaging	
150	99		
450	90	90 dB, noisy street	
950	83		
3950	71	70 dB, noisy talking	
28,950	54		
33,950	52	50 dB, meeting room	
		20–40 dB, whisper	
		0–20 dB, silence	

7.2.2 Electric Engine Is More Efficient

Compared to fossil fuel mechanical/thermal engines, the efficiency of electric motor is higher. First, there is no thermal energy loss in e-engines. Second, the structure of the e-motor is relatively simpler—there is no gearbox to shift speed and no energy loss in the transmission. The motor speed is continuously adjustable, which only depends on the input electric current. The efficiency of the motor is just to convert the input voltage and current into the mechanical torque and rotation. The conversion efficiency can reach 90% or higher.

The electricity itself can be generated more efficiently as well. Currently, more and more electricity are from sustainable resources such as solar and windmills. In China, green energy has already account for nearly half (more details are provided in Sect. 7.3.1). Such green electricity stored in batteries and supercapacitors can be used to drive electric engines, a very worthwhile approach compared to fossil fuel gas turbine engines. Of course, some electricity may also come from power plants, but its efficiency can be targeted at ~ 40%. Compared with many individual combustion engines running around the world, the centralized electric generation from large power plants still has higher overall efficiency and fewer emissions of pollutants/greenhouse gases.

Additionally, electric engines are more prompt because they are electric driven instead of going through a gradual thermal combustion ramping process. This is because the power output mode of the engine is different. Fossil engines need to reach a certain speed to output the maximum torque. The electric engine can reach the maximum torque in an instant when the maximum input voltage and current are applied. For example, Tesla Model 3 has 275 horsepower and 1745 kg vehicle in weight. Its 100-m acceleration has reached 5.6 s. The Cadillac ATS with 276 horsepower and 1550 kg weight needs 6.2 s for the same acceleration.

The electric engine is clean and not as "dirty" as the fossil fuel engine since there is no exhaustion involved. Electric vehicles have zero tailpipe emissions, leaving the air cleaner as you drive. EV cars produce no emissions as compared to fossil fuel cars, especially as the electric grid becomes more renewable. The same truth applied for the airplanes since the electrified takeoff reduces many airport contaminations (more in Sect. 7.3).

7.2.3 *Electric Flight Is Smarter*

The smart flight is the "medium scope" flight, both from the range and from the pax. The large flight such as a 400-seat overseas commercial flight is not as economical compare to a few flexible 20 pax flights. Besides, the majority of flights are from ~ 500 km in range with ~ 50 passenger. Electrification plays a more efficient role in this scenario.

Figure 7.6 shows the flight traffic versus the mileage, and an impressive share of the flights is taken for short distances with a flight journey of 500 km range and 1–2 h duration [6]. As seen from the chart, close to 50% of all domestic flights cover distances of less than 500 km. These distances correspond to the operational range where the fossil fuels planes are the less efficient due to the higher takeoff and landing portion compared to cruising. Figure 7.6 also implies that it would be easier to reduce CO_2 emissions by using electrified takeoff/landing by replacing or decarbonizing the conventional gas turbine airliners in these short flights. The potential savings of less than 500-km flights by electrified air vehicles represent most of the CO_2 reductions.

Another interesting example of short-distance flight is in China. There used to be an interesting question asked by Chinese: Why don't we build a bridge over Dalian and Yantai? Dalian and Yantai are two very important coastal cities in China (Fig. 7.7) belonging to two large provinces: Liaoning and Shandong. Interestingly, although they are so close in distance, they are not connected inland due to a short ocean channel in between such as the Suez Canal. Both Dalian and Yantai are node cities with diverse transportation hubs connecting other inland cities, major ports, etc. In fact, the Chinese did not establish a bridging to connect these two cities is not they are not capable build high and long oversea bridge nor under sea canal, but due to the massive big shipping traffic passing through this canal everyday, since there are many important cities inside the canal connecting the inland to the Pacific Ocean. Through the scientific analysis of the geological situation, it is just not as scientific logical to connect these two cities either by underground channels or via a bridge across the Yangtze River. Since it is just 100 miles distance, it is a very promising plan to use the aero-vehicles to connect these two cities considering the short-range

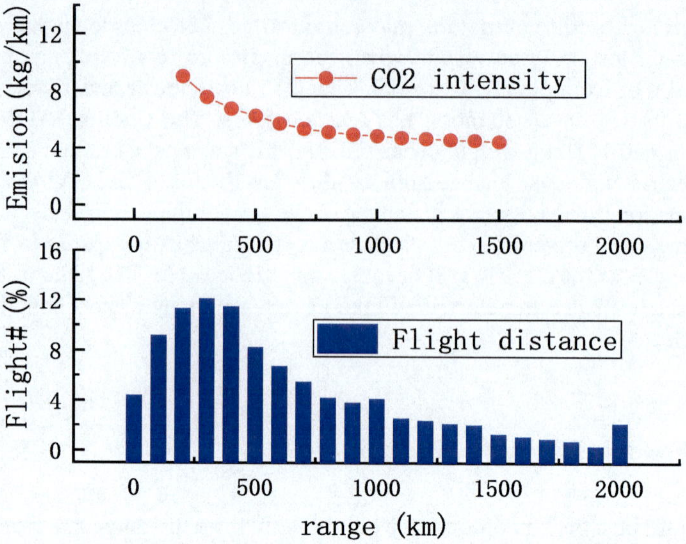

Fig. 7.6 The statistics of the most frequent flights and related CO_2 emission levels

feature of the electric aircraft, which happens to be in within 100 miles using the matured Li batteries. If one drives a car between these two cities, he has to drive around a big circle of over 1000 km (see Fig. 7.7 left). Not only in China, similar geographical situations worldwide can take a similar advantage by electric vehicles to connect two locations.

The Chinese philosophy is always seeking the "middle way", allowing more room to maneuver left or right, to shift either to the faster lane or slower lane on a highway. In this case, the middle way is the ~ 500-km flight with ~ 50 pax—the most frequent aviation traffic that the electric airplanes can fit in to maintain a proper balance between the performance versus efficiency. From the strategic level, small and medium-sized aircraft with medium thrust is always preferred for the best

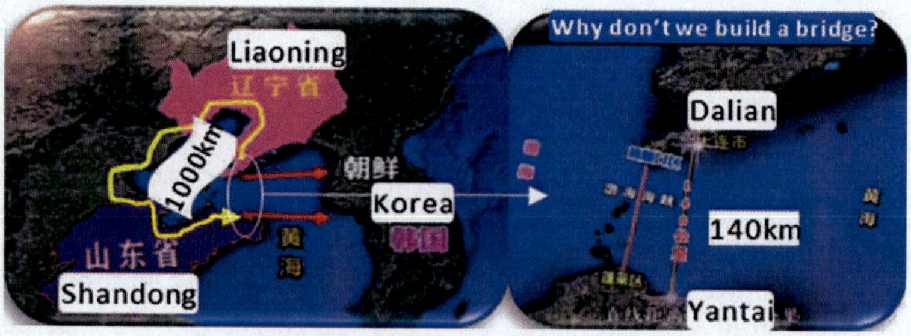

Fig. 7.7 Geographical site which is most suitable for electric aero-communications

economy. Most of the time, the 400 seats of large aircraft are not fully occupied and not as cost-effective as several adjustable-scheduled 20 seats van aircraft.

7.3 The Green Aviation—Less CO_2

What is "green"? "Green" refers to less carbon dioxide and methane in the air under the background of energy conservation and sustainability. There are two "greener" ways: one is passive, and the other is active. The passive way, for example, is to make CO_2 disappear by carbon neutralization, i.e., the focus is to solve a problem. The active way is to reduce the emission, i.e., to prevent a problem from happening. One is curing a disease while the other is to avoid it. The advantage of the second choice is obvious. Electrification is the active solution. Green electric aircraft technology has gradually become a hot topic f under the background of recent global warming, the abrupt climate change, and the reduction of fossil fuel resources [7, 8] Small and medium-sized aircraft and electric aircraft have low energy consumption and a high utilization rate, which is of great benefit to energy conservation and emission reduction. In addition, express drone aircraft is a hot market place in the near future, which will greatly enhance the IoT capability. It is especially cost-effective to use electric aircraft for fuel savings and emission reduction!

7.3.1 Greener Electricity

Electricity itself is becoming cheaper and more green in the coming century with the advent of more sustainable electricity generation such as solar, wind, and tidal. Figure 7.8 shows the electricity generation of China and America, the two largest vendors and users of electric power. The amount of electric generation has continued to grow since China opened its door to the outside world in 1978 by President Den XiaoPing. In addition, an increasing amount of electricity comes from renewable energy rather than from fossil. Renewable energy accounts for as much as 20% of total electricity generation in the States. The total power generation of the US in 2021 is ~ 4.1 trillion kWh, and the renewable energy power generation, including hydropower, solar power, wind power, and biomass power generation, is ~ 800 billion kWh. In particular, "solar power generation" is expected to become the main renewable energy source in the United States. In China, the total power generation in 2021 is approximately 8.112 trillion kWh, and the power generation from renewable energy is 2.485 trillion kWh, accounting for 30.6%. The power generation from wind power, solar energy, and biomass energy is 586.67, 300.9, and 148 billion kWh, respectively.

The Final Chapter of Green Energy

The ultimate green energy generation of mankind must be the full use of solar energy, tidal energy, and wind energy. The ultimate green energy storage should use the

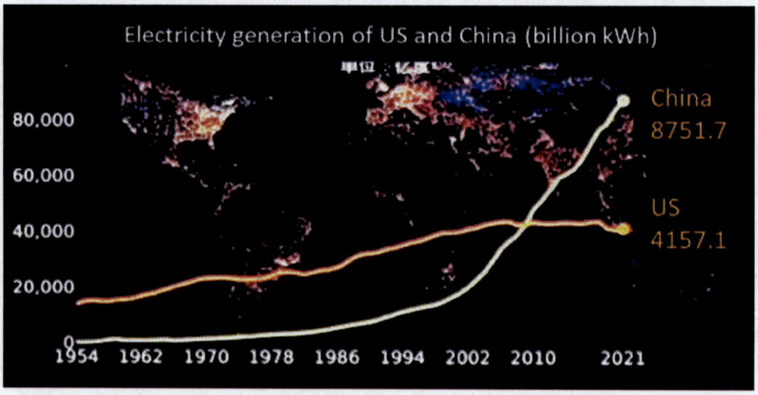

Fig. 7.8 Electricity generation of the US and China

above energies to hydrolyze water to extract hydrogen, storing hydrogen as fuel instead of fossil fuels, and using hydrogen fuel to create electricity from fuel cells. The byproduct is the H_2O, the water. That is the ideal way of green power. The real green energy solution is to use hydrogen as an energy storage fuel instead of using a Li battery and/or fossil fuel.

That is why Japanese people prefer to develop FC instead of Li battery cars although Toyota was the first to introduce the Prius hybrid car to the market as early as 1997. Japan has not participated in the development of Li battery EVs in recent years; part of the reason is due to Japanese understanding of clean energy and its own national conditions. Unlike the US or China, which have boundless resources, Japan is an island country and is seeking its self-feeding independence. According to Japan's philosophy, Li battery is never considered as clean energy because the battery production and recycling still cause a great deal of pollution to the environment. They firmly believe that Li batteries are only a temporary substitute to deal with the oil crisis. The hydrogen is the final chapter of the green energy, despite the current bottleneck of the hydrogen generation technique. Although there are boundless hydrogen in H_2O, it is just take too much electricity to separate it from water right now, and it is just a question of time when the breakthrough come into being.

7.3.2 Greener Flight

Greener Trip

Table 7.3 compares the cost of trips per capita by airplanes, gasoline cars, and electric cars in China and in the US as well as Chinese advanced CRH train, the cheapest and most effective cool public transportation taking full advantage of Chinese landscape geography. In China, taking CRH trains costs only ¥0.0022 per person per km, the cheapest long-journey travel in the world. Additionally, as seen in the table, it is more economical to drive electric cars in China than gasoline cars, and the expense of electric car traveling is only 1/9–1/5, because Chinese electricity is cheaper than that in the US and Chinese gasoline is more expensive than that in the US. Using EV car to replace fossil fuel car is partially driven by the government policies in China in the future, to shift from fossil fuels to purely electric vehicles. It is wise national strategy in the long run. Table 7.3 also contains many other useful info for future technology and policy judgment.

Greener Journey

One can also see from the above table that speed affects the efficiency of the high-speed trains—just reducing the traveling speed from 350 to 250 kmh saves 33% of the energy. This also applies for airplanes. Figure 7.9 shows the best navigation speed corresponding to the L/D_{max} ratios for airplanes traveling at different stages (with different weights of the airplane due to the kerosene consumption) [9]. It is thus concluded that more delicate tune-ups during the flight can save the energies to achieve a greener journey other than the above policies/strategic approaches.

Greener Strategy

The government is responsible for directing the overall orientation of the greener policy by taking full advantage of the nation's geographical and cultural situations. For example, in China, the CRH high-speed train is a wise decision for public traveling as well as for logistics deliveries in comparison to airplanes. It is both cheaper and green, just ¥0.0022/paxkm, less than 1% of traveling expense of the airplane and

Table 7.3 Comparison of the trip prices in the US and China by airplane, trains, gas car, and electric car

	E-motorbike	Motorbike	Tesla car	BMW car	China hi-speed train	Boeing 787	RR's e-plane
	Electric	Fossil	Electric	Fossil	Electric	Fossil	Electric
kWh/km, capita	0.048	0.110	0.072	0.193	0.003@250kmh 0.0045@350kmh	0.16	0.130
kWh/km, total	0.048	0.110	0.143	0.386	19.2@250kmh 27.4@350kmh	32	0.130
Passengers	1	1	2	2	600	200	1
¥/kWh/China	0.478	1.95	0.478	1.95	0.478	1.95	0.478
¥/kWh/US	0.86	1.34	0.86	1.34		1.34	0.86
¥/km capita/China	0.0229	0.2145	0.0342	0.3764	0.0022	0.3120	0.0621
¥/km capita/US	0.0413	0.1474	0.0615	0.2586		0.2144	0.1118
Notes	1 man	1 man	2 men	2 men	8 carriages, 600 passengers	200 passengers	1 man

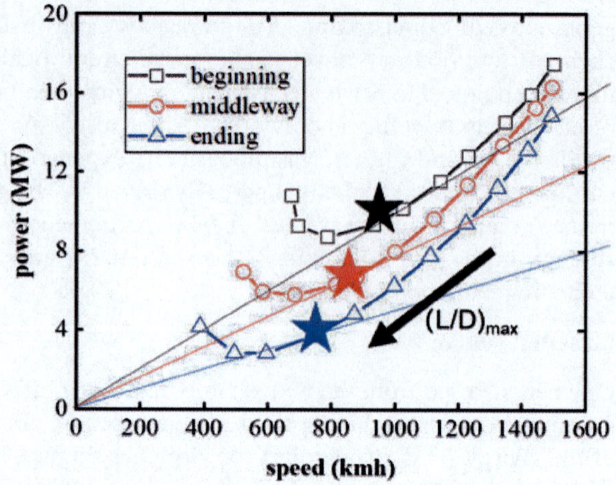

Fig. 7.9 Illustration of different $(L/D)_{max}$ ratios (at the points of tangency) when fuel is consumed in the journey

less than 10% of electric cars. The railway station security checking is also much easier than in the airport. The loading time is also much shorter—no-ticket/check-in is required.

From a strategic point of view, unless purely necessary such as continental long-range travels, large aircraft is not as efficient to operate as smaller aircraft. Large thrust consumes much more power than medium thrust, and continental flights are less frequent than domestic travels. Through the analysis of the markets and air flows, medium-scale local flight processes a large portion of air traffic, and electric planes fit into this scenario more efficiently.

7.3.3 *Greener Takeoff*

Background: Pollution from Takeoff

The air pollution during takeoff/landing is higher and more intense since the engine is in full thrust and emissions are on the ground with concentrated airplane traffic (Fig. 7.10). NO_x (nitrogen oxides) is the main pollutant from engine's full thrust emissions, a poisonous gas derived from nitrogen and oxygen combustion under high pressure and temperatures during takeoff. NO_x contains nitric oxide (NO) and a smaller percentage of more poisonous nitrogen dioxide (NO_2); both are poisonous gases that contribute to acid rain and suffocating smog. In the takeoff phase, during the low-speed coasting process after the engine is started, the NO_x emission index increases from 0.96 to 8.10 g kg^{-1}. When entering the takeoff process with full thrust, the engine temperature rises, and the NO_x emission rises. The instantaneous maximum value exceeds 114.00 g kg^{-1}, with an average of approximately 105.80 g kg^{-1}, which is much higher than the rest of the flight.

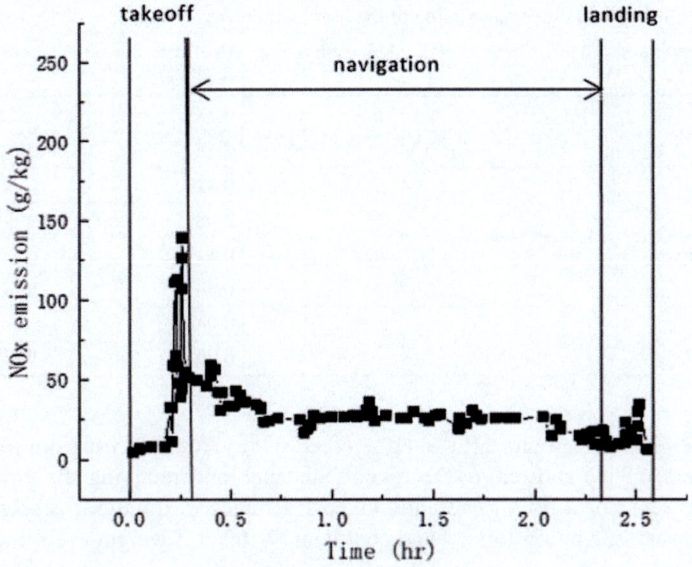

Fig. 7.10 NO_x levels in takeoff, navigation, and landing

The landing takeoff cycle (LTO) of the A320 series aircraft is 215 per day, accounting for approximately 40% of the total LTO of the airport. Based on this calculation, the NO_x emitted from the A320 series aircraft at Baiyun Airport in Guangzhou, China, is almost equivalent to the sum of 270,000 cars in a single day. Airport pollution is heavy and intense and usually requires a large space to digest other than its requirements on takeoff/landing distances.

Table 7.4 shows the details of fuel consumption rate of the engines together with the thrust and duration of the takeoff/landing of two typical engines, the small-scale AE3007 (32 kN thrust, 717 kg weight), and large jet engine CF6 (300 kN thrust, 4.9 tons weight) series. The proportions of aircraft engine NOx emissions are 28.16%, 38.60%, 8.46%, and 4.10% for the four modes of LTO: takeoff, climb, approach, and taxiing, respectively.

Greener Takeoff Methodology—VTOL and STOL

For larger aircraft, it is not practical for vertical takeoff since lifting up a 100-ton airliner aircraft requires 1000-kN engine thrust. However, shorter takeoff is feasible to offer an optimized and advanced takeoff process by reducing the amount of on the acceleration runway. Taking a 10-ton aircraft as an example, the friction loss accounts for 1/3 of the total takeoff energy. This takeoff loss can be avoided by shorter takeoff. Table 7.5 evaluates the takeoff distance savings for different planes by using RDF jets to leverage the weight of the airplane. For a 20-ton airplane equipped with 9 RDF jets, only, 500 m are needed for takeoff, half of the takeoff distance without vertical lift and half of the friction energy loss in takeoff.

Table 7.4 Throttle and fuel consumption of the four LTO stages

Mode of operation	Thrust throttle (%)	Duration (min)	Fuel consumption rate (kg/s)	Aero-engine
Takeoff	100	0.7	0.383	AE3007A1
			2.581	CF6-80C2A5
Ramping	85	2.2	0.318	AE3007A1
			2.082	CF6-80C2A5
Descending	30	4.0	0.113	AE3007A1
			0.687	CF6-80C2A5
Taxiing	7	26.0	0.046	AE3007A1
			0.207	CF6-80C2A5

The weight of large aircraft can be reduced with vertical thrust from rim-driven jet thrusters, hence shortening the takeoff distance and reducing the emission of harmful gases. For a rough estimate, takeoff distance *S*, duration, acceleration *a*, speed *V* thrust *F*, friction force *f* and coefficient *k*, takeoff weight *m* follow:

$$S = \frac{1}{2}at^2 \tag{7.1}$$

$$V = at \tag{7.2}$$

$$F - f = ma \tag{7.3}$$

$$f = mgk \tag{7.4}$$

We have

$$S = \frac{1}{2}\left(\frac{V^2}{F - mgk}\right) * m \tag{7.5}$$

For an ordinary commercial airplane, the takeoff speed is 270 kmh; the takeoff distance is 1000 m, and the takeoff time is 30 s. The thrust *F* must be large enough to accelerate the airplane to the takeoff speed within the runway distance *S*. That renders different aero-engines with different levels of maximum thrust (e.g., two engines with 290-kN thrust each). The takeoff distance *S* is proportional to the takeoff weight of the airplane. Using 9 RDF jets with a maximum vertical thrust of 10 tons, the weight is reduced by 1/5 for a 50-ton airliner. Ignoring the friction, the takeoff distance savings of the 50-ton aircraft are 1/5 * 1000 m = 200 m. In Table 7.5, we estimate the takeoff savings on distance for three weight airplanes under different thrust levels but assume the same takeoff distance of 1000 m including the friction loss at the friction coefficient $k = 0.01$.

Table 7.5 Reduced takeoff distance by vertical thrust for three airliners

RDF jet thrust (kg)	20 ton airplane			50 ton airplane			100 ton airplane		
	Reduced weight (kg)	Takeoff distance (m)	Takeoff saving (%)	Reduced weight (kg)	Takeoff distance (m)	Takeoff saving (%)	Reduced weight (kg)	Takeoff distance (m)	Takeoff saving
0	20,000	1004		50,000	1001		100,000	1004	
2000	18,000	901	10.3	48,000	959	4.1	98,000	984	2.0
4000	16,000	798	20.5	46,000	918	8.3	96,000	963	4.1
6000	14,000	696	30.7	44,000	877	12.4	94,000	942	6.2
8000	12,000	594	40.8	42,000	836	16.5	92,000	921	8.2
10,000	10,000	493	50.9	40,000	795	20.6	90,000	901	10.3
	Thrust	*58 kN*		*Thrust*	*146 kN*		*Thrust*	*290 kN*	

In summary, the STOL is the best approach for greener takeoff, saving airport space and time by using the vertical thrust to reduce the MGTOW. The Beijing airport traffic every day has ~ 22,920 takeoffs. A 10–50% takeoff distance reduction is a considerable amount of saving for CO_2 emissions and airport resources.

7.3.4 Greener Landing

Background: Pollution from Landing

The engine cannot be shut off during aircraft landing for safety concerns. Besides, many models still need the engine power to provide the reverse thrust to decelerate the plane during the landing. Moreover, the aircraft needs some power for taxiing and connecting taxiways.

When the aircraft is approaching the runway and preparing to land, the engine throttle is gradually reduced. In the final stage of landing, the engine throttle is at the minimum (idle) position. The combustion stability and economy of the engine are poor under the idle state. There are three main reasons for this: first, the small supercharging ratio of the engine at idle makes the thermal efficiency of the engine low and the economy poor. Second, the combustion chamber efficiency is low in the idle state. At idle conditions, the inlet temperature and pressure decrease, and the mixture is relatively lean, resulting in a lower flame temperature in the combustion chamber, which increases the emissions of carbon monoxide and hydrocarbons, both of which are products of incomplete combustion. Nearly, 50% of pollutants are discharged when the aircraft is in taxiing/idle mode. It is indicated that when the engine operates at low thrust, the emission index is relatively high, producing more emissions per unit of fuel consumed. Taking Beijing Capital Airport as an example, when the sliding time decreases from 26 to 22 min, the total amount of gaseous pollutant emissions is reduced by 6%. From an environmental point of view, the general approach is to reduce the duration of the on-state of gas turbine engines during landing.

CO is more prominent during landing, as shown in Fig. 7.11 although its content is not as high as NO_x in Fig. 7.11 during the takeoff. CO is more harmful than NO_x. The most common symptoms of CO exposure are fatigue, headaches, and dizziness due to inadequate oxygen delivery to the brain. This can lead to serious tissue damage or even death. In the landing phase during the approaching phase, the engine's low fuel flow operation and the reversed thrust cause the increasing CO emission index. In addition, the duration of this CO pollution is longer, causing more serious airport pollution than NO_x pollution during takeoff.

Greener Landing Methodology—Electric Reverse Thrust

Traditionally, the reverse thrust technique is applied for landing deceleration. The speed of the aircraft during landing is ~ 250 kmh, and the taxiing distance is between 2000 and 3000 m. This consumes many airport resources. Deceleration is the basic

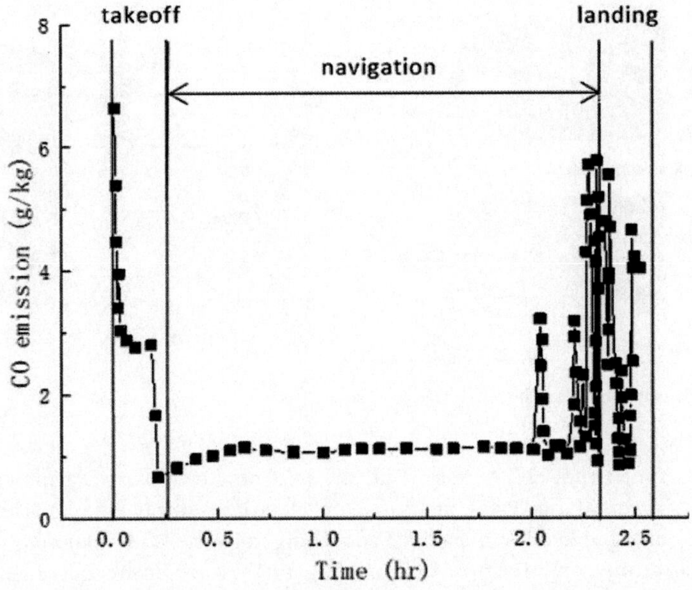

Fig. 7.11 CO levels in takeoff, navigation, and landing

solution to enhance the airport landing efficiency. The conventional deceleration methods are wheel braking and reverse thrust (Fig. 7.12), and the reverse thrust is the more effective and powerful and preferred. The drawback of this deceleration is its highest level of CO emission for landing compared to other LTO stages due to its low fuel rate combustion and reverse thrust. Such pollutant and noisy deceleration can be replaced by electric reverse thrusters. Multiple RDF jet can provide ~ 100 kN thrust just equivalent to a Boeing 787 reverse thruster deceleration force (Suppose the approaching speed is 250 kmh and the taxiing distance is 2500 m, the landing time is thereby 71 s, and the deceleration rate is approximately 1 ms^{-2}. Suppose the MGTOW of Boeing 787 is 200 tons, the fuel is almost exhausted after the journey; the aircraft weight at landing is 100 tons, and the decelerating force is close to 100 kN.)

Replacing the on-state gas turbine engine with an electric reverse thruster avoids much CO contamination from the commercial airliners—a large contribution to the greener landing.

In summary, green airports are more crucial since contamination is intense in the takeoff and landing periods. The peak and major levels of CO and NO_x are larger in the takeoff/landing stages than in the flight stage. In addition, airports are concentrated places for airplanes, and these accumulated effects exacerbate air pollution. Taking advantage of the rotatable wings to shift the electric jets vertically and backwardly can greatly enhance the airport takeoff/landing efficiency and reduce the associated airport pollution of NO_x and CO poisonous gases. The power to drive these electric engines may come from cleaner and quieter rechargeable electricity resources such

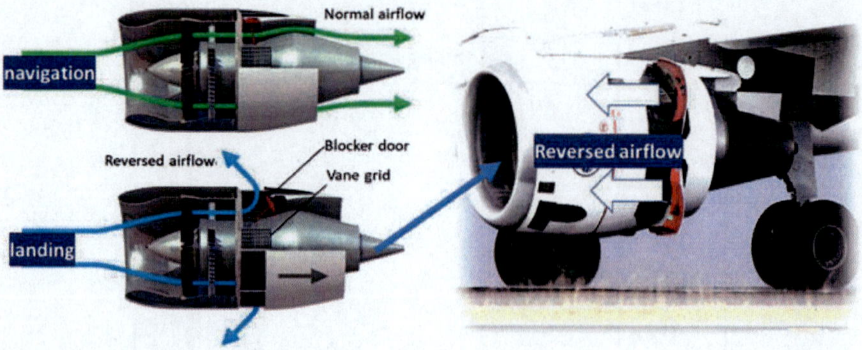

Fig. 7.12 Reverse thrust of the jet engine for landing

as Li battery and fuel cell battery. But, the best power is our 3D supercapacitor, which is lightweight, powerful, and safe aero-electricity storage media. The electric power is discharged in takeoff and will be recharged by the BTW generator from the existing gas turbine engine (more in Chap. 10) and then use this recharged electricity for clean and quiet landing. This clever arrangement of takeoff and landing is much more economical and much cleaner.

7.4 Case Studies/The Learnings

Table 7.6 compares the features, pros and cons of a few typical cases of the e-aviation together with our proposed e-aircraft/propulsion platform. There are two kinds of efforts on electric aviation: one is targeting the commercial market, and the other is for R&D; one is for short-term profit, and the other is long-term technology. The former is highly market driven, such as the twin-seat private airplane equipped with propeller and lithium battery—the two mature existing yet let advanced technology. Mature technology is easy to implement and simple to use, but its function is not superior enough. The latter is technology leading, which is the main interest of our AIAA/IEEE technical experts. It should be noted that while developing technology, we must have a strong sense of engineering infield experimentation instead of purely theoretical argument.

The value of the first three airplanes lies in its infield test data, which can be used as reference points for later development work. It borrows the matured Li battery achievements from the mature electric car industry to test the flight principles of electric propulsion and airplanes. Nevertheless, even these early testing are still preliminary—the immigration from the car industry to the aero-experiment is not as smooth or as up-to-date during the initial integration efforts. For example, the range of the infield test of the RR's SoI e-plane is still far less than the estimated value.

Table 7.6 Review of electrification on airplanes

Showcase	Moller VTOL Skycar	Rolls-Royce SOI	Eviation Alice	Yuneec E430	Boeing 787	Delta VTOLer	20-seater aero-van	3 × 3 STOL for C919
Driven by: shaft/rim	Shaft	Shaft	Shaft	Shaft	Shaft	Rim	Rim	Rim
Power: battery/generator/SC	Generator	Battery	Battery/generator	Battery	Generator	Battery/SC	SC/LTG	SC/LTG
Generator: ethanol/diesel/gas turbine	Ethanol	na	Gas turbine	na	Diesel	na	Gas turbine	Gas turbine
Fuel: fossil/electricity/hybrid	Fossil	Electric	Hybrid	Electric	Fossil	Electric	Hybrid	Hybrid
E-engine	Duct fan	Propeller	Propeller	Propeller	APU	RDF jet	RDF jet	RDF jet
VTOL	y	n	n	n	STOL	y	y	y
Rotatable wing	y	n	n	n	n	y	y	y
L/*D* ratio	n	y	y	y	y	y	y	y
Weight	325 kg	1250 kg	6668 kg	470 kg	250 ton	500 kg	7 ton	73 ton
Size meter ($W \times L \times H$)	6.5 × 2.5 × 2.25	7 × 8 × 2	19 × 17 × 3	13.8 × 12.5 × 2	60 × 63 × 17	3 × 2 × 2	7 × 6 × 2	36 × 38 × 12
Pax	1	1	11	2	200	3	20	168
Prototype	y	y	y	y	y	n	n	n

Sufficient battery efficiency data and airplane efficiency are still lacking. Stronger teams and sponsorship are urgently needed for future work.

In addition, from the first principle of physics, the Li battery is not an ideal choice for aviation electricity, both from its capacity and safety point of view. The intrinsic-specific power of the battery is just too low, and if operated in high-power output mode, its safety and reliability/lifetime (recharging times) will be greatly affected. From this perspective, new electric power supply resources must be developed that are more suitable for aviation electricity.

The value of the Moller's effort in Table 7.6 is its VTOL feature—it is the only manned VTOL infield test so far by successfully replacing the propeller with turbofan and with rotatable wings compared with other e-airplanes. This practical case, although far from being commercialized yet, serves as a good reference when developing the advanced eVTOL technologies. As for the Boeing 787's case, the value in a bold pace to adapt the new technologies, such as more electric engines, lightweight materials, and advanced aero-plane designs, with its superior R&D background support. These highly mature advanced technologies can be transplanted to electric engines, electric generators, and electric aircrafts.

In the following, we review the above electrification achievements from two perspectives. One is from the existing two mature cases from Boeing 787 and Honeywell, and the other is the three typical flying e-objects such as the Moller's VTOL carcopter, RR's SoI e-airplane, and DJI's agriculture UAS. The purpose of reviewing the existing process is for future advanced e-aviation technologies—to make traditional aviation more intelligent and effective. There are two key technology families involved in e-propulsion addressed in Chaps. 9 and 10 and advanced airplane/engine integration (DEP in Chap. 11).

7.4.1 Electrification of Boeing 787

Boeing 787 took advantage of new electrical engineering achievements to aid the existing aircraft (Fig. 7.13, from [10]). It is the first aircraft in history which makes the most electrification on the commercial airplane compared to its previous generations and other counterparts (Table in Fig. 7.13). In its first generations, it uses four 250 kW VSVF diesel generators to start the GEnX gas turbine engines during takeoff and two 225 kW engines from Honeywell for APU. The power over weight ratio of these generators is close to 1–2 kW/kg. For the recent generations, it uses four 250 kW generators weighing 95 kg (the specific power is 2.63 kW/kg).

What we learned from B787 electrification is although the power generation efficiency of diesel generator is lower than that of the gas turbine generator, its technology is simple, mature, and easy to implement. In B787's first introduction generation in 2011, the specific power is 1.25 kW/kg [11]. Currently, this ratio becomes 2.63 kW/kg (B787 in 2022). Boeing expects to achieve 5 kW/kg in 2030, and NASA has an even more ambitious goal achieve this specific power over 10 kW/kg [12].

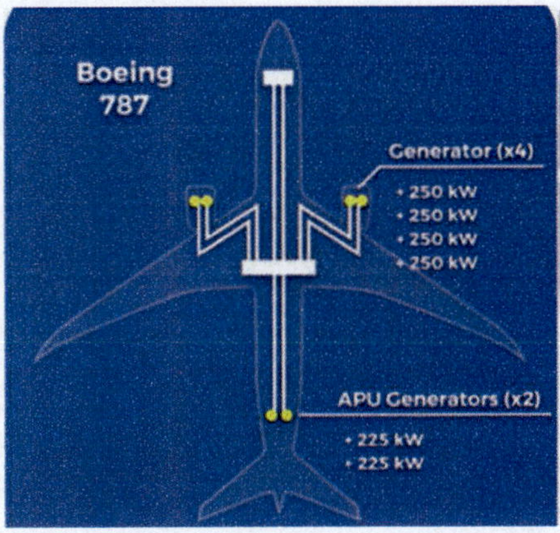

Aricraft	Electic power used	Year	MTOW	Power /MTOW	Engine
B787-8	1000kW	2011	228 tons	4.387	Trent 1000/GEnx
A380	600kW	2007	575 tons	1.043	GP7000/Trent 900
B737NG	180kW	1998	79 tons	2.278	CFM56
A340	360kW	1993	277 tons	1.302	CFM56/Trent 500
B777	240kW	1995	247 tons	0.971	GE90
B767	240kW	1982	204 tons	1.176	CF6/RB211/PW4056

Fig. 7.13 Six diesel generators equipped on Boeing 787 plane with the maximum of 1.45 MW power right: electric power amount of different aircraft with introduction year

7.4.2 Rotatable Wing VTOL Carcopter

In 1991, Paul Moller completed his M400 Skycar after 50 years' endeavor in VTOL [13]. Compared with other flying cars, Muller focuses on its technological aspect rather than the commercial market. His great achievement is that he completed the vertical takeoff and landing using the rotatable wings (Fig. 7.14). It is the first practical case thus far to use a rotatable wing to adjust small-scale aero-engines to adjust the horizontal and vertical propulsion. During takeoff, the four engines project to the ground for the VTOL and turn to the proper angles for hover and navigate—some propulsion power holds the airplane in air, and some power moves the airplane forward. Four simple ethanol engines drive ducted fans to provide the thrust. This is the prototype that is closest to the future VTOL concept electric aircraft. Skycar M400 is designed for 4 people for duration of 5.9 h, passenger capacity 325 kg, aircraft weight 1 ton, speed 500 kmh, ethanol as fuel, and maximum range 1206 km. Engine maximum power is 45 kW, weight 22 kg, size 25 × 28 × 28 cm, displacement 530 cc, 24.5 L per hour. The aircraft size ($L \times W \times H$) is 6.5 m × 2.5 m × 2.25 m [14].

The value of this prototype is that it is a proven case to verify the working principle of a rotating wing by attaching a compact duct fan to it to achieve the VTOL instead of using rotating propellers without proven cases. Although this prototype is far from being adopted by marketing, its main focus on technical aspects is more important to us compared to most of the other developers who are still using propellers and other matured technologies to design a product on a planned roadmap. Of course, the disadvantages were the lack of strong sponsorship, such as the big airliners, and lack of the perseverant efforts of follow-ups. Its technology is more advanced but needs further R&D investment. In addition, it is not an electric aircraft; it is noisy and only weighs 500 kg. Despite all these drawbacks, its trial-and-error experience is much

Fig. 7.14 Carcopter from Paul Moller

more worthwhile to study and follow-up when developing our e-propulsions using the RDF jet because they share many commonality.

7.4.3 Three E-plane Cases

There are many showcases of small Li battery-powered aircraft. The maximum takeoff weight is basically within 1.5 tons with less than 4 seats. Most of the time, people intend to identify electric aircraft with electric cars simply because they both use Li battery as electric power. However, there are still some differences in the battery features required for these two types of vehicles as listed in Table 7.7. The working conditions of existing Li batteries in electric airplanes are far more stringent than those of new energy automobiles. For example, the power output rate is much higher for electric aircraft, esp. during the takeoff. It is certainly the bottleneck of the battery from its first principle of physics—low specific power and lack of safety, esp. under high-power output. Li battery helps e-plane enters the e-aviation market faster, but it is not an ideal technology for aviation. In reality, as we see in the following real cases, the use of lithium battery technology borrowed from Tesla or BYD electric cars is not as professional and requires more delicate integration efforts. It is a big achievement of Rolls-Royce's Spirit of Innovation's test flight with its opened technical datasheet, where we can see that the actual flight mileage has not yet reached its expectation. Many tune-ups are still needed on the road of development.

Mobile aviation power is still the key bottleneck for electric aviation. In the long run, new electricity options must be developed other than Li batteries.

Real prototype test flights are still lacking to accumulate enough flight data to analyze for future R&D and commercial use. Here, we select only three cases—"Alice" from the Israeli/American Eviation Company, "Spirit of Innovation" from the British Rolls-Royce company, "E430" from a Chinese company. All cases have some infield flight data that can be used as reference for future studies. There are

Table 7.7 Li battery parameter comparison of EV car and EV plane

Parameters	Electric car	Flying car
Energy density (Wh/kg)	140–200	180–300
Discharge rate (C)	0.5–2	1–5
Peak power duration (S)	10–30	≥ 60
Number of cycles	1000–1500	≤ 500
Fast charging frequency (%)	10–50	≥ 90
Safety (%)	50	≥ 90

two prerequisites when we select the cases: they have deep research and development background support (for the perseverance and sustainability) and actual flight test prototypes (infield validation and technical datasheet). These models have good historical continuities and excellent supporting platforms, so they will not be like many "short life" cases. The battery is key bottleneck for all cases, and they need to work closely with concurrent EV vendors such as Tesla and BYD to catch up the newest battery developments (for example, 4680 battery in Tesla and Blade battery technology in BYD, more details in Chap. 10).

RR's Spirit of Innovation

Rolls-Royce's Spirit of Innovation [15] claimed in 2021 the first time that this SoI airplane flew more than 15 km at an average speed of 532.1 kmh and broke the world's highest electric aircraft speed record (600 kmh). It took 202 s to climb to 3000 m. The data of the RR aircraft are as follows: single-person aircraft, aircraft weight 1250 kg, 7 m (W) $\times$ 8 m (L) $\times$ 2 m (H), lithium battery weight 475 kg, power 74 kWh (= energy density 155 Wh/kg * 475 kg), power 237 kW (power density 500 W/kg * 475 kg). Three electric engines with a total power of 133 kW $\times$ 3 = 399 kW, 2200 rpm propeller, 750 V voltage.

The "Spirit of Innovation" is part of the ACCEL or "Accelerating the Electrification of Flight" project. Half of the project's funding is provided by the Aerospace Technology Institute (ATI), in partnership with the Department for Business, Energy & Industrial Strategy and Innovate UK. Its strong sponsorship background ensures that such an airplane plan has perseverant follow-up in the following years, and its real flight data provide valuable reference points for the following researchers.

Eviation's Alice

Unlike the RR's SoI, Eviation's Alice [16] electric airplane has not yet performed its infield flight and plan to make its first try from 2021 to 2022. It is a larger airplane with 11 seaters, all electric with 820 kWh lithium battery in total weight of 3720 kg. It covers an area of 19 m (W) $\times$ 17 m (L) $\times$ 3 m (H), with a maximum takeoff weight of 6668 kg. Table 7.8 lists its target parameters. It is our opinion that such plane has bigger ambition to achieve than RR's SoI which involves more technology challenges (much heavier aircraft which must involve the hybrid technology instead of pure Li battery). With such a goal in mind, the delivery year of 2022 is just too

Table 7.8 Key parameters of Alice's hybrid "van-aircraft"

Performance		Weights/power	
Max cruise speed	250 kts	MTOW	16,500 lbs
Max range	440 NM[a]	Payload	2500 lbs
Landing distance	2040 ft		
Take-off distance	2600 ft	Model	magni650
Climb rate	2000 ft/min	Max power	2 × 640 kW

[a]Target range, zero wind, no payload, IFR reserves

early considering the recent global situation. The more reasonable prototype delivery date shall be 2025–2030.

Chinese E430 Dual-Seat Electric Plane

Yuneec E430 [17] is a Chinese domestic two-seat electric aircraft, with a total takeoff mass of 470 kg (= battery 200 kg + two people 160 kg + aircraft and engine 110 kg). The lithium battery energy density is 200 Wh/kg, and the power density is 100 W/kg. The total energy carried with the airplane is 40 kWh with a maximum power of 20 kW. The cruise speed is 90 km/h, and the range is 227 km, so the flight time is 2 h and 30 min. The ordinary e-plane market in China is not as prominent compared to the unmanned e-aircraft for various mobile scenarios such as camera, delivery, and agricultures.

7.4.4 Unmanned Aero-system

UAS is one of the air mobilities in advanced air mobility (AAM). UAS opens a broad playground of market which brings about the vast versatile aerospace usages and needs the proper aero-administration rules when there are too many aero-vehicles in air. The keyword of UAS is the "unmanned" which implies much less safety concerns which involved stringent technical requirements such as robustness and reliability. There are two families of UAS. One is the vertical hovering ability, such as camera drones. The other is horizontal aero-machine, such as the Switch Blade series. From the thrust perspective, the first type focuses on the vertical lift ability similar to a helicopter, and the design rules are focusing overcoming the gravity. The second type is focusing on the acceleration for horizontal navigation as ordinary planes. The first type UAS seems to have more market places. In both UAS cases, the weight is the key factor to leverage the load of the engine as well as the battery. Such UAS is usually below 100 kg. At the current stage, none of them is capable of rotatable wings to accommodate both VTOL and horizontal navigation. Below, we analyze these two typical UAS. As a rough estimate:

- The horizontal speeding capability. The power need in the flight must match the power of the battery as well as the engine. For example, a UAS carrying a 0.25 kg

Li battery (150 Wh/kg and 400 W/kg) can provide a maximum 100 W power and 37.5 Wh energy. With a power/weight ratio of 1 kW/1 kg, a 0.1-kg electric motor can provide a power output of 100 W. The power needed to fly at 120 kmh (=33 ms) speed is $P = TV$, where the thrust T = mg (L/D). With a designed thrust/weight ratio of 10, $P = 82.5$ W Therefore, 100 W of power from both the electric motor and battery is sufficient to support the navigation of this horizontal UAS airplane.

- The vertical hovering capability. With the special design of the propeller together with enough power output, the vertical thrust should overcome the gravity of the UAS. The power/weight ratio should be greater than 3.75 kg/kW, a typical value for a helicopter. For an electric UAS, this value can reach 7 kg/kW or higher. For example, to lift an agriculture drone of 90 kg, one needs at least 13 kW of power from the battery, i.e., at least 32.5 kg of Li battery with a power density of 400 W/kg is necessary to lift this 90-kg T40 agriculture UAS. This is the proven data from DJI.

Horizontal UAS

The two main characteristics of a horizontal flight UAS are the range and duration. The range R follows this formula, and the duration t is the energy E divided by the output power P

$$R = E\,\eta \frac{1}{g} \frac{L}{D} \frac{m_b}{m_a} \tag{7.6}$$

$$t = E/P \tag{7.7}$$

For example, the energy carried in SB300 with a 0.1-kg Li battery is $E = 15$ Wh. The power was assumed to be $P = TV$ = mg (L/D) $V = 125$ W. Then, $t = E/P$ = 7 min.

As discussed earlier, for a certain airplane, there exists a best navigation speed for the best efficiency with a max lift/drag ratio L/D. Therefore, the best design of the lift/drag ratio and optimized flight velocity is the key. It is noted that the L/D ratio is different during takeoff compared to the $L/D_{\max}$ in navigation. For example, B787 takeoff requires 60 tons of thrust to lift 240 tons of weight at a speed of 270 kmh, and L/D = 1/4. During the flight, the thrust used is 8 tons with a speed of 900 kmh to keep the airplane at 10,000 m altitude. The L/D ratio is 8/240 = 30.

Table 7.9 lists the three typical UASs from AeroVironment unmanned aircraft systems [18]. It is noted here that only the switch blade 300 and 600 use electricity as the power; JUMP 20 UAS uses gasoline engines/fossil fuels as navigation power and MOGAS 190 cc electronic fuel injection (EFI) as its engine [19]. Unlike the DJI, this type of UAS is still a horizontal airplane which need some range to take off.

Vertical UAS

Table 7.9 Three typical small UAS (two one-time use)

	Fuel	E (Wh/kg)	Eta	mb (battery) (kg)	ma (MGTOW) (kg)	Range (estimated) (km)	Range (data) (km)	Endurance (min)
Switch blade300	Li battery	150	0.8	0.1	2.5	10.4	10	6.0
Switch blade600	Li battery	150	0.8	10	54	40.0	40	40
JUMP 20	Kerosene	2000	0.3	17	100	183.6	185	

DJI (Da Jiang Inc. [20]) is one of the leading Chinese drone companies founded in 2006 focusing on versatile drones. There are two typical application scenarios: camera drone for geometrical studies and agricultural drone spraying fertilizer and sowing seeds (Table 7.10).

Table 7.10 Small weight drones for agriculture and for camera shooting

	Parameters	DJI agricultural drone	DJI camera drone	
Battery	mb (battery)	12 kg	0.12 kg	
	P-density	1112 W/kg	1112 W/kg	
	Max power	13,344.0 W	134.6 W	
	Max lifted weight	93 kg	1 kg	
	E-density	130 Wh/kg	130 Wh/kg	
	Total energy	1560 Wh	16 Wh	
	Duration	7 min	33 min	
Airplane	I	30,000 mAh	3850 mAh	
	V	52.0 V	7.4 V	
	Used power	1560.0 W	28.5 W	
	Duration	7 min	30 min	
	ma (airplane)	50 kg		
	mp (payload)	40 kg		
	MGTOW	90 kg	0.25 kg	
	Weight (ability)	93 kg	1 kg	

DJI Agras T40. T40 drone is DJI's new flagship for *digital agriculture*. This high-endurance drone is optimized for agricultural application scenarios such as precision spraying fertilizers, spreading the seeds, and aerial surveying and mapping. Equipped with a coaxial twin-rotor design, the T40 farming drone boasts its spreading capacity of 50 kg and a spraying capacity of 40 kg. Effectively, this drone can spread 1.5 tons spray pesticides on a 320-acre field in an hour. The drone also features omni-directional radar and binocular vision to detect obstacles at a distance of up to 50 m. As such, DJI T40 can also be flown in complex terrain, such as orchard hills, by spraying up to 60 acres of fruit trees in an hour.

T40' has max MGTOW 90 kg (40-kg payload), size 2.8 m × 3.15 m × 0.78 m, can hover max 7 min with 1560 Wh energy and max battery power output 13.4 kW (full load). There are eight 18 cm propellers, each driven by a rotor with 4 kW, and the max electric engine power is 32 kW. To lift the 90-kg T40, the weight-to-power ratio must be higher than 90 kg/13.4 kW = 6.7 kg/kW. This ratio is higher than that of a typical kerosene-fuelled helicopter (3.75 kg/kW), indicating that the electric engine used in DJI possesses higher lift-to-power ratio than an ordinary machines.

DJI Mini 3 Pro. This tiny camera drone weighs only 0.25 kg, with 30 min hovering time, 18 km range, 21.6 kmh speed. The size is $L \times W \times H = 251$ mm × 362 mm × 70 mm. Li ion battery 3850 mAh, 0.121 kg, 7.4 V, 28.4 Wh, using slightly tilted rotating wings or a dedicated propeller to provide horizontal thrust.

What we learned from the above examples is that (1) they are unmanned flying objects, (2) they are within 100 kg weight, and (3) they are matured products.

7.5 The Distinct Topics of E-propulsion

The electronic propulsion for a VTOL aero-machine is a pretty comprehensive technology and process. We use Fig. 7.15 to illustrate their complex connections. A VTOLer must fly horizontally fast enough (speed) as well be able to vertically take off (weight). The engine/the battery must be good enough to provide enough thrust/power. Both the energy density and the amount of fuel should provide enough energies for a long-range flight.

7.5.1 Power Versus Thrust, Engine Versus Battery

First of all, both the engine and battery must be able to support enough power (in kW) as well as thrust (in kN). The power and thrusts are two different concepts. Enough output power is needed from the battery, and the engine must be able to accept this power, digest it, and convert it to propulsion energy either as a propeller or a turbofan. This acceptability is power of an engine. Therefore, both the battery (or other electric power supplier) and engine must be powerful enough. Besides, vertical takeoff thrust

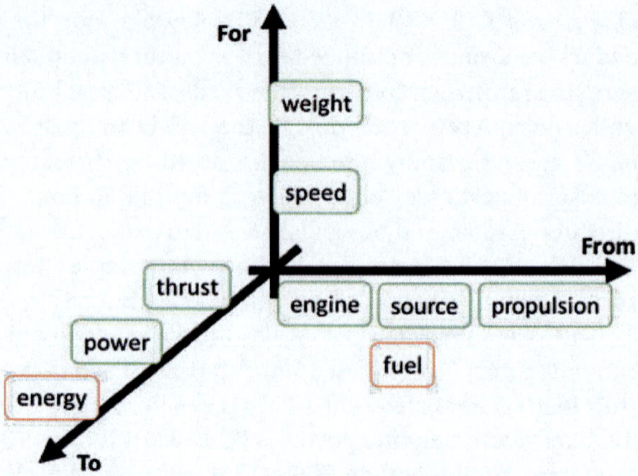

Fig. 7.15 E-propulsion system—from, to, and for

requirement is different from the horizontal thrust. The horizontal thrust is used for acceleration and then to use the wing to take off an airplane. The key ability is for the horizontal acceleration. The vertical thrust is to conquer the weight of the airplane, i.e., to overcome the gravity "mg". The latter needs more thrust and is harder to achieve than the former. For example, two GEnX engines can horizontally take off a Boeing 787, but it is almost impossible to lift this 200-ton airplane machine. As an advanced electric airplane, we prefer to call it a "VTOLer", which must possess both qualities: vertical lifting itself and enough horizontal speed.

VTOL is a pretty comprehensive technology, which involves both the power and the thrust from both engines and batteries for both horizontal speeding and for vertical lifting. As outlined in Fig. 7.16, the new e-aircraft is different from the previous generations in its integrated capabilities of both VTOL (e.g., ~ 7 tons weight) and horizontal flight high speed (e.g., ~ 600 kmh) via a rotatable wing.

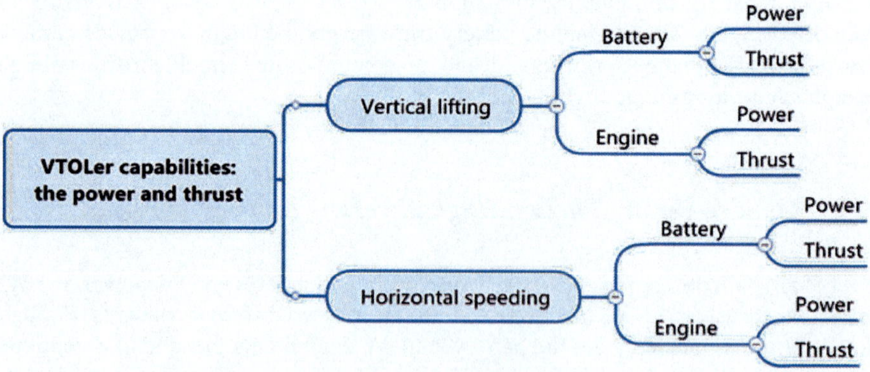

Fig. 7.16 A VTOLer must have both vertical and horizontal capabilities from both engine and battery for both power and thrust

Below, we discussed this matter one by one.

(1) **Vertical lifting**

The vertical takeoff requires both power and thrust. To raise the 700-kg aircraft with a speed of 1 m/s, the power needed is P = FV = mgv = 7 kW. On the other hand, the helicopter has a conventional rule that under enough helicopter thrust like a long propeller, the lift-to-weight ratio is 3.75 kg/kW. From this perspective, 187 kW of power is needed to raise a 700-kg VTOLer. On the other hand, the estimated lift-to-weight ratio of DJI's UAS can reach 7 kg/kW. Considering the thrust privilege of the RDF jet, this lift-to-weight ratio can be even higher. The needed power if using the 7 kg/kW is just 100 kW. In terms of lift-up strength, the vertical thrust must overweight of the 700-kg aircraft, i.e., 7 kN. Enough power/thrust must be provided both by battery/engine:

Battery

Assuming the battery weighs 400 kg in the 700-kg plane equipped with three ϕ30 RDF jet engines (Chap. 9), the max power of the battery is 160 kW (the power density of the Li battery is 400 Wh/kg), which meets the minimum requirement of 100 kW.

Engine

(1) Power: The power-to-weight ratio of the e-engine is 1–7 kW/kg. Three ϕ30 RDF jets weigh 60 kg which shall deliver 60–420-kW e-power, which can may meet the minimum requirement of 100-kW power margin provided enough power from battery/other electric suppliers.
(2) Thrust: Three ϕ30 RDF jets provide 3–18 kN depending on the amount of the feeding power from the battery. With enough battery power, the thrust can be greater than 7 kN weight, fulfilling the need for VTOL.

Therefore, the power/thrust from the battery/engine is good enough for vertical takeoff. After that, the RDF jets are rotated horizontally by rotatable wing to speed up the airplane, and eVTOLer relies on the lift-to-drag ratio to maintain its altitude.

(2) **Horizontal speeding**

The power needed to fly at 900 kmh (=250 ms) speed is $P = TV$, where the thrust T = mg $(L/D)^{-1}$. For the 700-kg airplane with a thrust/weight ratio L/D of 30, the thrust needed is T = 233 N, and the power is P = 58 kW. Both these thrust and power shall be provided by battery and engine with enough amount.

Battery

The power of 400-kg battery (400 W/kg) is 160 kW, which is greater than need (58 kW). Assuming the energy density is 200 Wh/kg, then total energy of 400-kg battery is 80 kWh. If the airplane keeps flying at 900 kmh, the time span is 80/160 = 0.5 h; the range is 450 km.

Engine

Three ϕ30 RDF jets can deliver greater than 60 kW e-power, which can also fulfill the need of the minimum requirement of 58 kW. Besides, three ϕ30 RDF jets provide greater than 3-kN thrust, also greater than the thrust requirement.

Of course, the previous calculation is just a rough estimation of the ideal case, such as assuming the best level *L*/*D* ratio ~ 30 and no other energy losses. Yet, it is already seen that the main bottleneck for a VTOLer is the vertical takeoff, the potential powerfulness of engine and power when outputting the vertical thrust and related power in the lightweight. In our sample case above, this VTOLer potential is fairly easy to achieve since the airplane is still not heavy, just 700 kg. Besides, we carry a lot of battery (400-kg Li ion battery with the best level of energy density (200 Wh/kg) as the year of 2022 with the power density of 400 W/kg) in just 700-kg airplane, with a big assumption of lightweight aircraft and payload (just 300 kg of airplane, engine, and pilot).

7.5.2 Specific Power of E-engine

The specific power of an engine refers to the ratio of the power output versus the weight of an engine (kW/kg). It used not to a big deal for the vehicles traveling on the ground, but this parameter is very crucial for the aero-engines. Beside, even for ground vehicles like the modern EV cars, seeking the higher P/W ratio is also a key parameter when developing the advanced e-engines. Table 7.11 compares the power/weight ratios of various engines of fossil versus electric, car versus airplane.

Table 7.11 Specific power of typical engines

	Motor	Para			
		M (kg)	*W* (kW)	Power/weight (kW/kg)	Year of introduction
Cars	EA111	128	100	0.781	1974
	EA211	106	100	0.943	2012
	BMW740i's B57	334	294	0.880	2015
Air-planes	Diesel generator/B787	95	250	2.632	2021
	Honeywell aero-generator	127	1000	7.874	2021
	Gas turbine/CFM56-2-C	2104	13,000	6.179	2010
Electric engine	E-engine-old	7	1	0.143	1960
	E-engine-Tesla	134	340	2.537	2020
	E-engine-Lucid Air	73	502.5	6.884	2021

Most of the traditional fossil fuel car engines do not have a good PWR performance. In comparison, the ordinary aero-engines for aviation have a much better performance by considering the weight factor in engine design. Most of the gas turbine machines can reach the power-weight ratio about 7 kW/kg. In Boeing 787 first generations of diesel generators, the PWR is just 1 kW/kg. For the most recent B787, four 250 kW generators weighing 95 kg have been implemented with an improved PWR of 2.63 kW/kg. In the future, much better PWR electric engines will emerge according to their roadmap of development [21].

What we learned from B787 electrification is that when advanced aero-electric power generation technology is ready, diesel and other mature generators can be used with lightweight modifications to generate electricity on a small and medium scale. Although the power generation efficiency of a diesel generator is lower than that of a gas turbine, its technology is simple, mature, and easy to implement. In B787's first introduction generation in 2011, the specific power was 1.25 kW/kg [11]. Currently, this ratio will become 2.63 kW/kg in B787 in 2022. As seen, Honeywell already claimed its lightweight gas turbine generator (in 2021). Other companies are also working on developing large PWR machines such that Boeing expects to achieve 5 kW/kg in 2030, and NASA has an even more ambitious goal of achieving this specific power over 10 kW/kg [12, 22].

Big PWR E-motors

The early power-to-weight ratio of traditional electric motors is very low. About 7–12-kg heavy-duty motor is often required to output 1 kW of power (0.08–0.14 kW/kg). Driven by electric cars and other advanced robotic mobile devices, more and more advanced motor design technology are adopted to enhance the power-to-weight ratio (PWR in kW/kg). For example, the Tesla Model 3 Motor weighs 92 kg with 2.35 kW/kg PWR, and the weight of the Lucid Air motor is just 73 kg reaching a high PWR of 6.8 kW/kg. Lucid recently claims it permanent magnet motor integrated with an inverter; differential and transmission system with an ultra-high voltage 900 V power unit weighs only 74 kg with a max of 9.45 kW/kg PWR.

As seen, adopting a rare earth permanent magnet into electric motors is the key technology to increase the PWR. Rare earth permanent magnet motor systems are already in broad use in Chinese high-speed railways (CRHs). The manufacturing cost of rare earth permanent magnet (PM) motors is reduced by more than 20%; the weight is reduced by 50–80%; the volume is reduced by 80%, and the energy conversion efficiency is increased by 10–15%. Rare earth PM most often use Nd-Fe-B composites, and LaSeFeB can also be added into PM family to leverage the need for pure NdFeB rare earth materials. PM is an indispensable material for building rim-driven aviation engines (in Chap. 9). As a future roadmap for advanced electric propulsion of electric aviation by improving the power density and efficiency of motors, the realistic near-term target is to develop 50–300 kW electric motors with PWRs of 5–7 kW/kg and weights of 8–40 kg. For example, for an 8-kg rare earth PM electric engine with a PWR of 5 kW/kg, 48 such engines can deliver up to 2.24 MW for the DEP airplane with a total thrust of 5 tons. To achieve this goal, the integrated design of materials, structures, and systems is the key instead of the individualized efforts of the professionals in different fields in AIAA and IEEE. The mutual cooperation is highly necessary to achieve at highly interdiscipline goal of electrified aviation.

7.5.3 Torque, Power, and Speed of Electric Motor

Figure 7.17 compares the engine behaviors of the famous and matured EA 211 gasoline engine and an ordinary electric engine which is still on the road of developing for better PWR. Here, we just use this chart to compare the different performance behavior of the fossil engines versus electric engines. There are two distinct difference between two engines. For the electric engine, the maximum torque can be achieved even with the low rotating speed. In other words, the torque is only dependent on the amount of applied electric powers (the input current and voltage). With enough power supply, we can maintain the enough strength of the engine. The second difference is that the electric motor has a maximum constant power limit which is the product of the torque and speed. In other words, if the torque is too big, the engine cannot maintain the speed. Such a phenomenon will be elaborated in Chap. 9 that in the RDF e-engine; if the resistance level is too high (the torque is too big), the turbofan will slow down the speed and cause the lower air intake rate and then reduce the thrust. Such a phenomenon does not occur for gas turbofan with stronger power level. Due to the limited electricity output from battery, its power is just not enough to maintain the speed under high torque when driving a bigger fan size, i.e., the thrust cannot reach its expected value even with a larger fan size (Sect. 9.6, Table 9.3).

In summary, driven by the strong need of advanced and smarter electric-driven e-aviation industry, with the help of the most advanced e-generator and e-engine, the vertical VTOL aero-machines will soon become possible to fit in the vast air mobility needs in the very near future.

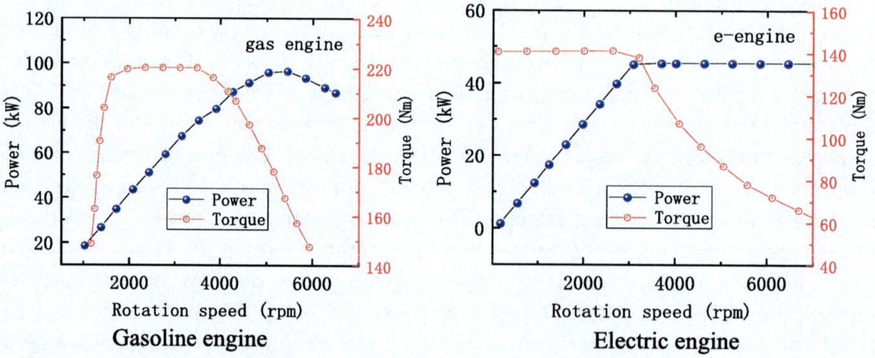

Fig. 7.17 Engine performance of typical gas engine and electric engine: torque, speed and power. Left: 220 Nm, 110 kW, 106 kg fossil engine. Right: 135 Nm, 50 kW, 200 kg, electric engine

References

1. AAM—Advanced Air Mobility. https://www.nasa.gov/aam/
2. North, D.D., Busan, R.C., Howland, G.: Design and fabrication of the LA-8 distributed electric propulsion VTOL testbed. In: AIAA SciTech 2021 Forum, p. 1188 (2021)
3. Moore, M.D.: The third wave of aeronautics: on-demand mobility. SAE Trans. 713–722 (2006)
4. Boom's XB-1 supersonic high-speed flight. https://boomsupersonic.com/xb-1XB1
5. Han, B., Liu, Y.T., Tan, H.Z., et al.: Emission characterization of civil aviation aircraft during a whole flight. Acta Sci. Circum. **37**(12), 4492–4502 (2017)
6. Murphy, A., Simon, V.: In-House Analysis by Transport & Environment. Transport & Environment (2021). https://www.transportenvironment.org/wp-content/uploads/2021/08/2021_05_private_jets_FINAL.pdf
7. Huang, J., Yang, F.: Development and challenges of electric aircraft with new energies. Acta Aeronaut. Astronaut. Sinica **37**(01), 57–68 (2016). (in Chinese)
8. Saite, W.U.: Feasibility analysis and prospect of electric aircraft application. Energy Conserv. **39**(4), 57–60 (2020). (in Chinese)
9. Lyu, Y., Liem, R.P.: Flight performance analysis with data-driven mission parameterization: mapping flight operational data to aircraft performance analysis. Transp. Eng. **2**, 100035 (2020). ISSN: 2666-691X. https://doi.org/10.1016/j.treng.2020.100035
10. Thomson, R., Sachdeva, N., Nazukin, M., et al.: Aircraft Electrical Propulsion—The Next Chapter of Aviation, pp. 1–32. Think: Act (2017)
11. Whyatt, G.A., Chick, L.A.: Electrical Generation for More-Electric Aircraft Using Solid Oxide Fuel Cells. Office of Scientific & Technical Information Technical Reports (2012)
12. Tom, L., Khowja, M., Vakil, G., Gerada, C.: Commercial aircraft electrification—current state and future scope. Energies **14**, 8381 (2021). https://doi.org/10.3390/en14248381
13. The video of Muller's rotatable wing—Moller Skycar M400. https://evtol.news/moller-skycar-m400/
14. The 2018 data of Skycar® 400 datasheet from: https://www.moller.com/
15. Rolls-Royce's spirit of innovation. https://careers.rolls-royce.com/usa/stem/spirit-of-innovation
16. Alice: Future of Aviation: Battery-Powered Planes Produce Zero Emissions. https://www.eviation.co/, https://www.eviation.co/aircraft/
17. Surhone, L.M., Tennoe, M.T., Henssonow, S.F., et al.: Yuneec International E430. Betascript Publishing (2010)
18. AeroVironment unmanned aircraft systems. https://www.avinc.com/
19. MOGAS EFI Engine and Power Generation. https://www.mogas.com/
20. Da Jiang UAS Inc. https://www.dji.com/
21. Singh, M., Yadav, K., Singh, S., Chumber, V., Chavhan, H.: Optimization and analysis of an elite electric propulsion system. Int. J. Aviat. Aeronaut. Aerosp. (2019). https://doi.org/10.15394/ijaaa.2019.1419
22. Madonna, V., Giangrande, P., Galea, M.: Electrical power generation in aircraft: review, challenges, and opportunities. IEEE Trans. Transp. Electrif. **4**, 646–659 (2018). https://doi.org/10.1109/TTE.2018.2834142

Chapter 8
The Evolution of Aero-engines

Engine is the heart of an airplane. In fact, in the aero-industry, the significance value of the engine is almost equivalent to the plane. That is why in the four AIAA Forums, there is a dedicated forum specifically for the aero-engine—the AIAA Forum on Propulsion and Energy. In the 2018 P&E Forum, the electric aero-engine was the hottest topic which attracts both the AIAA and IEEE people. Since then, electric propulsion keeps shedding new light to the next generation of aircraft propulsive force:

> *In terms of thrust, the power ranking is:*
>
> *Rocket > Turbine jet > Turbine fan > Electric aero-engine*
>
> *In regard to intelligence, the ranking becomes:*
>
> *Electric propulsion > Turbine-fan > Turbine jet > Rocket.*

An electric propulsion versus gas turbine engine is analogous to efficiency versus power. The value of developing electric aero-engine does not lie in chasing the powerfulness but in seeking for intelligence. Let's review the development of the aero-engines starting from the early days in order to get some insight from past to future.

The Evolution

As shown in Fig. 8.1, the first generation for aero-propulsion is a propeller driven by piston engine, such as the Spitfire fighter and Merlin engine in WW II. The second generation is the gas turbine and electric-driven propeller. Both technologies are matured and are being used right now. The future aero-engine shall be the clever combination of the re-engineered gas turbine (for example, big BPR gas turbofan GEnX) plus the advanced electric propulsion technology (RDF jet in Chap. 9).

© The Author(s), under exclusive license to Springer Nature Singapore Pte Ltd. 2023
F. L. Duan, *When AIAA Meets IEEE*,
https://doi.org/10.1007/978-981-19-8394-8_8

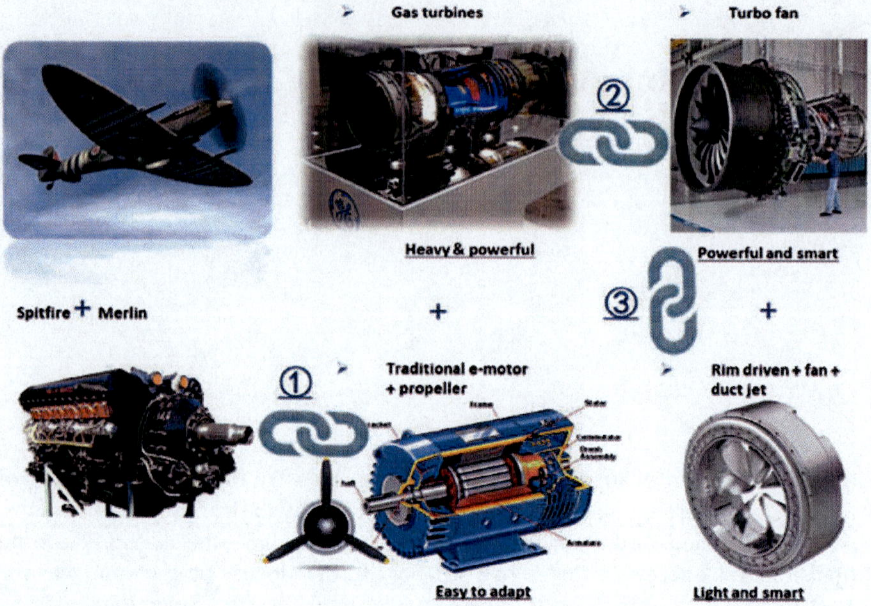

Fig. 8.1 Evolution of the aero-engine from past to future

The Road Map

A roadmap of future aero-engine is shown in Fig. 8.2. There are two separate paths: stronger/more efficient gas turbine engines and smaller/smarter electric aero-engines. Gas turbine engines are highly dependable on advanced technologies in mechanical, thermodynamic, and material engineering and stringent working ambient conditions (high temperature, high pressure, and intense stress). In contrast, electrified propulsion may circumvent these tough requirements. In these two paths, the first engine serves big size aircraft for continental and official purpose. We use larger bypass ratio to enhance the efficiency of gas turbofan engine and to use the electric propulsion to shorten the takeoff distance. The second aero-engine provides the propulsion for small aircraft serving the local, private, and versatile purposes. We use electric propulsion to achieve the integrated design of multiple engines/airplane (DEP) and to serve as the driving engine for 300-kg–10-ton VTOLers (aero-vehicles capable of both vertical takeoff and horizontal navigation with rotatable wings.

As to future electric power, there are three separate road maps (Fig. 8.3),

1. Li battery and diesel generator are two matured technologies which can be quickly used in propeller-based private airplanes and to help the electrification for big airplane as in B787. Battery-driven propeller plane is still in R&D stage and not yet fully commercialized. The main challenge is the range is too short and the less satisfied safety when adopting the Li battery from EV cars to airplanes. Even the EV cars are still in the developing stage of enhancing the Li battery efficiency

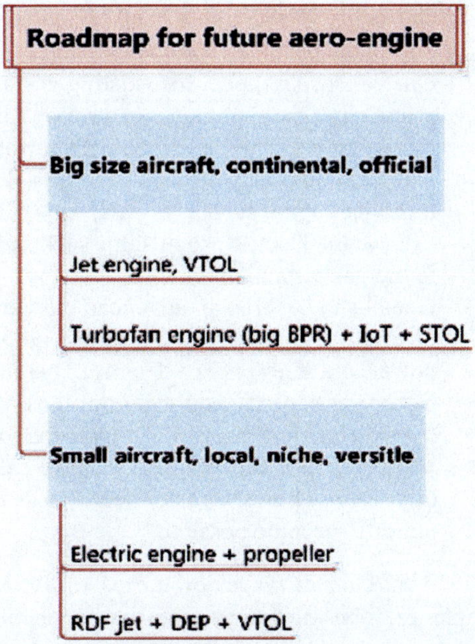

Fig. 8.2 Roadmap of future airplane propulsion

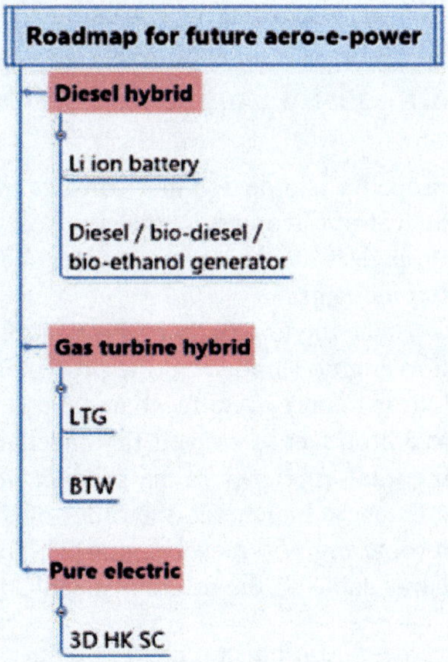

Fig. 8.3 Road maps for future aviation electric powers

and safety. These issues shall be maturely addressed before Li battery-powered e-airplane becomes commercialized.

2. One can also remodel the existing gas turbine machines to generate the aviation electricity. One approach is to develop advanced lightweight gas turbine generator by re-engineering the existing gas turbine engine and generator. The other is to make a by-the-way electric generator just behind the big fan in commercial gas turbofan engine such as GEnX. This BTW electricity can be used as the power to drive the electric propulsions such as the RDF jets as a supplement in addition to the gas turbine thrust.
3. One has to develop an advanced pure electric power source for VTOL purpose. Such electric power supplier must be purely electric and possessing both high power and high energy density. The current pure electric suppliers such as Li battery or supercapacitor do not have both qualities at the same time, and the hybrid electric generator is not pure electric and cannot meet the need of a green VTOL. A new supercapacitor as we called as "3D HK SC" can fulfill both the high energy/power density needs so as to make the "over 300 kg weight VTOL aircraft" become possible.

In this chapter, we review the existing technical details of propeller and gas turbine engines, focusing their technology connections with electrified propulsion and try to make use them to inspire us a new generation of electric aero-engine and aero-power.

8.1 Pistol Engine and Propeller

Propeller is a device that consists of ① a central hub with ② radiating blades with helical surface and ③ that is used to propel a vehicle (such as ship or airplane). Propellers are driven by a center shaft, powered by diesel engine/electric motor/gas turbine engine.

Since the first flight of the Wright Brothers, airplanes have used internal combustion engine as power drive propellers for thrust. The engine converts the chemical burning energy into mechanical energy on the shaft, and the shaft is then connected to a propeller to convert this mechanical energy into propulsion energy. Therefore, the total efficiency is the product of the combustion efficiency and the propulsion efficiency. In general, the order of the efficiency is as follows: electric motor > gas turbine engine > diesel engine [1]. The efficiency of the propeller propulsion is highly dependable on the rotating speed [2]:

$$\eta = \frac{\text{propulsion power output}}{\text{shaft power input}} = \frac{\text{thrust} * \text{axial speed}}{\text{resistance torque} * \text{rotational speed}} \tag{8.1}$$

Therefore, when we were talking about the propellers, it actually includes both the propellers and engines (Fig. 8.4). We must realize that the propeller and engine are two separate units in propeller propulsion, and they are fairly independent. For

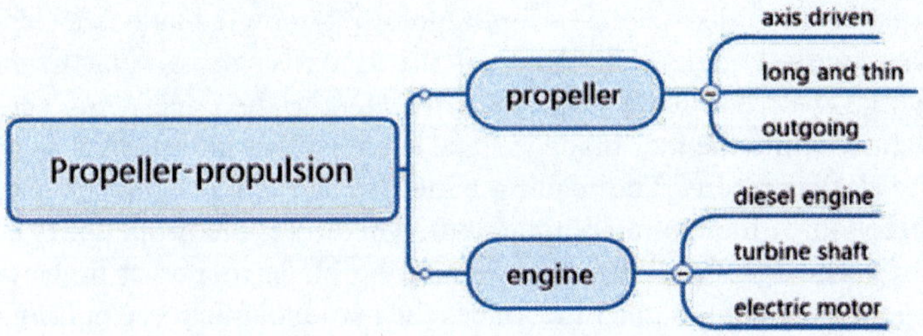

Fig. 8.4 Propeller propulsion has two distinct and separable components: propeller and engine

example, when we say that propellers are most suitable at subsonic speed range, we are discussing the characteristics of propellers; when we say the thrust to weight ratio and the fuel consumption rate, we are actually talking about engine. This is very important, especially when we are comparing advantages and disadvantages of different thrust systems; we should discuss engine and the propulsion separately—which is different in jet engines.

8.1.1 The Pistol Engines

Traditionally, the engine to drive the propellers is the piston engine. The working principle of this engine is that the piston in the combustion chamber makes repeated movement in the cylinder, transforms this movement into shaft rotation, and then, the shaft drives the propeller to rotate to create the thrust. Since the first flight of the Wright brothers, the thrust power has increased from less than 10 kW to approximately 2500 kW, and the fuel consumption has been reduced from 0.5 kg/(kW h) to approximately 0.25 kg/(kW h). After the end of the Second World War, the invention of the turbojet engine gradually replaced the piston engine in the main aviation field, especially for the big continental commercial airliners. Currently, the pistol engine can be easily replaced by electric engine to drive propellers for low-speed local aviation purpose. This is actually the main market place for the concurrent electric aviation industry.

8.1.2 The Propeller

The first air propeller originated as a "bamboo dragonfly" in the Chinese Ming Dynasty (1368–1644). It refers to the device that generates propulsion by rotating the blades in the air. It is composed of several blades and a central hub. When the propeller is rotating with a center hub, the inclined plane of the blade generates dragging force and pulls the aircraft forward. The mechanism by which a propeller creates thrust is due to its special design topology. A single blade is a slender wing blade with a twisted angle. The shape of the blade is very similar to that of the airplane wing. The front blade surface is equivalent to the upper wing surface with large curvature, while the rear blade surface is equivalent to the lower wing surface with nearly flat curvature. The leading edge of each blade is consistent with the rotation direction of the engine output shaft. Therefore, the propeller is equivalent to a pair of vertically installed wings. When the blade rotates at high speed, two forces are generated at the same time: one is the aerodynamic force pulling the blade forward, and the other is the reaction force generated by pushing the air backward by the blade torsion angle. Because the curvature of the front blade surface is different from that of the rear blade surface, when the blade rotates, the pressure of the air flow on the front blade surface with large curvature is small, while the pressure on the rear blade surface with a nearly straight curve is large. Therefore, the pressure difference between the front and rear blade surfaces is formed, resulting in an aerodynamic force pulling the blade forward—the power to pull the aircraft forward.

Another reaction force is generated by the twisted blade pushing the air backward. The blade is installed at right angles to the engine shaft and has a torsion angle. When the blade rotates, the air in front is sucked in by the torsion angle of the blade, and a backward pushing force is added to the inhaled air. At the same time, the air flow also gives the blades a reaction force, which also provides additional power to pull the aircraft forward. The aerodynamic force generated by the blade special-shaped surface and the reaction force generated by the blade torsion angle pushing back the air occur at the same time. The resultant force of these two is the total aerodynamic force that pulls the aircraft forward.

The pulling force of the twisted angle variation of the propeller with rotating speed and horizontal flight speed is as follows: due to the increase in engine output power, the propeller speed (tangential speed) rapidly increases to a certain value, and the propeller tension increases. As the flight speed increases, the blade angle of attack begins to decrease gradually. This pulling force also decreases gradually. The aircraft resistance increases gradually, and the increasing trend of speed also slows down gradually. When the pull is reduced to a certain extent (i.e., the pull is equal to the resistance), the speed of the aircraft will no longer increase. At this time, the flight speed, rotating speed, blade angle of attack, and propeller pull remain unchanged, and the aircraft will fly at a saturated speed.

The pull (T) of the propeller, the power (P), and the efficiency required to overcome the resistance moment of the propeller (η) are in the following formula:

$$T = C_t \rho n^2 D^4 \tag{8.2}$$

$$P = C_p \rho n^3 D^5 \tag{8.3}$$

$$\eta = J \frac{C_t}{C_p} \tag{8.4}$$

where C_t and C_p are the force and power coefficients, respectively, ρ is the air density, n is the rotating speed, and D is the propeller diameter. C_p and C_t depend on the geometric parameters of the propeller, and their values vary with propeller design (coefficient J). The characteristic curve determines the propeller pull coefficient and power coefficient, and this is the key technology for designing propellers and aircraft performance.

When the propeller diameter increases, the towing force increases tremendously, but the power increases even higher, which eventually limits the size of the propeller. In general, the propeller cannot be too long due to the power limitation of the engine. In addition, when the linear speed on the tip of the propeller reaches the sound speed (340 m/s), shock waves occur, which greatly reduce the propeller efficiency. Below is the linear speed at the tip of the propeller for the ship, aircraft, and helicopter ($v = r$ * rpm * $2\pi/60$):

- ship $r = 0.5$–6 m, 100–400 rpm, $v = 20$–60 m/s;
- helicopter $r = 3$–6 m, 300 rpm, $v < 180$ m/s.
- aircraft $r = 1.5$ m, 3000 rpm, $v = 470$ m/s.

The famous Spitfire [3] used in World War II has a propeller approximately 1.5 m long. Its linear speed on the tip of the propeller already reaches the sound speed where the shock wave occurs. More detailed discussions regarding the shock wave effect will be addressed in a later section.

8.1.3 The Privilege of Propeller Aircraft

The analogy of the propeller versus the turbofan is as diesel versus gas turbine. Technology-wise it is much less advanced, structure-wise it is much simpler, light, and low-cost. Although jet propulsion has much more advanced performance than traditional pistol propeller propulsion, propeller aircraft still occupies an important market place, especially in general aviation, which are small weight and size, low flight speed at low altitude. Propeller aircraft can better fit these scenarios than gas turbine engines. The disadvantages of turbojet engines are becoming increasingly prominent mainly due to high fuel consumption and low efficiency at low speed. The horizontal cylinder piston engine with power less than 370 kW is still widely used in small general-purpose aircraft in these market places such as administrative aircraft, agricultural and forestry aircraft, exploration aircraft, sports aircraft, private aircraft, and various UAVs.

More importantly, piston engines can be easily replaced by electric motors to continue the legacy of the above features. Propeller aircraft is still the most welcome choices for flight enthusiasts, meeting the low-end requirements. As a matter of fact, as shown in Fig. 8.1, replacing the diesel engine with an electric motor powered by a Li battery is the quickest and most straightforward way to make the airplane commercially electrified.

8.1.4 *The Drawback of the Propeller*

There are three distinct features of the propeller machine: ① axis driven, ② very thin and long blade instead of the fat and wide blade in turbofan, and ③ exposed outstretching compared to the compact turbofan inside a shell. These three features render the intrinsic drawbacks of the propeller, and one of the effective solutions is the rim-driven motor with enclosed fan jet (RDF jet) that we proposed in Chap. 9.

Drawback of Axis Driven

The axis-driven machine uses the shaft rotation which uses small radius shaft to drive the larger radius blade to rotate. As shown in the analogy chart in Fig. 8.5, it is like to lift a heavy good from the shorter end on the level which is against the Archimedes principle. For shaft-driven rotation, most of the twisted tension is on the root of the axis connecting the blade. The centrifugal force of the propeller at the blade root can reach 200 kN (20 ton force). In comparison to the rim driven, to drive, the rotation is much easier. The blade is fixed on point A and the tip of the blade on the point B (Fig. 8.5). The rotation mechanisms are very different from the axis driven, and much easier torque efficiency can be obtained for rim-driven machines. This principle will be fully elaborated in the RDF jet session in Chap. 9.

Bottleneck of the Propeller—High Speed Limit

When the airplane speed with the propeller engine exceeds 700 km/h, the propulsion efficiency decreases tremendously. The power needed to drive the propeller is in proportion to the V^3. When the speed V of the aircraft is too high, the engine cannot provide enough power to drive it rotate at high speed. Besides, shockwave in front of the propeller causes the efficiency dropping sharply. No matter how powerful the engine is, the plane still cannot fly fast.

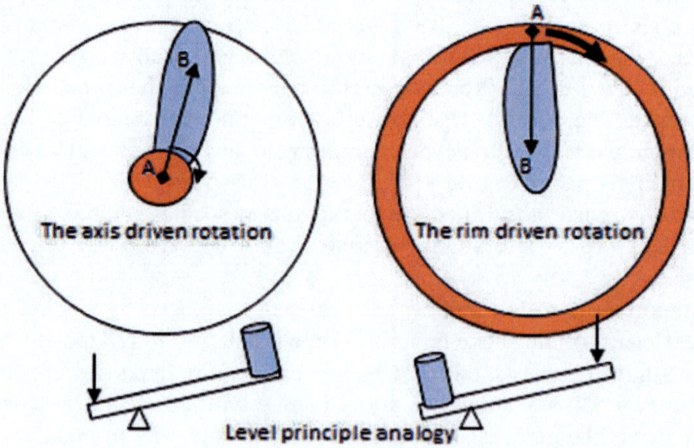

Fig. 8.5 Shortcoming of axis-driven rotation with the analogy of the Archimedes level principle

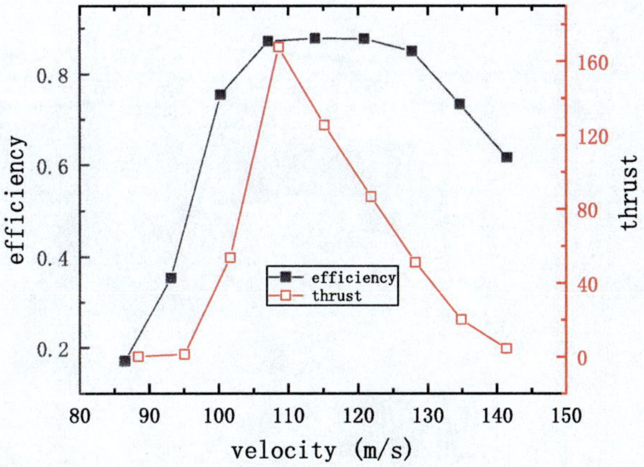

Fig. 8.6 Shock wave: both efficiency and thrust are dependent on the speed

The Shockwave Effect

A higher rotating speed usually enhances the efficiency and thrust. However, when reaching to a certain limit, the flow in front of the aircraft and the tangent speed of the propeller will add up, and the compressed air will cause a shock wave and form a great resistance. In addition, the shock wave propagates over root, which may cause the damage of the propeller system.

When the propeller is at high speed, both the efficiency and the thrust of the propeller decrease sharply due to the shock wave at the tip and will exhibit a peak value with blade velocity due to the shock wave (Fig. 8.6 [4]). Therefore, there exists a velocity range that will produce the highest efficiency for the propeller [5]. When above and below this range, the propeller efficiency drops off. There is an optimum cruise velocity that will produce the maximum thrust for a given propeller [6]. Rim-driven propulsion can circumvent this issue since the tip of the blade is at the center point, which has a nearly zero liner speed when rotates.

Propeller Versus the Turbo Fan

One of the biggest differences is propeller is open and outstretching, while the turbo fan is concealed in a compact shell. When the size is too long, the propeller becomes not as environmental-friendly and has the risk of hurting others as well as itself when it rotates. This disadvantage is more distinct when we are using the bigger rotating wings during the VTOL operations when shifting the vertical propelling to horizontal.

The propulsion principle of propeller and turbofan is also different. In the propeller, the force is directly applied on the blade, and it is this pushing force from repelling the air that gives propulsive force of the airplane. The force is intensified on the blade itself. For the turbofan, the propulsion force is not from the fan blade itself. The fan blade sucks in as much air as possible, and the projection force is from the exhaust jet. This is the intrinsic advantage of the turbofan versus the propeller. In the very early days when jet engines were first introduced, the propeller fellow

Fig. 8.7 Camera UAS, which uses multiple short propellers and avoids using long-stretching helicopter propellers

could not determine how this "compact object" can amazingly create such a strong propulsion because they cannot see the rotating or jet inside shell as they easily see the visible rotating propeller blades. This same compact shell concept will be used in our RDF jet electric—aero-engine. The RDF jet has all the rotating parts within a compact shell together with many other advantages.

Currently, the general approach to overcome the drawback of the long and outstretching propeller is as follows:

(1) Use multiple small radius propellers as in the current drone market for small UAS systems, such as hovering camera drones (Fig. 8.7). The approach of multiple small radius propellers is more suitable for lightweight USA drones but not as ideal for passenger aircraft;
(2) Use a duct fan jet with an enclosed shell, such as the RDF jet. This is a better approach, especially for heavy payload that needs a larger thrust. The other advantage is its rotatable wings which enables both the vertical thrust for VTOL and horizontal propulsion for fast navigation.

8.1.5 From Propeller to Jet Engine

In 1930, Sir Frank Whittle and the other German fellow von Ohain coinvented the first jet engine, which produced thrust by the high-speed gas ejected from the nozzle. At the initial stage of the jet development, Whittle faced quite a lot of challenges to persuade the traditional propeller fellows to accept this new idea of jet propulsion, considering that at that time the propeller fighters have a brilliant contributions in the war between the British and Germans. Most engineers believe that propellers are amazing machines and are not ready to shift their enthusiasm from propeller to jet. It was not until the late 1950s that the gas turbine jet engine replaced the old piston engine plus propeller configuration due to its obvious advantage of more powerful and excellent performance and became the preferred power unit of fighter planes and commercial aircraft replacing ships for continental traveling.

Jet engine follows totally different principle to propellers. The thrust of a propeller occurs via the intimate interaction between the blade and the air; the turbo jet works like a rocket which the exploded air exhausts via the outlet nozzle to create the accumulated jet effect. The first one is the intimate individual force, and the latter one is accumulated airflow momentum. The engine and the propeller are separated in the propeller system, and the energy is transferred from the shaft. In contrast, the turbojet machine is a consistent flow from power to thrust, i.e., from compressing the air to combustion and to jet. It is more consistent energy flow.

8.2 The Gas Turbine Machine

Different from propeller propulsions that the engine and propeller are two separate units, a gas turbine engine combines the power unit and propulsion unit in one. The thermal and mechanical processes are unified in one unit as shown in Fig. 8.8.

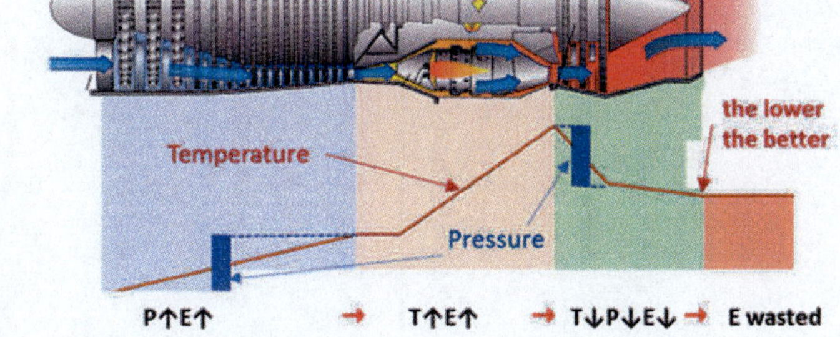

Fig. 8.8 Gas turbine operations (*T*: temperature, *P*: pressure, *E*: energy)

There are 5 main components of the jet engine: inlet, compressor, combustion chamber, turbine, and jet nozzle. Among them, the compressor, combustion chamber, and turbine form the **core**—the key thermodynamic performance of a jet engine. The core engine determines the total power and thermal efficiency. It can be transformed into an electric generator if it used on the ground. It can also be transformed into an engine if adding the front fan duct and jet nuzzle. The distinct examples are GE's LM2500 gas generator and GE's GE9X aero-engine which are all derived from a core of GE9.

8.2.1 The Core

The core is the essential part of a jet engine to convert combustion thermal power into mechanical power. The thermodynamic process of the core is shown in the Brayton cycle in Fig. 8.9.

For a jet engine core, point 1 indicates the initial state of the surrounding atmosphere (altitude and speed). The air entering the inlet and being compressed by the compressor is an isentropic compression process from point 1 to point 2. Ideally, the total entropy of air remains unchanged, and of course, in real world, there will always be some loss. At point 2, the gas is compressed after passing through the compressor, and its temperature rises to 600 K from point 2 to point 3 and is sent to the combustion chamber to mix with fuels. The temperature at point 3 is also called the temperature in front of the turbine, which can reach more than 1400 K. From point 3 to point 4, the gas expands through the isentropic expansion via the turbines. In this thermal to mechanical energy transfer process, the turbines gain energy and rotate at high speed, and the internal energy decreases, which is reflected in the decrease in temperature. Usually, the turbine and the front compressor are connected by a shaft. The shaft transmits the turbine energy to the front compressor. For this reason, the jet engine cannot start by itself. When starting, something like an electric motor must bring it to a certain speed before ignition and start the combustion process. After the

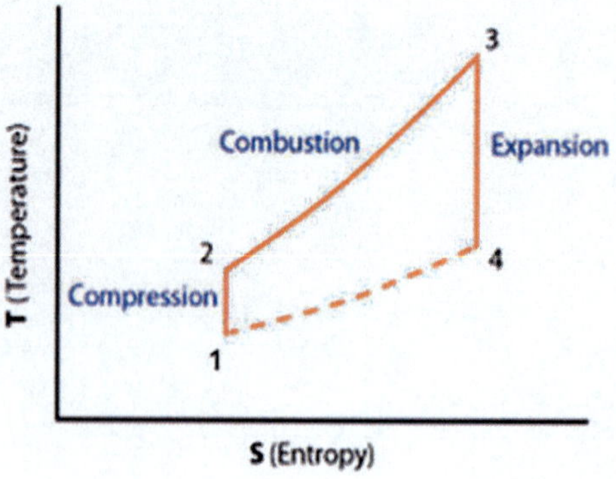

Fig. 8.9 Thermal dynamics of gas turbine core operation

gas is discharged from the engine, it is cooled at constant pressure in the atmosphere, which is represented by the dotted line from point 4 to point 1 in the figure.

By analyzing this thermodynamic cycle, it can be seen that the higher the temperature at point 3 is, the higher the internal energy of the gas in front of the turbine. The more work done when passing through the turbine, the more energy transferred from the turbine to the compressor, and the better the core's performance. Therefore, the development of core engines mainly focuses on trying to improve the temperature in front of the turbine and the compressor boost ratio.

8.2.2 The Inlet and Outlet

The jet engine uses the inlet to inhale lot amount of air and compress them in the compressor and then send to combustion. A large fan is used to suck in the air, and the fan is usually driven by the turbines via the center shaft. When the aircraft is flying at high speed, the air flow speed is very fast, and this high-speed low-pressure air flow must be converted into low-speed high-pressure air flow through the inlet, since the compressor cannot directly absorb the high-speed air flow. Usually, the front air flow is 0.5–0.6 local Mach velocity.

As to the outlet, the tail nozzle creates the thrust by controlling the ejection speed of air flow and makes the outlet pressure as close to the outside atmospheric pressure as possible. Obviously, there is energy loss due to the high temperature/pressure of jet air compared to its ambient environment. For this reason, a large bypass ratio (BPR ~ 9) engine is emerged by using a big front fan and the outer duct jet to create the main thrust instead of using inner jet nuzzle.

8.2.3 The Working Process

The air first enters the inlet of the engine. When the aircraft is flying, it can be regarded as the air flowing to the engine at the flight speed. Because the flight speed of the aircraft is variable and the incoming flow speed adapted by the compressor has a certain range, the function of the inlet is to adjust the flow to the appropriate speed through the adjustable pipe. When flying at supersonic speed, the inlet decreases the air flow velocity to subsonic speed and then directed it to the compressor. The compressor behind the inlet is used to increase the pressure of the air flow. When the air flows through the compressor, the rotating blades of the compressor raise up the pressure and temperature. The high-temperature and high-pressure gas flows to the combustion chamber and burn up the mixed kerosene fuel to generate a combustion. After combustion, the fuel gas energy in front of the turbine increases greatly. The high-temperature and high-pressure gas flowing out of the turbine continues to expand in the tail nozzle and is discharged backward from the nozzle along the axial direction of the engine at high speed. This speed is much higher than the speed of air flow into the engine, so that the engine obtains reactive thrust. This is the basic principle of the turbojet engine. The combusted air pushes the turbines machines, and part of the internal energy of the gas that expands in the turbine is converted into mechanical energy to drive the compressor, forming a consistent cycle of compression-combustion-turbine-compression with the constant kerosene fuel feed-in.

8.2.4 Turbojet Engine

From the late 1930s to the early 1940s, the birth of jet engines in Britain and Germany created a new era of jet propulsion and a new era of aviation. The turbojet engine is still a kind of heat engine. Therefore, in terms of the principle of generating output energy, the jet engine and piston engine are the same. They all need four stages: intake, pressurization, combustion, and exhaust. In the piston engine, these four stages are carried out in a time-sharing sequence with two sequences wasted, and the pistols are moving back and forth. However, in a gas turbine engine, these four stages are carried out continuously. The gas flows through each part of the engine in turn, creating a consistent flow without back and forth movement. The working principle of the gas turbine is certainly more advanced than the pistol engine operation. The prominent disadvantage of the turbojet engine is that the jet temperature and pressure are too high, resulting in a waste of energy and high fuel consumption. For this reason, a new engine has been derived as a solution—from turbojet to turbofan.

8.2.5 Turbofan Engine

There is a saying in the gas turbine industry

The inner culvert provides power;

The outer duct provides thrust.

The advantage of the turbojet engine is its high speed and power, which is very important for fighters. However, it is unacceptable to use it on civil commercial aircraft with stringent economic requirements. To improve the thermal and propulsion efficiency, the turbofan engine appears. This engine adds several stages of turbines on the basis of the turbojet engine. These turbines do more work from the combusted gas and reduce the outlet jet energy waste (lower temperature/pressure outlet jet). Moreover, the energy created from the turbine is transferred to the front compressor as well as the front fans to suck in more inlet air. The air flow behind the fan is divided into two parts: one part enters the compressor (inner channel), and the other part is directly discharged into the air (outer duct) without combustion. The exhaust energy from the high temperature/pressure jet is reduced, and the propulsion efficiency is improved. With a large BPR (by pass ratio), nearly, 80% of the thrust comes from the outer duct air flow such as GEnX commercial engine for Boeing 787. This principle of outer duct thrust is greatly made use of by our RDF jet, which is elaborated in Chap. 9.

8.2.6 Turboshaft Engine

The turboshaft engine uses a turbine machine to replace the piston engine to rotate the shaft. The turboshaft engine is composed of a propeller and a gas turbine machine. The main engine structure is basically the same, except an intermediate deceleration transmission system is needed between the turbine and propeller, since the speed of the propeller is much lower than that of the turbine. The structure of the reducer is heavy, complex, and has a high manufacturing cost—its weight is almost equivalent to the total weight of the compressor and turbine. As a part of the whole engine, the reducer plays a very important role in design, manufacturing, and testing. The efficiency of the turboshaft engine is higher than that of the turbofan engine at low speeds less than 900 km/h. This kind of engine is used in medium- and low-speed aircraft or patrol, anti-submarine, or fire-fighting aircraft with strict requirements for low-speed performance.

8.2.7 Gas Turbine Efficiency

According to the first principle of physics, the energy conservation law tells us that the less energy wasted, the more efficient the machine has. From this perspective, the piston engine in the combustion process has two cycles wasted; therefore, it has less efficiency than the gas turbine. Yet, even with gas turbines, the maximum efficiency is also limited due to the thermal energy loss.

The energy conversion efficiency η (ETA) of thermal engines is as follows: ① with certain amount of fuel, combustion energy is generated and then converted into mechanical energy; ② mechanical energy is transformed into propulsion energy, ③ what is the energy conversion rate from the fuel energy to propulsion energy, i.e., the efficiency η or ETA. Aero-engine efficiency is the product of thermal and mechanical efficiencies. The thermal η_{th} is 0.25–0.35, where most of the thermal loss is the exhaust thermal loss ~ 55–75%. The propulsion η_p is 0.5–0.75. The total Eta $\eta_0 = \eta_{\mathrm{th}} * \eta_{\mathrm{p}} = 0.3–0.4$. The total efficiency is directly related to the fuel consumption rate (specific fuel consumption, sfc). When the flight speed is constant, the total efficiency is inversely proportional to sfc. The following table lists the fuel consumption rates and related parameters of four generations of aircraft engines from propeller to jet

Table 8.1 Few generations of aircraft engines since World War II

Takeoff SFC (kg/Nh)	Takeoff thrust (N)	Bypass ratio	Pre-turbine temperature (K)	Compression ratio	Age/aircraft	Engine
0.349	2133 kW	–	1088	7.2	60s Y7	WJ5
0.51	94,308	1	1500	16	70s B707	JT3D
0.35	89,172	6.1	1588	24.3	80s B737	CFM56-3
0.278	382,500	8.4	1703	39.3	90s B777	GE90

and to turbofan. The general trend of the aero-engine development is the improved sfc or η as shown in Table 8.1.

Compared to mechanical/thermal engines, electric motor efficiency is much higher (~ 85%) since there is no thermal loss. People might argue that the electricity source is from the fossil fuel generator which still has low efficiency. Yet the overall energy generated from electric power plant with the dedicated gas turbine generator should still be more efficient and more economical than the individual aircraft/car. In addition, electricity may also come from sustainable resources such as solar, wind, and tidal forces. In the long run, electric engines are certainly preferred as aviation engines wherever replaceable.

8.3 The Electric Aero-engine

The electric aero-engine is an integrated engine with either propeller or turbofan to create the aero-thrust. The e-propulsion combines the previous engineering achievements (diesel and gas turbine) and the associated propulsion (propeller and turbofan). In the following, we are focusing on what we can learn from them to form our new theme of electrified aero-engines by analyzing their pros and cons. In general, propellers are suitable for easy-simple and small-scale propulsion, fitting for AAM and the quick private personal airplane market. Gas turbine engines and generators are good for middle- and large-scale commercial airliner services or power plants.

8.3.1 Analysis 1: Propeller Versus Duct Fan

Propulsion and power are two separate technologies. When we compared the gas turbine machine versus the propellant machine, we had to compare the propulsion and power separately. Propulsion refers to the thrust; the power refers to the engine. As a matter of fact, in the gas turbine machine, those two are combined. However, for the propeller machine and electric propulsion, those two units are separated.

(1) Comparing the engine—pistol versus gas turbine

The gas turbine has an advanced working principle compared to the pistol engine. The energy flow in a gas turbine engine is more fluent. A gas turbine engine follows a consistent, unified, one-directional energy flow. In diesel engine, there are four working cycles in which two cycles are wasted during the energy flow. In addition, the directions of the force applied on the rotor shift back and forth in pistol engine. In gas turbine machines, the air flows in the same direction. From the first principle of physics, gas turbine engine is more ideal than pistol engine.

(2) Comparing the propulsion—propeller versus turbofan

Propeller and turbofan have different interactive forces between the blade and the air, and their aerodynamics are very different, serving different purposes, possessing different shapes, and rotating at different speeds.

(a) The purpose is different:

- Turbofan: the air intake rate, the more it takes in, the higher the jet thrust.
- Propeller: the speed and the thrust.

(b) The topology is different:

- Turbofan blade: wide and fat, sucking in a large amount of air flow, enclosed with the high-pressure environment;
- Propeller blade: slender narrow and long blade, sparsely distributed, high-speed air flow, open environment.

(c) The rotating speed is different:

- Turbofan, 3000 rpm, compressor 6000–10,000 rpm, turbine, 10,000–20,000 rpm,
- Propeller: 300 rpm for ship, 3000 rpm for propeller aircraft, 100 rpm for helicopter.

The Electric Aero-engine Should Learn from the Propeller + Jet + Gas Turbine

- Propeller: replace diesel to electric to fit in a quick market place on 2-men private airplane and UAS (drone). Lessons learned: engine, motor, propeller.
- Jet: modern turbofan engine with bigger BPR which mainly uses turbofan plus ducted jet for thrust. Lessons learned: fan and duct nuzzle designs and apply them to our RDF jet.
- Lightweight generator: adopt the technology from high power-to-weight ratio of the gas turbine machine. Lessons learned: transform the lightweight engine into lightweight gas generator.
- The by-the-way generator: modify the commercial gas turbine to generate a BTW electricity and use this electricity to drive electric aero-engines or recharge the batteries/supercapacitors. Lessons: integrate turbine energy transmission and gas generator technology inside the existing GEnX.

New Technology Generated from the Above Analysis

- Engine: RDF jet by rim-driven motor + shiftless fan + duct shrink jet for improved mass intake rate and enhanced jet speed (more details in Chap. 9).
- Energy: ① medium power (1–10 MW) and lightweight (1–2 ton) gas turbine electric generator by reconfiguring GE's LM6000 gas generator and CF700 lightweight jet engine. ② Embedded gas generator inside the existing commercial GEnX engine to provide additional BTW electric power to drive the RDF jet propulsion and recharging the 3D HK SCs.

8.3.2 Analysis 2: Shaft Driven Versus Rim Driven

Conventional commercial engines use shafts to drive the large fan in the front to create the outer duct jet thrust. This driving method requires very large torque and power. In comparison, rim drive requires much less power and torque. Let us take the GEnX engine as an example to compare the disadvantages of shaft drive versus rim driven. As seen in Fig. 8.10:

$$F = m * w^2 * r = m(2\pi * \text{rpm}/60)^2 * r \tag{8.5}$$

$$t = Fr = m * w^2 * r^2 = m(2\pi * \text{rpm}/60)^2 * r^2 \tag{8.6}$$

$$P = t * \text{rpm}/60 = m(2\pi)2 * (\text{rpm}/60)^3 * r^2 \tag{8.7}$$

where the rpm and ω is the rotating speed, r is the radius, m is the mass, τ is torque, F is the centrifugal force, and P is the power.

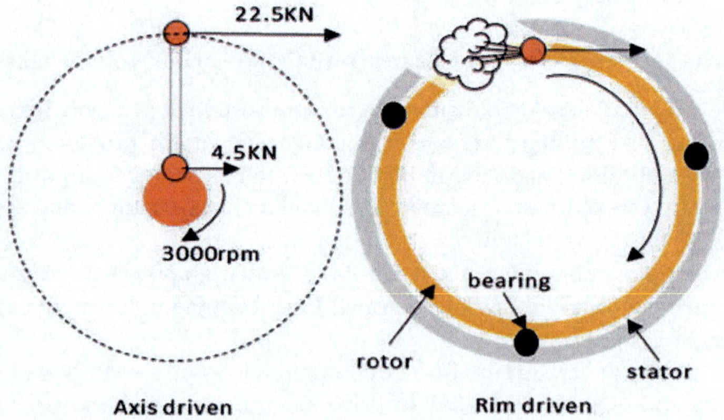

Fig. 8.10 Chart to explain the torque difference of the shaft driven versus the rim driven

The GE9X engine has a large fan that is 1.5 m long. Suppose the shaft is 0.3 m in diameter, and the value of the centrifugal force per kilogram is just 4.5 kN when rotating at a speed of 3000 rpm. If this shaft drives a 1 kg mass at a distance of 1.5 m, its centrifugal force is 22.5 kN, which is nearly 2 tons and five times larger. The rotating power is 25 times larger. Since the GE9X engine is a very powerful machine, it can still provide enough power to drive this fan to create the thrust when rotating at 3000 rpm. In addition, this rotating speed is its maximum speed since the tip of the large fan blade already reaches the ultrasonic speed at 3000 rpm, which renders the shock wave effect. That is why in large BPR commercial gas engines, the reducing gear is necessary to reduce the turbine speed from 10,000 to 3000 rpm.

For the rim-driven machine, there is no such problem (Fig. 8.10). The force of the rim motor is derived from the tangential magnetic force of magnets. The acting point to the blade from the motor is from the edge, and the torque is almost 0. The rim-driven machine is similar to a rocket or a bullet; the acting point is the same as the thrust force. Using the rim-driven motor to drive the inner rotation fan blade is much easier than using the shaft to drive a large fan. Rim driven has more privilege to offer larger torque as compared to ordinary shaft-driven machines.

8.3.3 Inspiration 1: Fan-Ducted Jet

The idea of high thrust from rim-driven fan was first inspired by big BPR commercial aero-engines such as GEnX. Seventy to eighty percent of the thrust comes from the outer duct jet. The thrust is merely proportional to the area ($\pi(D/2)^2$ (D is the diameter) of the turbofan. Based on the extrapolation from GEnX, a 50-cm-diameter fan should deliver 1250-kg thrust.

The most advanced gas turbine engine, such as GEnX as shown in Fig. 8.11, converts the turbine energy to the fan energy for a large air intake mass rate to create the outer duct thrust. The large bypass rate of the outer duct air flow already reaches 90%, contributing 80% of the total thrust. This smart architecture inspires us to make use of the electric-driven fan to create the large intake air mass and use the inner shrinking duct to enhance the jet speed. The air intake and compressing process occurs in the compact shell, which brings no outer disturbance like the propellers. We use an electric engine to replace the turbines to drive the front fan, and we use the culvert to compress and direct the airflow. The compressed jet at high speed creates the thrust. The advanced lightweight material from GEnX to build a 3-diameter fan blade can be reused to make our electric jet machine, rendering a more efficient lightweight compact electric jet engine. The application of carbon fiber resin matrix composites and CMC materials (the density is only 1/10–1/3 of that of Ni–Cr super alloys in gas turbine machines) technology will also greatly enhance the advantages of this closed-shell electric-driven aero-engine.

In addition, in our shaftless design, since there is no central shaft to block the airflow, air inhalation is more sufficient. Moreover, the sucked air is compressed to the tail spray culvert through the streamlined designed closed shell. By optimizing the

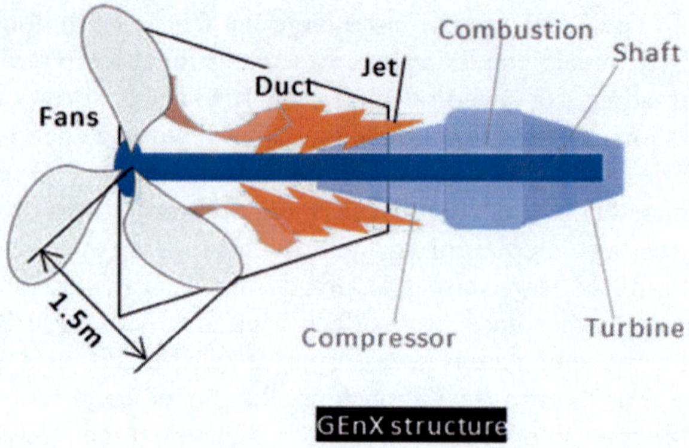

Fig. 8.11 General configuration of the GEnX aero-engine, i.e., to use the shaft to convert the turbine energy into rotating fan energy

geometric dimensions of the streamlined connotation channel and tail spray culvert to form the best propulsion capacity and efficiency, the RDF jet may offer an efficient and feasible smart small and lightweight electric aero-engine.

Specifically, the advantages are ① easier for vertical takeoff operation, low requirements for VTOL site due to its small volume and compact size, lightweight, and better flexibility. It can hover and fly by rotatable wings for a flexible shift between horizontal and vertical propulsion and maneuver the movement at a large angle. ② The compact compartment shell protects the rotating blade from damage, reduces the aerodynamic noise of the rotor itself, and has better concealment. ③ The more concentrated air flow and more ducted flow improve the propulsion efficiency; the air flow can be consciously induced, and the air velocity on the upper surface of the ring wing is greater than that on the lower surface. Compared with the helicopter, the wake effect of the ducted rotor is more significant, which can provide additional thrust for the aircraft [7]. ④ The efficiency loss at the end of the blade tip is small. A very important energy loss of open rotorcraft and helicopter lies in the blade end loss, which is caused by the different pressure at the bottom and top of the rotor and shock wave effect. At the edge of the wing, the air flow under the rotor will produce eddy current. Compared with the isolated propeller, the vortex and tip loss of the RDF jet can be ignored [8].

8.3.4 *Inspiration 2: Lightweight Aero-generator*

Fossil fuel-based electrical generators are usually classified into two categories: the diesel generator family and the gas turbine family, as illustrated in Fig. 8.12. The first type of electric generator is cheap and simple but much less efficient and advanced compared to the second family. It still has its own marketplace due to its distinct advantages. The biodiesel and bioethanol in replacement of the traditional fossil fuel to feed the generator have many more technology advantages [9]. A diesel generator can be used as a supplement to larger aircraft, such as the Boeing 787, to help the gas

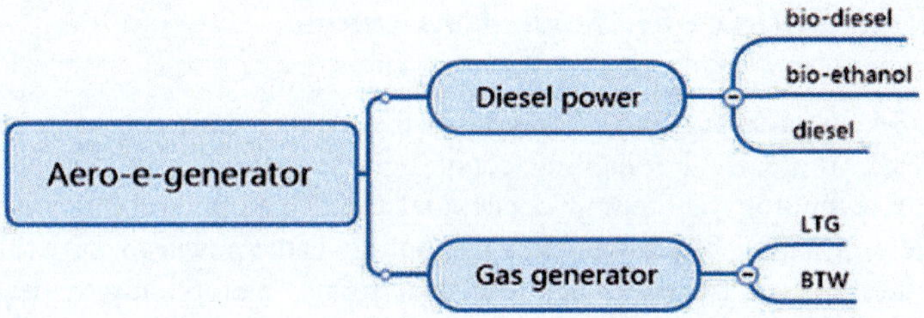

Fig. 8.12 Types of different electric generators as aviation power

turbine start and for the initial airplane operation, such as air conditioning. However, its low power/energy efficiency may not fit the need as a driving power supplier for aviation purpose.

A gas turbine generator is a promising choice for the aviation power supply. LTG (lightweight gas generator) and BTW generator (by-the-way generator) are the two options for aviation electric power generation. The BTW generator makes use of the existing commercial gas turbine machine with just a small modification. The main marketplace is for the STOL, i.e., to shorten the takeoff distance for the big commercial airliners and leverage the burdens of the airport. LTG is a newly developed power source suitable for small and mid-sized electric aircraft (2–20 passengers or 500–5000-kg drone delivery craft). The details of the BTW and LTG generator are discussed in Chap. 10. Different from ground and submarine gas generators, aviation gas turbine generators must be lightweight.

8.3.5 Inspiration 3: Edge-Driven Turbine Machine

The existing shaft-driven turbine engine can be refitted into an edge-driven aero-engine to effectively transfer the turbine power from back to the front for rim-driven fan and for compressor. In this new configuration, the rotating rim is connected by CMC bearings with the outer ring fixed on the shell. The rim rotation consists of three components, the front fan, the compressor, and the turbine, which share the same rim. The speed is set at ~ 5000 rpm instead of 10,000 rpm, considering that the momentum generated by the rim turbine work is larger than that driven by the shaft. The rotating speeds of the front fan, compressor, and turbine are the same. This is very different comparing to the current shaft-driven turbine machine, which requires reducing the rotating speed of the turbine before transmitting to the front fan and compressor because the efficiency of the compressor and fan at high speed will be greatly reduced. The combustion chamber is fixed by a center shaft, which is fixed on the frame on two ends to the shell. Bearing cooling together with high-temperature bearing will be the key technologies of this new flange-powered aero-engine.

8.3.6 R&D Matrix of Electric Aero-engine

Figure 8.13 illustrates the "XYZ" matrix of the electric aero-engine development. The "XY" axes represent the enabling technologies, and the Z is the marketplace. The first driving technology is the new conceptual electric engine/propulsion system—our RDF jet in Chap. 9. The second key technology is the power to drive this engine, including the electric generator and electric storage methodologies described in Chap. 10. Both of them should be lightweight and possess both high energy and power density for both VTOL operation (need high vertical thrust and power) and enough energy for long-term navigation (need more energy density from kerosene fuels). Electric storage is necessary for vertical takeoff, while the lightweight generator provides long-term navigation energy. Fossil fuels still possess the unbeatable energy density privilege compared to other existing energy media. In addition, biofuels such as biodiesel and corn ethanol [10] are also on the way of development intending to replace unsustainable fossil fuels. On the *Z* axis, the two major markets are located. One is the VTOL for small- and medium-sized planes. The other is the STOL for larger aircraft.

X axis—motor:

(a) the principle and new technology of the duct fan jet in modern gas turbine engines such as GEnX will be borrowed in the new electric engine design,
(b) the old-fashioned propeller will be replaced by the RDF jet concept: rim driven+ fan + duct + jet + compact size,
(c) compared with the propeller, gas turbine, and conventional axis-driven motors, the RDF jet is good at ① lightweight and multiple. Multiple engines can be integrated in the airplane design for more efficient aerodynamic control and redundancy for reliability and ② small and compact and easy to operate. The compact enclosed cylindrical design of the RDF jet allows easy manipulation

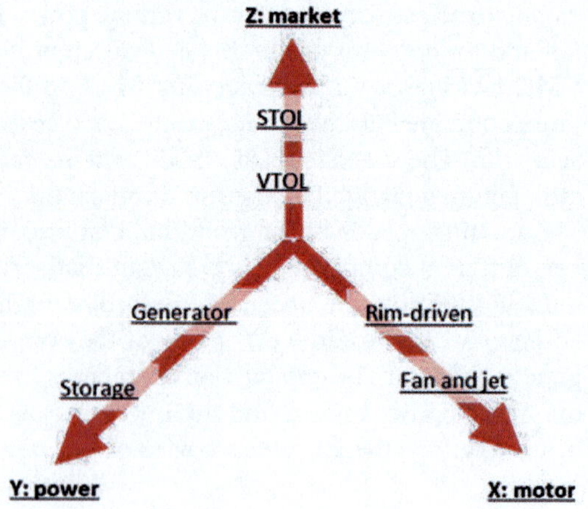

Fig. 8.13 R&D metrics to build the electric propulsion for electric aviation

of various angles. Another advantage is to reduce the incidence of bird collision since the fan of the RDF jet is facing upward during takeoff.

Y axis—power:

(a) Taking advantage of both the high energy and power density of kerosene and other biofuels, LTG specially designed for lightweight aviation electric generator will still play a major role as long-term aero-electricity energy provider. BTW e-power is another electric generation that takes advantage of the existing gas turbine machines to create a by-the-way electric energy during the normal gas turbine operation.
(b) In the meantime, special high power density battery/supercapacitor are needed to drive RDF jet for VTOL operation. This green and convenient takeoff/landing is one of the key reasons to develop electric aero-engines among others. The existing Li-ion battery can be used as a media for electricity storage, yet its safety issue must be properly addressed as an aviation power source. Very importantly, Li-ion batteries are intrinsically short of the power density since their electricity is created by electrochemical processes in nature. A new pure physically operated supercapacitor—3D HK SC—is proposed and discussed in detail in Chap. 10, which can both fulfill the need for high power and high energy density. Both the Li battery and 3D HKSC can be recharged by the LTG or BTW generator during navigation, and the recharged energy can be used for vertical landing purposes. For this reason, fuel cell batteries are not rechargeable batteries to serve this purpose.

Z axis—marketplaces:

(a) VTOL. Basically, there are two types of aero-vehicles depending on the range and weight. The first is the UAS market, such as DJ's T40 agriculture fertilizer sprayer or the camera drone. In general, their takeoff weight is less than 100 kg. The second type is aero-vehicle with takeoff weight in the range of 500 kg to 10 tons, corresponding to the 2-men car airplane and 20-seater van aircraft, or serving as a delivery drone for Amazon or Taobao. VTOL techniques from advanced electric aero-engines can make such planes to achieve pure electric vertical takeoff and landing, but hybrid power is still needed for long-range navigation and recharging purposes.
(b) STOL. For larger continental commercial airliners, it is not as practical to achieve vertical takeoff and landing due to its heavy weight, but the vertical thrust from e-propulsion technique can leverage the weight of the airplane to achieve shorter distance takeoff and landing, which greatly leverages airport resources, reducing airport pollution and noise.

In the following Chaps. 9–11, this "XYZ" will be discussed in detail from their application scenario and technological/economic perspectives.

References

1. NASA: Propeller propulsion. https://www.grc.nasa.gov/www/k-12/airplane/propeller.html
2. Spakovszky ZS: Archived 2012-06-28 at the Wayback Machine. 11.7.4.3 Efficiency Archived 2015-02-26 at the Wayback Machine. MIT turbines, 2002. Thermodynamics and Propulsion, main page Archived 2010-02-17 at the Wayback Machine
3. Spitfire and Merlin. https://www.quora.com/How-did-the-Spitfires-propeller-change-during-its-time-in-production?top_ans=220936742, 10 ft 9 in (3.27 m) diameter, 3 bladed de Havilland constant speed units, with narrow metal blades. Castle Bromwich with a wide bladed Rotol constant speed propeller of either 10 ft 9 in (3.28 m) diameter, with metal blades, or (on late production Spitfires) 10 ft 3 in (3.12 m) diameter, with broader, compressed wood blades.
4. Colozza, A.J.: High Altitude Propeller Design and Analysis Overview. Federal Data Systems, NASA Glenn Research Center, Cleveland, OH (1998)
5. Morgado, J., et al.: High altitude propeller design and analysis. Aerosp. Sci. Technol. **45**, 398–407 (2015)
6. Monk, J.S.: A Propeller Design and Analysis Capability Evaluation for High Altitude Application. Ph.D. dissertation, Wits Institutional Repository on DSPACE (2011). https://core.ac.uk/display/39668395?source=3
7. Choi, Y.H., Hong, S., Suk, J., Jang, J.S., Lee, D.: Mathematical modeling of a small scale ducted-fan UAV. In: AIAA Modeling and Simulation Technologies Conference, pp. 6519–6532 (2011)
8. Williams, M.H., Chot, J., Dalton, W.N.: Unsteady aerodynamic analysis of ducted fans. Propulsion **7**(5), 800–804 (1991)
9. Altun, Å.: Emissions from a diesel power generator fuelled with biodiesel and fossil diesel fuels. Energy Environ. **26**(4), 563–572 (2015)
10. Mumm, R.H., Goldsmith, P.D., Rausch, K.D., Stein, H.H.: Land usage attributed to corn ethanol production in the United States: sensitivity to technological advances in corn grain yield, ethanol conversion, and co-product utilization. Biotechnol. Biofuels **7**(1), 61 (2014). https://doi.org/10.1186/1754-6834-7-61

Chapter 9
The RDF Jet—A New Electric Aero-engine

The Rim-Driven Fan (RDF) jet is a small, light duct fan jet that serves as the electric aero-engine for various aircrafts with flexible vertical/horizontal propulsions. The RDF jet has a simple structure and is easy to install on the plane. This $\phi 30 \times 30$ cm cylindrical 20 kg compact jet is expected to provide 100–600 kg thrust. The RDF jet uses three special techniques plus the high-electric power to achieve this goal: rim-driven motor for bigger torque, inner embedded Tai Chi fan for more air intake, duct + shrunk outlet for higher jet speed. An electric power supply with both high specific energy and specific power, provided by lightweight gas turbine generator together with a 3D high-K supercapacitor, is needed to achieve such a high thrust, although conventional lithium and fuel cell batteries can still support RDF jets with lower thrust for small aero-vehicles in various UAS and AAM scenarios.

Most of the current electric aviation still use traditional propeller technology, that is, to replace the old diesel pistol engine with electric engine and to use the shaft to drive the propeller for thrust. It has two weaknesses: ① inferior propulsive efficiency compared to jet propulsion, ② inconvenient to shift VTOL to horizontal propulsion for large blades, and ③ shock wave stall at high speed. The replacement of the propeller engine by turbojet in late World War II is based on its substantial improvement in the overall propulsion efficiency.

On the other hand, jet engines' driving principle is to convert thermal energy into mechanical energy and then use the compressed jet as propulsion energy. The turbofan engine is an improved version of the jet engine that makes use of the turbines to convert the combustion energy into shaft rotating energy and then uses this energy to drive the turbofan in front of the engine. This turbofan sucks in a large amount of air, compresses it, and then ejects it via the outer duct as a jet energy. Limited by the transmission efficiency of the shaft drive, the rotation speed of turbofan must be reduced, and the driving efficiency is also limited.

In view of the shortcomings in the previous engines/propulsions, an RDF jet is proposed—a shaftless electromagnetic rim-driven turbofan plus a ducted jet nuzzle. The RDF jet integrates the two technologies in one: propulsion and energy—both are highly interdisciplinary projects involving both electric and aeronautical engineering.

© The Author(s), under exclusive license to Springer Nature Singapore Pte Ltd. 2023

F. L. Duan, *When AIAA Meets IEEE*,
https://doi.org/10.1007/978-981-19-8394-8_9

It is inspired by the following engineering achievements: the jet thrust of a duct fan in a commercial aero-engine, the electric power from a supercapacitor and gas turbine generator, the rim-driven propulsion in a submarine. The RDF jet has a few advantages: small in size, light in weight, rotatable wing for flexible propulsion, no thermal combustion involved.

In this chapter, we first describe the advantage of the RDF jet. Then, its operating principle is explained. Relevant new theories are explored regarding to the magneto-reaction force inside the RDF jet system. Then, we decompose the RDF jet structure, including the rim drive engine, Tai Chi fan, duct, and jet and explain them in details.

The outline of this chapter is:

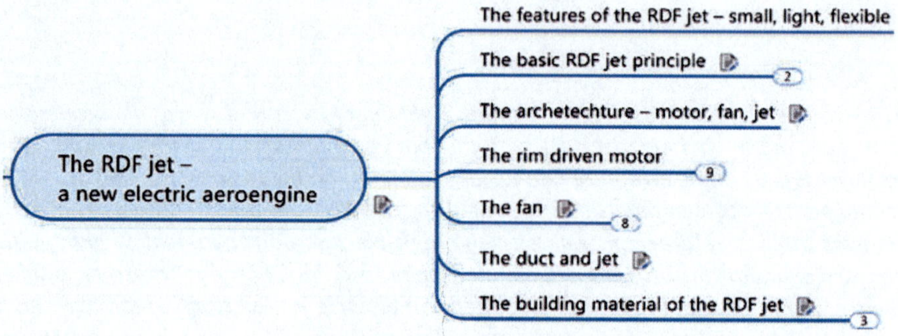

Outline of the RDF jet engine

9.1 The Features—Small, Light, Flexible

(1) **Small**. An RDF jet is a compact cubic cylinder with two small sizes: either Φ50 cm for up to 900 kg thrust or ϕ30 cm for less than 600 kg thrust to fit in smaller aircraft. It is very easy for installation and for orientation adjustment. The distributed electric propulsion (DEP [1]) can be readily achieved with a multi-RDF jet installed at various locations of the aircraft. Its compact design is attributed to the enclosed turbofan rather than the outstretching long propeller. The RDF jet is the skillful integration of the shaft less motor, embedded turbofan, inner-duct channel, and tail jet propulsion inside an enclosed cylinder. The RDF jet has good concealment since it has no exposed propeller and no high-temperature jet flow.

(2) **Light**. The RDF jet engine is an electric motor which does not involve the complicated thermal/mechanical systems in gas turbine or diesel engines. Rim-driven motor adopts the newest brushless electric motor technology which greatly enhances its power over-weight ratio. In addition, lightweight materials to build the GEnX front fan can be reused to build the RDF jet. 3D nested FRC (fiber-reinforced composite [2]) and CMC (ceramic matrix composite [3])

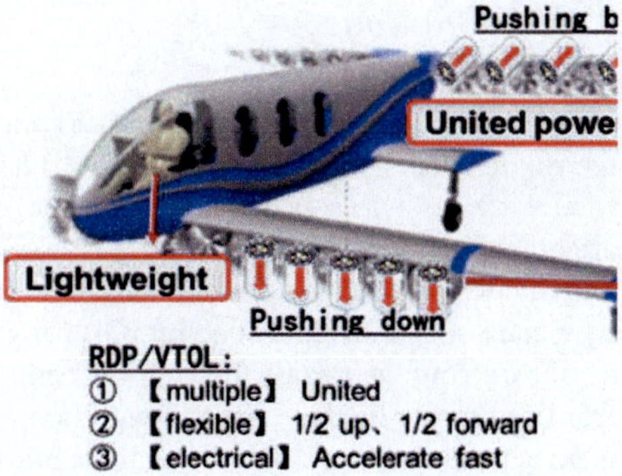

Fig. 9.1 The multiple engine arrangement during takeoff

materials are two favorite candidates for building shells, rims, and fan blades. Combining these advanced achievements, the RDF jet can be built very light. A rough weight estimation is 20–60 kg with expected thrust in the range of 100–900 kg. A detailed evaluation of the weight is at the end of this chapter. A small engine is suitable for the 3-passenger rescue aircraft, and a larger engine is suitable for the larger thrust to drive 20-passenger van aircraft or for STOL of commercial airliners.

(3) **Flexible**. Multiple engines can be integrated in airplane design for more efficient aerodynamic performance. As shown in Fig. 9.1, half of the engines are direct to the ground and half parallel during takeoff to shorten the takeoff distance, and then all turned parallel for horizontal propulsion. This configuration is very different either from helicopter or from the ordinary jet engine, which cannot accommodate both the vertical and horizontal propulsion for the best performance.

9.2 RDF Jet Principle

Most of the current e-plane still uses traditional propeller technology, which suffers its intrinsic drawbacks. On the other hand, gas turbine engines have limitation of the energy transmission efficiency by using the shaft. The RDF jet engine circumvents these shortcomings by adopting their pros and avoiding their cons. An RDF jet uses the new rim-driven principle instead of the shaft driven, allowing much bigger torque when driving the big fan blades and much higher energy transfer efficiency. RDF jet adopts the duct fan idea of the turbofan engine and adepts some advanced achievements such as Li battery and high-efficient e-engine in the current e-plane industry to fit in its own engine design, such as to integrate the motor, fan, duct, and jet together with the battery or the 3D high-K super capacitor in one 30 × 30 cm cylinder shell. An RDF jet can be built as an independent separable unit of ϕ30 cubic cylinder with 20 kg weight and up to 600 kg thrust.

9.2.1 The Concept

Different from the conventional axis-driven propeller or turbofan, the rim driven uses the rim to drive the fan blades to rotate. The rim rotates under the dynamic magnetic force between two sets of magnets of the inner rim and outer ring/shell. The rotors are embedded in the inner rim with NdFeB high-intensity permanent magnet plates which does not electric connections. The outer rim is embedded with stators made by electro-magnets controlled by IGBT high-power IC circuitry. In this way, instead of transferring the torque through the central shaft, RDF directly delivers the power from rim to the blades. The driving is much more efficient. In addition, since there is no central shaft to block the airflow, allowing more air to flow in.

The air is compressed and then leads to the shrunk outlet of the tail jet culvert through the streamlined duct inside the closed cylinder casing. The jet speed V_2 is increased by the narrowed outlet. The thrust of the duct fan follows:

$$F = m(V_2 - V_0) \tag{9.1}$$

where m is the mass flow rate (kg/s), V_2 is the outlet jet speed and V_0 is the airplane speed. In the RDF jet, the high mass flow rate m plus the high-speed jet V_2 can greatly enhance the propulsion with an expected thrust of up to 900 kg.

The larger intake mass flow rate m is achieved by the strong torque of the rim-driven strength together with the fat blade topology. An analogy of rim driven versus shaft-driven is to lift a heavy load from the long end at Archimedes' level. As shown in Fig. 9.2, it is much easier to conquer the large torque by using rim driven instead of axis-driven method.

Another difference of the rim driven vs. axis driven is the "direct power transfer". In the conventional turbofan, the original driving force is from turbine, then to the

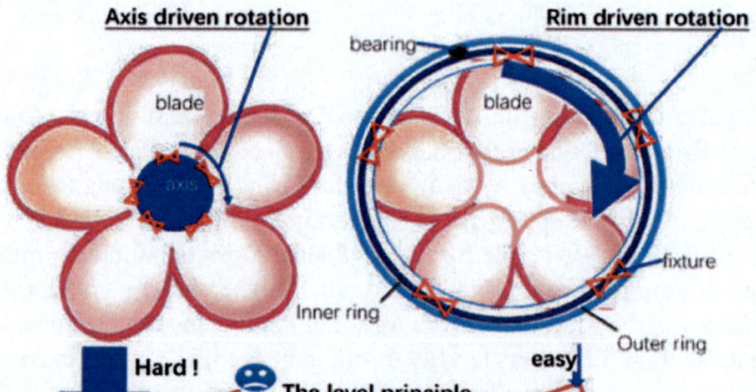

Fig. 9.2 Axis-driven rotation versus rim-driven rotation—Archimedes' lever principle

shaft, then to the fan blade-it's a two-step indirect power transfer. In RDF, the power of the rotor is directly delivered from the rim to the fan blade. This higher drivability allows a fatter blade to rotate faster and easier to concur the resistant torque from the high thrust force of the RDF jet engine.

Regarding to the fan blade of the rim-driven machine, the shape the rotating mechanism is different from ordinary shaft-driven fan or propeller—the center is hollow and the driving force is from the edge. This allows a more spacious fan blades be designed. Inspired by the ancient Chinese Wisdom logo—the Tai Chi, a Tai Chi fan blade is designed according to its shape and is manufactured by the 3D-printing additive method. CFD is used to fine tune the curvature for the maximum air intake rate.

9.2.2 The Thrust

The feasibility to achieve 900-kg thrust using the RDF jet is from engineering extrapolation of the existing turbofan data of commercial aero-engines by GE, (GE90, GEnX, GE9X from 1995, 2010, 2020[1]) CFM56, and PW products. In modern large BPR (bypass ratio) gas turbine engines, 80–90% of the thrust comes from the outer jet. One of the important methods to increase the BPR value is to enlarge the fan size. Figure 9.3 compares the outer fan thrust vs. the fan diameters. The thrust is nearly proportional to the area $(\pi(D/2)^2$ (D is the diameter) of the turbofan. Based on the extrapolation, the 50 cm diameter RDF jet should deliver 1250-kg thrust; therefore, the estimated 900-kg thrust target of our RDF jet is a possible expectation. It is also seen that with dedicated fan blade design, the vertical thrust can be higher than expected (8000 kg vs. extrapolated value 5277 kg). This further encourages us the feasibility to achieve the maximum 600 or 800 kg thrust by using the RDF jet

[1] **About GE90, GEnX,GE9X from 1995, 2010, 2020.**

GE90 engine is the originator of many modern jet engines' cutting-edge standard technologies. The legendary engine GE90 came out in the 1990s. GE90 engine is of great significance. The structure and mechanical design affect GE 's' and CFM's every turbofan engine in the next 20 years, including the popular GEnX engine and the record selling CFM leap engine.

The GEnX engine is used by Boeing 787 dream liner, the best engine in the world. The blade tip diameter of the GEnX fan blade is 2819.4 mm, and the fourth generation CMCs are used.

GE9X's fan blade has the largest diameter in the world, with a tip diameter of 3.4 m. It is the largest fan at present, and the largest engine in the world is GE9X. In 2019, the thrust of GE9X engine reached 134300 pounds (about 61 tons), breaking the record of 127900 pounds (about 58 tons) of GE90-115B engine in 2002. GE9X is a new large bypass ratio turbofan engine developed on the basis of GE90 and genx. Boeing announced in March 2013 that GE9X was the only engine choice for B777-x series.

The naming of GEnx-1B54 ① indicates the fan diameter: "1" is 111In and "2" is 104 in; ② Indicates the aircraft used: "a" refers to Airbus; "B" refers to Boeing 301.9 kn (66,500 lbf); ③ Indicates the thrust value: engine thrust = this value × 1000 lbf.

GEnx-1B54, thrust level 53,200 lbf (24100 kgf, 236 kn), for Boeing 787–3;
GEnx-1B64, thrust level 63,800 lbf (28,960 kgf, 284 kn), for Boeing 787–8;
GEnx-1B70, thrust level, 69,800 lbf (31,680 kgf, 310 kn), for Boeing 787–9;

D (cm)	Thrust(kg)		Vendor
300		22720	GEnX
184		10300	CFM56-5C1
174		8560	CFM56-2
152		7000	CFM56-3C1
135		5600	JT-3D
	real	extrapolated	
127	8000	5277	F35
50	900	1227	RDF Jet1
30	600	800	RDF Jet2

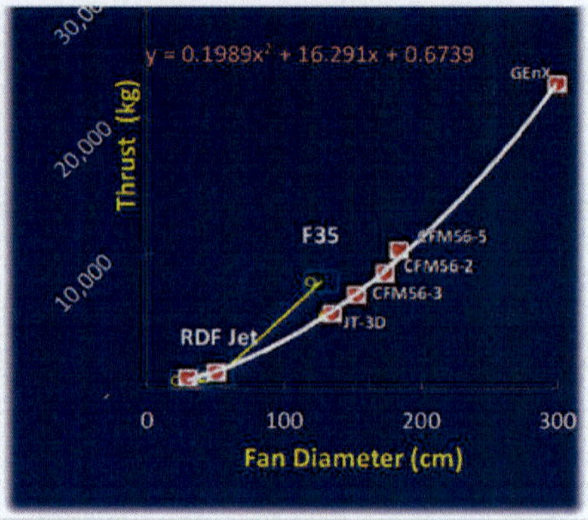

Fig. 9.3 Thrust versus the diameter of the turbofan

engine. Of course to achieve this goal, not only the engine should possess this power, but also the electric power is compatible with the gas turbine machines.

9.3 The Architecture—Motor, Fan, Jet

An RDF jet is a compact electric engine consisting of a motor, fan, and jet. The front half portion inside the cylinder is the rim-driven motor and fan, and the latter half portion is the duct and jet outlet. The compact enclosed cylindrical design of the RDF jet allows easy manipulation of various angles by using rotatable wings for horizontal/vertical propulsion transformation.

The components of the shaftless rim-driven turbofan ducted jet are:

- Permanent magnet rotor provides a rotational driving force;
- Fan blade which is connected to the rim sucks in a large amount of air;
- Inner duct directs the airflow toward the shrunk nuzzle;
- A high-speed jet from the nuzzle creates the thrust;
- The power to drive the motor is either from the power cable or from the embedded supercapacitor/battery inside the RDF jet compartment.

The RDF jet integrates the motor, fan, duct, and jet into one. As an option, the battery can be integrated inside the compartment if the jet nuzzle is designed as a ring-the battery can fit in the hollow space on the rear portion of the RDF jet.

The rim-driven permanent magnet motor includes:

- Outer ring, which can be integrated with the shell embedded with electromagnets built by flat-shaped electric coils;
- Inner ring embedded with evenly arranged flat-shaped permanent magnet plates;
- Bearings linking the outer ring and the inner ring.

The duct and jet include:

- The inner duct comprises an annular inner wall and an outer wall from front to the back, and the diameters shrink gradually and smoothly to form a streamlined air flow;
- A conduit and cable are buried in the outer shell to connect to the electric power;
- The shrunken nozzle jet has ~ 10% of the inlet area. In this way, the jet speed will be enhanced over ten times of the intake air flow speed (discussed later).

For VTOL, the jet nuzzle is facing the ground to provide the off-ground thrust; for navigation, the RDF jet is laid horizontally for parallel propulsion. In the following, the detailed mechanisms of the RDF jet are described, including the motor, the fan, the duct, and jet.

9.4 The Rim-Driven Motor

9.4.1 A Quick Review of the E-motor

In fact, an electric motor includes both engine and generator, since they follow the same basic principle in physics, which is the induction of electricity and magnetism. It's important to understand this first principle since the rim-driven electric motor in this book refers to both rim-driven engine as well as rim-driven generator-both play an important role for electric aviation: one generates electricity, the other creates the thrust, both use the rim-driven idea. To understand this first principle, let us quickly review the history of electricity and magnetism from the root in its early days.

Coulomb—the electricity. Although the acting force of magnetic and electric charge is very similar, when Coulomb invented the law, he asserted that there are no connections between electricity and magnetism, and it is impossible to convert each other. Because electric charges can be separated, positive and negative charges can exist independently, while the magnetic poles are inseparable. Regardless of how many segments a magnetic rod is folded into, each segment is a new magnetic rod with North and South poles.

Oersted—the electromagnetism. Later, Auster discovered the relationship between electricity and magnetism through a coincidental event. When he was doing experiments for students in class, he turned on the power and accidentally found that the magnetic needle parallel to the thin platinum wire swung a bit. Only Oersted knew that this was not an accident but a new science discovery in the classroom. It was the first time that human beings consciously discovered the relationship between electricity and magnetism! After three months of in-depth research, Auster finally found that there was indeed a magnetic field around the electric wire. In 1820, Auster first discovered the subtle relationship in electromagnetism.

Faraday—the electric motor. Michael Faraday completed the invention of an e-motor. In 1821, Faraday was inspired by Oersted's discovery that if a magnet was fixed, the coil might move. Based on this assumption, he succeeded in inventing a simple device. In the device, as long as there is current through the line, the line will rotate around a magnet. This is the first electric motor. In 1831, Faraday found that when a magnet passes through a closed line, there will be current in the line. This is the principle of the generator. Faraday invented both the e-generator and e-engine-the e-motor.

Gramme—the electric motor can be either an engine or a generator. It is a "comedy of errors" to convert an e-generator into an e-engine. Gramme, a Belgian, discovered during the exhibition in Vienna in 1873 that when accidentally connected the wrong wires, he found that when the motor rotates, the current output appeared at the input end. The electric engine becomes an electric generator if the wires are connected in different ways. In fact, both the electric engine and the electric generator originate from the same "first principle of physics" the electromagnet interaction when there exists a mutual movement between the electromagnet and the permanent magnet (or electromagnet).

Our point is to use this story to explain that a rim drive motor (RDM) can be used both as engine and generator for the next generation of aero-motor. The first is the RDF jet, the second is the rim-driven generator to provide aero-electricity. Both have their playgrounds in the electric propulsion industry.

Electric Generator

The induced electromotive (EM) force is the rate of magnetic flux change with time, i.e.

$$\varepsilon = \Delta\varphi/\Delta t \tag{9.2}$$

This EM is also called the electric voltage V. The basic equation of the electric current is (R is the internal resistance)

$$I = BLV/R \tag{9.3}$$

As long as the magnetic flux passing through the circuit, the induced electric voltage will be generated in the circuit. The generation of an induced EM force is the essence of the electromagnetic induction phenomenon. The magnitude of the induced EM is related to the speed of the change of the magnetic flux through the closed circuit. The internal resistance depends on the size of the wire diameter in the coil; the internal resistance is smaller for thicker wires.

A by-the-way electric generator can be formulated in other ordinary gas turbine engine. The gas turbine shaft can drive a solid plate of permanent magnet (PM) rotating in a series of radiant electric wires to create extra electricity in additional to drive the front fan for thrust-the BTW generator. The EM force and current in the coil are:

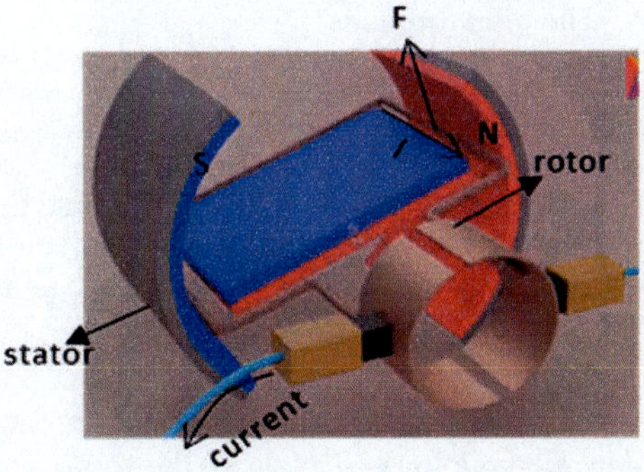

Fig. 9.4 The principle of an electric engine

$$\varepsilon = \frac{BR^2\omega}{2} \tag{9.4}$$

where the B is the magnet flux intensity of the PM, ω is the rotating speed, R is the radius of the PM plate. The operation principle of the rim-driven motor as an electric generator is as follows: when the internal rim containing PMs is rotating, while this rotation can be driven by the gas turbine machine or by some rotating engines, this rotation will create magnetic flux changes, and generate electricity inside the coils located on the outer rim. Connecting these electric coils will gather the collected EM into proper voltage output as an electricity supplier-the inner rim electricity generator. In Chap. 10, we will discuss the rim-driven generator in more detail.

Electric Engine

The traditional principle of an electric engine is explained in Fig. 9.4. An electric wire receives a Lorentz force under a magnetic field, and this force makes the coil rotate. The electric current must be alternated to receive this force consistently; therefore, a brush is needed in this alteration. There are many ways to make it happen, which renders various AC and DC electric engines.

The force of rotation is expressed by the Lorentz Law: $F = BLI$, where B is the magnetic field, L is the length, and I is the current. For a DC motor, when parallel currents flow in armature windings, the current flowing in each conductor is $I/2a$. If the length of the conductor is l, the magnitude of the ampere force is

$$f_i = BIl/2a \tag{9.5}$$

Assuming that the conductor is away from the armature axis in r, the torque that can be generated on a single conductor is

$$T_i = f_i * r \tag{9.6}$$

The total torque is

$$T = \sum Ti \tag{9.7}$$

The formula is sufficiently to evaluate the torque for an ordinary DC motor when the rotor is the electric coil. However, in the case of the rim-driven engine, the rotor is the permanent magnet. This methodology may not be as applicable. The interactive force between the rotor and stator is actually the interactive tangential force between two magnets with interlacing magnetic poles during rotation. This is the magnet-to-magnet interaction instead of the Lorentz force. Therefore, we have formulated a new methodology to evaluate the force, speed, and torque for RDM which will be elaborated in the following session.

9.4.2 Rim-Driven Principle

As shown in Fig. 9.5, an RDM consists of two rings connected by bearings. The outer ring is a stator embedded with electromagnets, and the inner ring is a rotor embedded with permanent magnets.

(1) The permanent magnet inside the inner ring moves counterclockwise under the repulsive force from electromagnet 1 and the attractive force from electromagnet 2 in the outer ring;
(2) When the permanent magnet is about to pass over electromagnet 2, the Hall sensor detects the incoming magnet and reverses the pole of electromagnet 2 so that the permanent magnet receives a repulsive force from electromagnet 2 while receiving the attraction from electromagnet 3—both forces keep making the permanent magnet move forward.

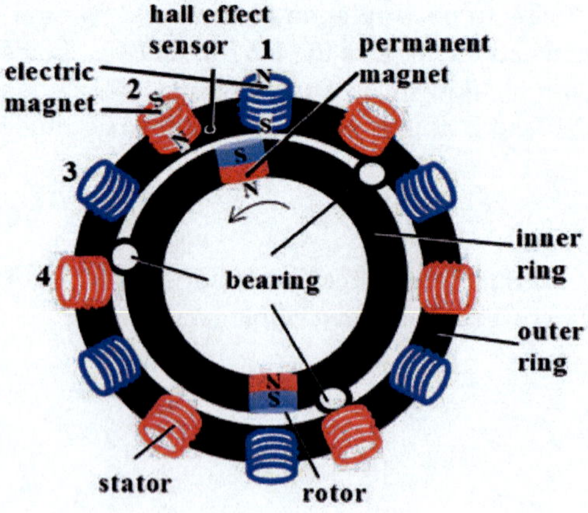

Fig. 9.5 Diagram of rim-driven magnetic propulsion

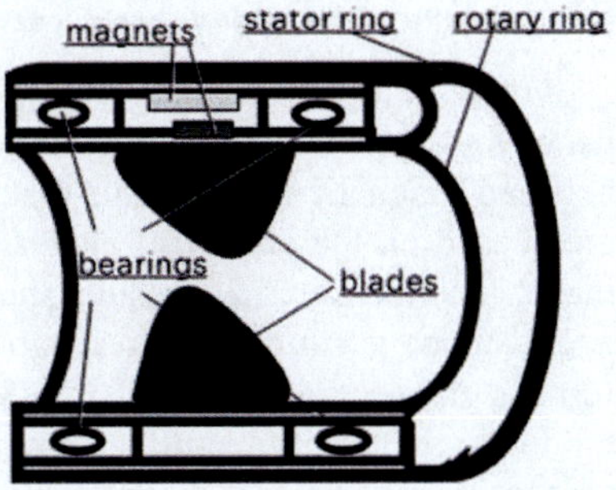

Fig. 9.6 3D configuration of the rim-driven motor

(3) In this way, the permanent magnet of the inner ring keeps rotates continuously with the dual alternating repulsion and attraction of the outer electromagnets.

When we attach the fan blades on the inner rim, the fan will suck in the air flow and output the jet backward for the thrust-which forms an RDF jet machine. The magnetic force conquers the resisting force from the thrust plus the friction of bearings/machines parts, and eventually the inner rim rotates at an equilibrium speed. The magnitude of the torque and the rotating speed are controlled by the electric power. Since the inner ring is rotating, it should be built by a permanent magnet without any wires, while the outer ring can be made by electromagnets driven by electric power. The neodymium permanent magnet flat plate rotor together with the flat electromagnet stator is the best combination to build a lighter-weight RDF machine. The power of the engine is under the dynamic control by IGBT devices from the electricity of battery/supercapacitor/hybrid generator. On the other hand, the power of the engine is the product of the torque and speed. In order to deliver enough thrust, one needs to match both ends well.

The 3D arrangement of the rim drive motor structure is shown in Fig. 9.6. Electric power is supplied to the stator to enable the magnetic force onto the inner rotor to make the inner ring rotate [4]. The blades attached to the inner ring generate the thrust to water for marine or the air for aviation.

RDM eliminates the center hub, resulting in unobstructed fluid flow through the central region, leading to a potential increase in the volume flow rate and therefore an increase in thrust. Unlike the traditional propeller, the free end of the blade (corresponding to the tips of the propeller) extends toward the center, which may avoid the shock waves when rotating at a very high speed (discussed later).

9.4.3 Rim-Driven Theory

The purpose of rim-driven theory is to derive the torque and speed of the RDM since the conventional Lorentz force method is no longer applicable. For this purpose, we have to evaluate the magnetic force between the two magnets, especially the tangible magnetic force interactions.

(a) *Coulomb Law of magnetic force*

A new quasi-Coulomb relation of the magnetic force versus distance is established for the first time quantitively with empirical coefficient. This is an initial study between magnetic force versus distance, magnetic flux density, and magnetic pole area based on the empirical analysis on the magnetic force vs. distance measurements. The measurement results show that the relationship between magnetic force and distance is similar to Coulomb's electrostatic law, but the attraction magnetic force is greater than the repulsion. Similar to the Coulomb force be proportional to the total amount of electric charges, the magnetic force is proportional to the total amount of magnetics, i.e., the integral of the magnetic flux density in the areas of the magnetic pole. Basic research work was just started, such as the magnetic force between two magnets vs. distance, the area of the pole and the magnetic flux density. It should be mentioned that such knowledge is still missing, although it seemingly exists in traditional physics textbooks [5].

The quantitive result of magnetic force is crucial to evaluate the speed and torque. The rotating speed and torque of the rim-driven motor are the two important parameters to estimate the thrust of the RDF jet. As shown in Fig. 9.7, under the magnetic forces of M_1 and M_2, the speed of magnetic M_0 is:

$$v_p = \sqrt{\frac{2}{m} * \int_O^P [(F_{M1} + F_{M2} - F_R)]dx} \tag{9.8}$$

where F_{M1} and F_{M2} are the cosine portions of the repulsive and attractive magnetic forces applied onto M_0. F_R is the dragging force, which can be equivalent to the RDF thrust, and $\int_O^P F dx$ represents the work done from the starting point O to point P.

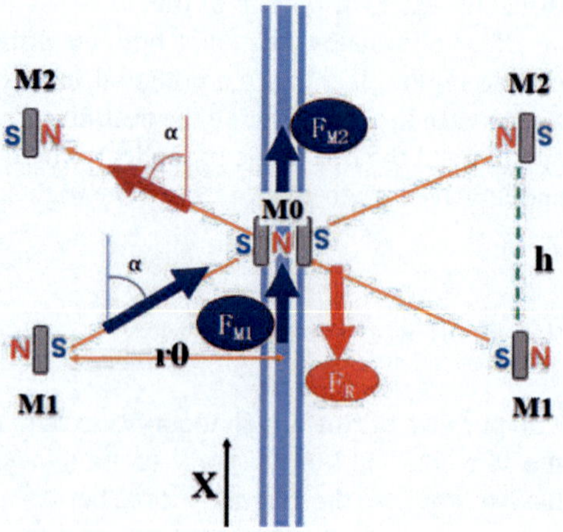

Fig. 9.7 Movement of permanent magnet M_0 in magnetic matrix of M_1 and M_2

Table 9.1 Coefficients of K and n obtained from the experiment

	ϕ15 mm	ϕ35 mm
	Solid round magnet	Hollow ring magnet
Attractive	$1.9 * 10^6$	$6.7 * 10^6$
Repulsive	$1.3 * 10^6$	$4.3 * 10^6$
Attractive	0.85	0.89
Repulsive	1	1.04

The magnetic force versus the spacing is measured using two neodymium magnet plates with different dimensions (Ø12 mm and Ø35 mm, thickness 5 mm). We first built a Coulomb-like formula for the attractive and repulsive magnetic forces as:

$$F = K * \frac{\int B_1 dS_1 * \int B_2 dS_2}{r^n} \tag{9.9}$$

All parameters follow SI units and B in Tesla. According to the data obtained by the experimental measurement, the values of the parameters K and n in the magnetic formula are explored and listed as follows (Table 9.1).

The force between the two magnet plates is approximately inversely proportional to the distance, where n is close to 1, and K is close to 10^6 [N/m Tesla]. It is already seen that the attractive force is bigger in this empirical formula. Unlike the Coulomb Law, the attractive magnetic force is stronger than the repulsive force as indicated in the experimental curve in Fig. 9.8. In general, the coefficient K in the Eq. (9.9) represents the difference between the attractive and repulsive forces with a relative ratio 1.5–1.8. For the Ø35 neodymium magnetic disk, the attractive magnetic force can be 5 N more than the repulsive. During the measurements, its magnetic force is too strong to control the distance when the two disks are too close to each other. (In fact, it is almost impossible to human hand to separate two Ø35 neodymium magnetic disks once they are attracted together, and special equipment is needed.)

For the repulsive magnetic forces, the north-to-north and south-to-south magnetic forces are the same.

Our tentative explanation that the attractive force is larger than the repulsive force is as follows:

> *The magnetic dipole tends to align together when the North Pole is facing the South Pole. However, when the North Pole is facing the North Pole or the South Pole is facing the South Pole, the magnetic dipole intends to repel each other to make the permanent magnet lose its permanent magnetism. The United States is always more powerful than the divided states fighting each other during the civil war. (The word "United States" is a plural such as "The United States are divided ...").*

A more vivid explanation is provided in Fig. 9.9. The magnetic dipole may exhibit angles along the magnification direction, and these angles increase when the same poles (N to N, or S to S) approach each other, while these dipoles become more aligned when the different poles (N to S) approach together. More aligned dipoles are more unified, more unified efforts possess more strength.

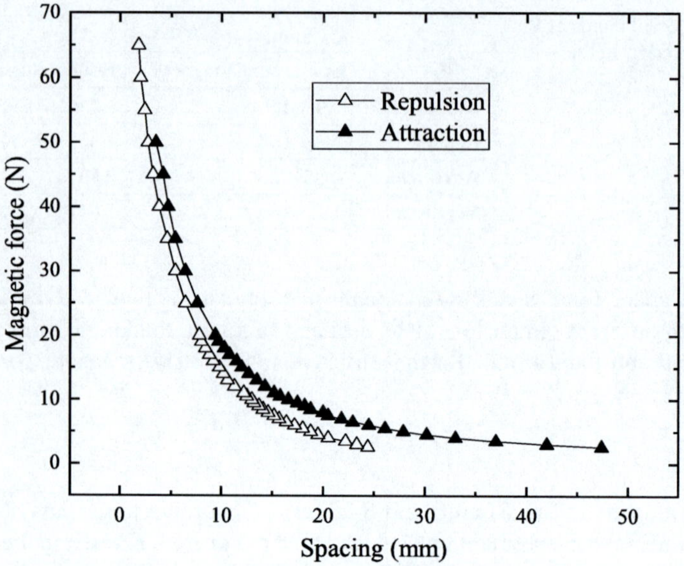

Fig. 9.8 Magnetic force of the Ø35 neodymium magnetic disk

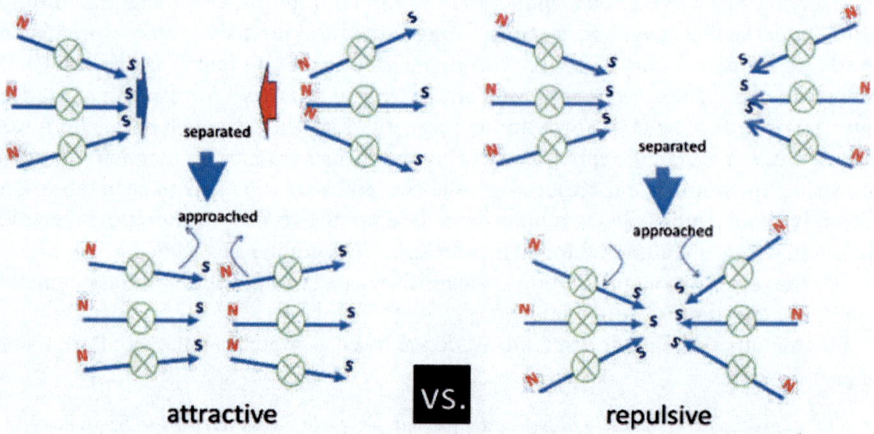

Fig. 9.9 Illustration of the more unified force of attractive magnetism versus divided

In summary, the attractive magnetic force is greater than the repulsive magnetic force because of the forward enhancement and reverse torsion of the magnetism. For electric charges, Coulomb's force only depends on the amount of charges. This is the key difference between the interactive forces of the mechanism and electronics.

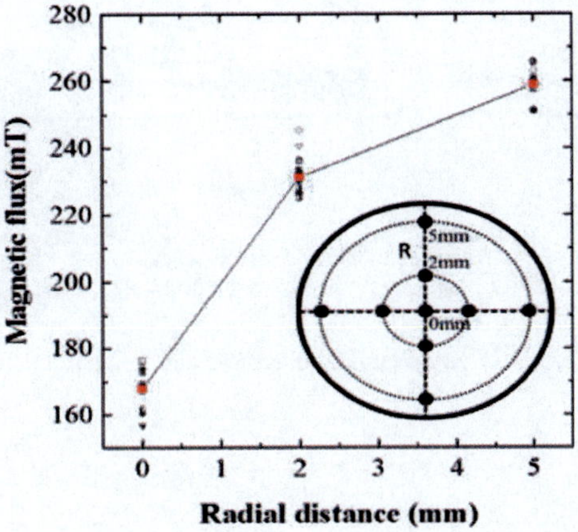

Fig. 9.10 Distribution of the magnetic flux on the surface

The Experiment

A series of tests were conducted to establish a Quasi-Coulomb relationship of magnetism. First, the numerator in formula (9.9) is the product of the magnetic field intensity and pole area. As the magnetic field distributes unevenly along the surface of the disk, the integration of the magnetic flux over the surface is used as the numerator, and the distribution of magnetic flux on the surface is characterized with the nine-point measurement. We measured 14 disks, and all measured data are shown in Fig. 9.10. The magnetic flux values at the same distance from the center are similar, and the variation is within 3%. The magnetic field is stronger at the edge.

Then, we carefully characterized the distribution of magnetic flux for different shapes of magnets-a solid magnet with a diameter of Φ12 mm and a ring neodymium permanent magnet of Φ35 mm. we have observed that:

(1) *For a solid magnet disk, the magnetic flux density is greater at the edge;*
(2) *For ring magnet, the distribution is different on the two surfaces,*

- *For convex surface, the minimum magnetic flux is at the inner edge and then reaches the maximum value at the outer edge, then gradually falls down along the radial direction until the edge.*
- *For the flat surface, the minimum magnetic flux is at the edge and then gradually increases along the radial direction until the edge.*

A test apparatus was established to measure the magnetic force, magnetic flus, distance, etc. The schematic is shown in Fig. 9.11a and the test system is fabricated as shown in Fig. 9.11b to measure the magnetic force vs. distance. One magnet is fixed on the wall, while the other is connected to the gauge. The distance between the two magnets is controlled by the rotation of the long screw at the rear end. The spacing between the two magnets is measured using a plastic caliper.

Based on the test results, we obtained the Coulomb-like formula of the magnetic normal attractive force (N) can be expressed as:

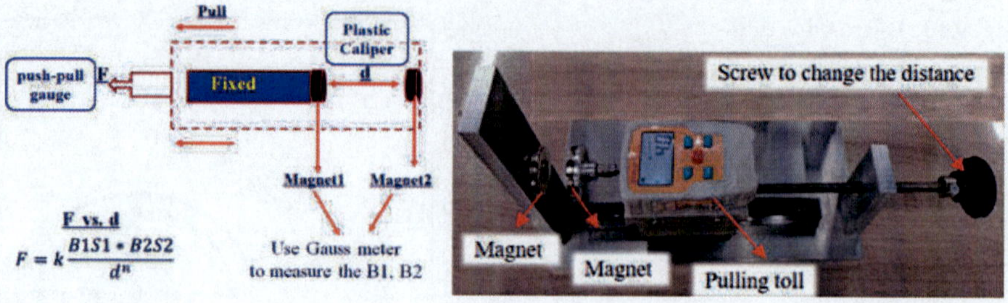

Fig. 9.11 **a** Schematic diagrams and **b** apparatus to test the magnetic force versus distance

$$F = 1.9 * 10^6 * \frac{\int B_1 dS_1 * \int B_2 dS_2}{r} \tag{9.10}$$

The Coulomb-like formula of the magnetic normal repulsive force can be expressed as:

$$F = 1.3 * 10^6 * \frac{\int B_1 dS_1 * \int B_2 dS_2}{r} \tag{9.11}$$

According to the obtained Coulomb-like law of normal magnetism, we concluded that the force between magnets is inversely proportional to its distance, which does not seem to be the same as Coulomb's law. However, considering that this magnetic force is between the two plates instead of two point charges, we conjecture that two line charges or surface charges may also exhibit the same behavior as magnetic force that we measured, i.e., the inverse-square relationship between forces versus distance in Coulomb's law will no longer hold even for electric charges. To further verify the correctness of the conclusion, we simulated the line charges and surface charges. The formula for Coulomb's law is

$$F = \frac{k * q * q}{r^2} \tag{9.12}$$

where $k = 8.9876 * 10^9$; this formula is applicable to the force between point charges with a distinct distance r. To study the force between line charges and surface charges, the simulation model of two line charges is established and a C++ program is written to solve the equations using the same procedure as above, set n to 11, $l = 0.001$ m, $L = 0.01$ m, and $d = 0.001$ to 0.1 m. The fitting line us shown in Fig. 9.12 together with the simulated data points:

From this, we can conclude that the force between the surface charges is also inversely proportional to the distance when the two surface charge is close enough and surface density is high enough, which is generally true in real world. Indeed, this is consistent with our magnetic experimental results.

In summary, under certain conditions, the electronic force between the line and surface charges follows the $1/r$ relationship instead of the $1/r^2$ of the pure Coulomb

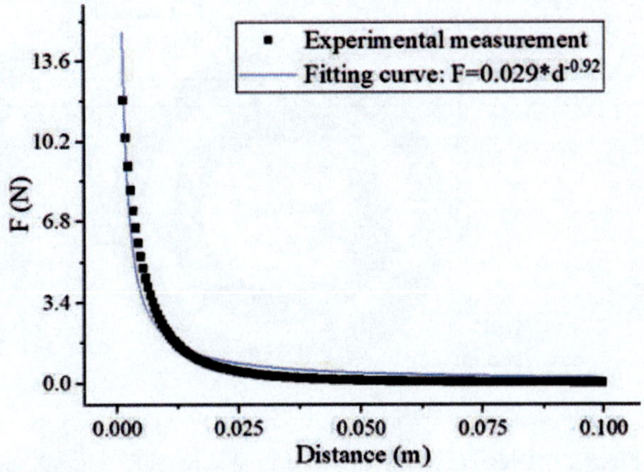

Fig. 9.12 Relationship between the surface charge F and d of the experimental (simulation) data and the fitting line

Law. From this perspective, the magnetic force between two magnetic charges may follow similar law as Coulomb's Law, but the attractive magnetic force is larger due to the slightly difference in the between the electric charge and magnetic dipole.

(b) *Study of tangent magnetic force*

As shown in Fig. 9.13 in a rim-driven system, the interactive force between the electromagnetic stator and the permanent magnetic rotor is not the normal force of the two facing magnets but their tangential force. Therefore, it is very important to study the tangential magnetic force between two magnets in the RDM system. Under the magnetic forces of $M_1$and M_2, the speed of the magnet M0 increases until the positive force and negative force is balanced:

$$V \infty (F_{M1} + F_{M2} - F_R)^n \tag{9.13}$$

where V is the velocity of magnet M_0, FM_1 is the repulsive force from M_1 and FM_2 is the attractive force from M_2 and F_R is the dragging force applied onto M_0. F_R can be treated as the propulsion force of the RDF jet.

To measure the tangential force between magnets, we designed the following experimental apparatus as shown in Fig. 9.14. The magnet on the right is moving with a screw in parallel to the magnet on the right. The tangential force between the magnets is measured by the gauge with varying distance between the two magnets. When facing directly, the tangential force should be zero.

The tangential forces of the attractive and repulsive forces were tested with two Φ50 strong Nb-Fe-B permanent magnets. The magnet fixed during the experiment is an N pole with BS integration of $1.76 * 10^{-4}$ in Eq. (9.9) For S pole the integration of the magnetic flux is $1.92 * 10^{-4}$. Experiments were carried out by changing the tangential distance L for two attractive and repulsive magnets. The relationship between the tangential force of the magnetic attractive force and the tangential distance with opposing distance of 5/6.5/8/11/13 mm are shown in Fig. 9.15. It is

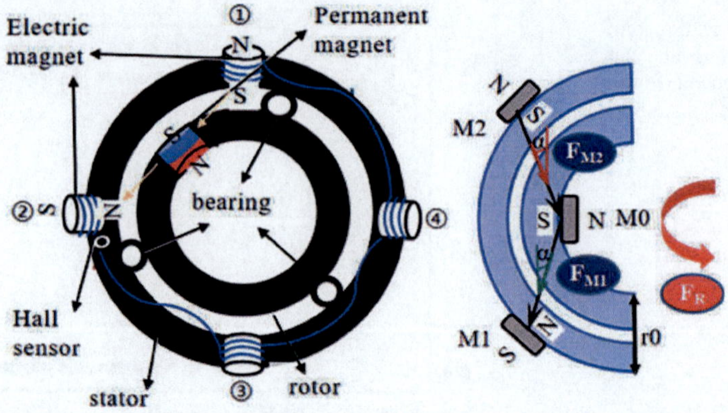

Fig. 9.13 Diagram of rim-driven magnetic propulsion (left), the movement of permanent magnet M_0 in magnetic matrix of M_1 and M_2 (right)

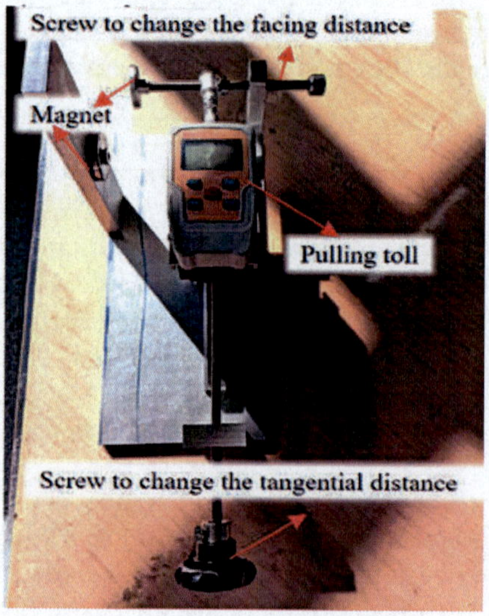

Fig. 9.14 Tangential force experimental device

expected to see that the magnetic force increases as they approaching closer, then the tangential force sharply decrease when they are facing straight. But it is also interesting to see there seems exist two peaks of tangential force as the two magnets moving together when their facing distance is very close.

A simulation using the surface charge was performed to study this phenomenon. We put four charges located on the four corners on the surface. The 16 forces generated by a total of 8 charges on the two surfaces are superimposed to obtain the total forces. When the two sheets of surface charge are closer enough we also observe two peaks as shown in Fig. 9.16. This simulation clearly indicates that even for two electric surface charge plates such phenomenon is an also expected result.

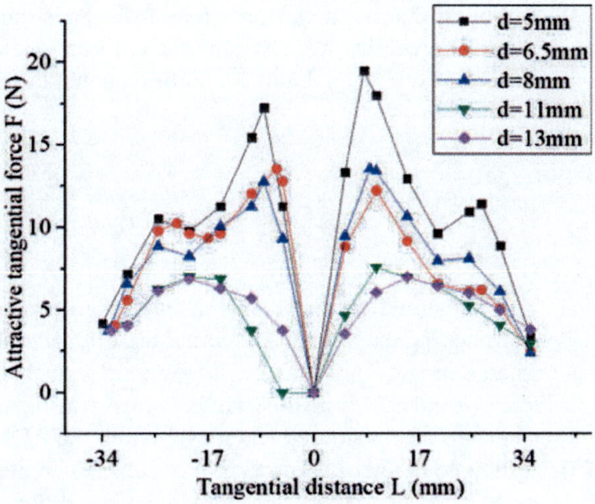

Fig. 9.15 The relationship between the tangential force and the tangential distance of the attractive force

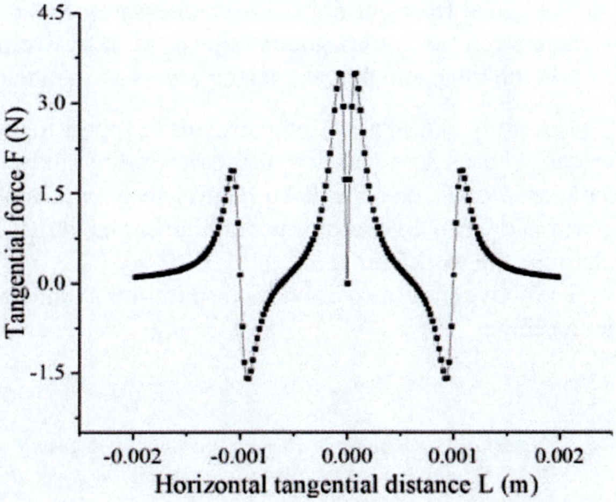

Fig. 9.16 Tangential force simulation of surface charges

We can draw the following conclusions based upon our studies both from simulation, curve fitting, and experimental results:

(1) The magnitude of magnetic force is proportional to the magnet area and the magnetic intensity. Therefore, large flat NdFeB permanent magnet plates are favorable to build strong permanent magnet rotor
(2) The Coulomb-like relationship to evaluate the force between two magnets could still be valid, since the behavior of the electric force of the line and surface charges are similar to our experimental observation on the magnetic force. That is, the relationship of the normal force of two surface magnets is inversely proportional to the opposite distance just as two electric surface charges.

(3) Although the magnetic force may follow a similar law as Coulomb's law for electric force, the attractive magnetic force is always larger than the repulsive force for both normal and tangential magnetic forces.

For this, our vivid explanation is that united force is more powerful than deviated force, i.e., "Together We Can Make a Difference".

(4) The tangential force vs. the distance has two-peak behavior when they are approaching each other and when the opposite distance of the two magnets is small enough. This "dancing" magnetic force behavior may bring some turbulence or add noise into the rotor movement during the rim-driven motor operation and shall be considered when evaluating the RDM engine performance.
(5) It may be neither necessary nor feasible to create an "exact Coulomb Law" in magnetism since there are no independent magnetic charges but rather densely populated magnetic dipoles. However, it is still useful to use the coefficient obtained from our normal force measurements to evaluate the magnetic forces between the two magnetic objects, such as two magnetic plates, as a function of the magnetic flus, the facing area, and the facing distance.

The study of magnetic force provides support for evaluating the magnetoelectric torque of brushless rim-driven motors in the study of next-generation electrified aviation propulsion. It will be helpful to build a model to evaluate the rotational speed of the rim drive motor under high engine thrust. More work is still ongoing to continue the work initiated.

The above study also provides inspirations to build a lightweight and smaller RDF jet machine.

***large flat** NdFeB permanent magnet plates are favorable to build a strong "permanent magnet rotor + electromagnet stator = rim driven motor" for a **big power-to-weight** ratio rim driven machine*

9.5 The Tai Chi Fan

In this section, we describe the special design of the Tai Chi blade for the RDF jet-its advantage, structure, and theory.

9.5.1 The Future of Rim-Driven Fan

A rim-driven fan is different from a shaft-driven fan or propeller, and there seems to be no historical reference to follow. Rim-driven fan has the inner rotating blades, and the driving force comes from the edge, which exhibits different behaviors as conventional shaft-driven rotations.

No Shock Wave Issue

The use of a propeller to generate thrust dates back to the beginning of human flight history of Wright Brothers. In the late 1940s, the notorious shock waves [6] were observed as the limiting factor on propelling speed. When rotating and navigating at high velocity, the tangent speed near the tip reaches supersonic range, and the shock wave occurs, which greatly reduces the propeller efficiency. The intense vibration caused by shock waves may also damage the propeller and propagate to its root. In contrast, the blade tip of the rim-driven fan sits inside, rotating at a nearly zero velocity which circumvents the shock wave issues of the propeller.

Bigger Torque

Most of the motor uses the shaft to drive wheels, propellers, and fans to rotate, i.e., to use the small radius axis to drive the larger radius machine part. Larger power and torque are needed in this energy transmission. As illustrated in Fig. 9.17, for shaft-driven rotation, most of the twisted tension is on the root of the axis connecting the rod. The force needed to drive the rod is $F = m\omega^2 r$. With the faster rotation ω and longer arm r, this torque force increases tremendously. In comparison, the torque force and power to drive the rim-driven fan is much less, allowing the bigger and fatter fan blade topology for larger air intake rate. In other words, the rim-driven motor exhibits more efficient drivability for bigger torque with less power needed. The drawback of the rim-driven machines is its less flexibility compared with the shaft driven machines such as the speed transmission and directional adjustment. However, rim-driven machine has an intrinsic advantage for the powerful drivability which is the most important for aviation thrust force.

9.5.2 The Design of Rim Driven Blade

DOE and CFD are used in combination for different trial-on-error iterations for the best air-sucking performance of the blade. The Tai Chi-shaped topography is the prototype for this design iteration cycle.

The Tai Chi fan

The Tai Chi fan integrates the Chinese Tai Chi logo (Fig. 9.18) into the rim-driven blade design. Tai Chi represents "the One", and it is Chinese belief that the One in Heaven is split into Two when down to the Earth: the Yin and Yang.

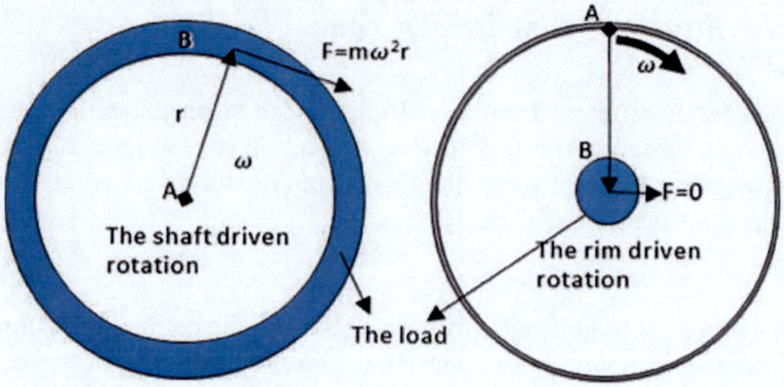

Fig. 9.17 The shortcoming of axis-driven rotation by the analogy of the Archimedes level principle

The concept of Yin and Yang in Chinese philosophy inspired Leibniz, a German philosopher/mathematician to initiate the "0 and 1" binary system which renders the resolution of the twenty-first century information technology together with a MOSFETs. Leibniz is therefore known as Aristotle in the seventeenth century for his discovery of the binary system based in Yin (symbol "–") and Yang (symbol "—"). The basic philosophical principle of "Yin and Yang" is that they must coexist, they can exchange each other and their connection is "round" instead of "straight". The Tai Chi logo refers these three relationships of Yin and Yang. The Tai Chi blade follows these principles in the blade design. For example, the transition of the Tai Chi fan blade is balanced, round, and smooth, representing the balanced smooth harmonic flow of the thrust force inside the RDF jet.

Geometry DOE of the Blade Design

The main object of a rim-mounted fan blade is to use the Design of Experiment (DOE) method (as shown in Fig. 9.19) to obtain the best input combination of length, width, thickness, and angle, together with the curvature to achieve the largest mass flow rate like we did in GEnX front fan.

The inner spiral blade using DOE design follows the "IPO" plan:

Input → process → output

- to compare the different propulsion forces (*O*)

Fig. 9.18 The Tai Chi logo

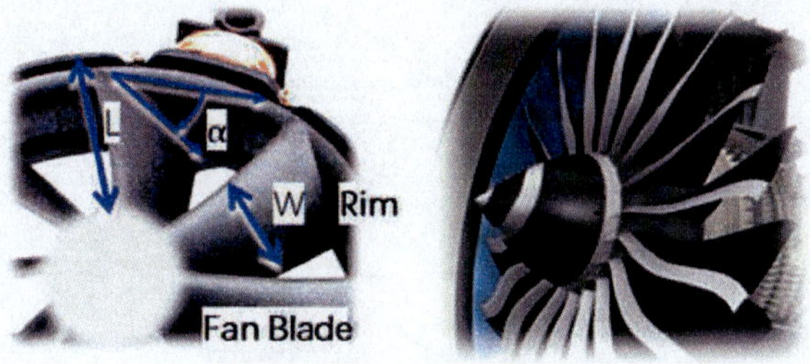

Fig. 9.19 Illustration of the DOE approach for the best angle, length, width of the rim-driven fan design (left), as compared to the best blade design of GEnX front fan (right)

- by using the CFD simulation tools (*P*)
- with variant blade parameters width, length, angle, etc. (*I*).

As an example, the plan of the DOE table is shown in Table 9.2:

The Curved and Twisted Design of the Blade Cascades

Figure 9.20 shows an example of the CFD design of the blade from the root to the tip. Curve, twist, and cascades are the three key parameters for the blade topology for the best aerodynamic performance (thrust force).

Different from the previous shaft-driven blades (turbofan, electric fan, and propeller), the rim-driven blades can be designed as a wholesome spiral plate with various inclination angles α and degrees of twist. Previous experience with airplane air foil design can be used as a reference. The lift and resilient force are both proportionally related to the speed, area, density, and attacking angles. For a flat plate moving at speed V, the lifting force is $\rho V^2 S\, k_1 * \sin(\alpha)\cos(\alpha)$ which is at max when α is 45°. However, the resistance is also high at this angle, causing a lower efficiency. The best angle of attack is usually at the lift/drag = minimum, which is close to 20°.

Making Use of the Existing Experience in Marine and Gas Turbine Industry

In addition to the previously approaches, the following two existing experiences are also helpful to design the RDF jet: the marine RDP and the turbofan of the commercial aero-engines.

(1) **The marine experience**. The original RDP was mainly for marine thrust dated back to 1970's since William Webster proposed rim driven concept in 1977 [7]. Therefore, RDP technology is not just an advanced concept but also a mature

Table 9.2 DOE plan of variables of RDF blade

Input parameters	Values
W (cm)	0.2, 1, 2
L (cm)	1, 5, 10
Angle (°)	22, 45
Blocking (m)	0.1, 0.5

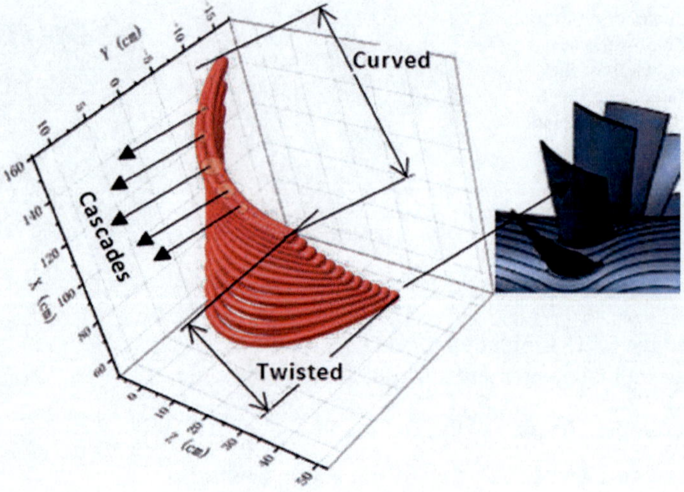

Fig. 9.20 Curve, twist, and cascade of the blade design

practice in marine industry, which took three decades of trial-and-error development [8, 9]. Its 30 years' marine career proved that the rim-driven concept is a doable practice and could serve as good reference when developing the aviation RDF jet machine. Although the flowing medium for the aircraft is different from that in a submarine, such as the rotating speed in water is much slower than that in air, and the energy exchanging pattern and velocity diagram of the blade for aircraft propulsion is also different from the water thrust for the ship, the commonality is both propulsion need to conquer the large torque from the thrust. Some design and manufacturing experience in marine [10–13], especially the machines and mechanics, can be borrowed as a reference for aviation RDP engines.

(2) **Experience in gas turbine machines.** The advanced design and manufacturing of the fan blade such GEnX is the main part that the RDF jet development can borrow from the previous efforts. The main goal of rim driven fan is the same, i.e., for the maximum massive intake air flow as in GE9X turbofan engine. The design and manufacturing experience can be borrowed from the existing GEnX engineering to build rim-driven fans. Although there is some preliminary work done on the simulation [14], more intensive and comprehensive simulation work is still needed to achieve an ideal rim-drive machine. In addition, experimental verification is also lacking and is highly needed. Blade manufacturing is fairly easy since there is no high temperature-related thermo-exchange involved as in turbine gas engines.

9.6 The Duct and Jet

The large amount of air intake by RDF is compressed and then leads to the shrunk outlet of the tail jet culvert through the streamlined duct inside the closed cylinder shell. The projection force of the jet is as follows:

$$F = m(V_2 - V_1) \tag{9.14}$$

m is the mass flow rate (kg/s), and V_2 is the outlet jet speed. At the takeoff stage, the airplane speed V_1 is close to 0. In addition to increase the "m", the other factor to enhance the "F" is "V_2".

The jet speed V_2 can be improved by shrinking the outlet area. V_2 is inversely proportional to the area of the jet outlet,

$$V_2 = \frac{V_1 S_1}{S_2} \tag{9.15}$$

This effect only becomes more prominent when the jet outlet is narrow enough. For example, in our RDF jet, the front inlet diameter is 50 cm. Assuming the outlet ring is 1 cm wide (2%), then the outlet area is ~ 1/10 of the inlet. Assuming the inlet air-sucking speed V_1 is 40 m/s, then the outlet jet speed V_2 becomes 400 m/s, even greater than the duct jet speed of 200 m/s in the GEnX commercial aero-engine.

> The GEnX *engine's total thrust is 310 kN. When BPR = 9, the outer duct thrust F = 0.8 * 310 = 252 kN. With suck-in air amount of 1200 kg, from F = m (V2-V1) we get jet speed* V_2 *~ 210 m/s.*

On the other hand, one also has to consider the power capable of driving these turbofans to achieve this high thrust. Imagine you hold a blade and force it not rotating, there exists a big torque applied on the motor. This is equivalent to shrink the jet outlet as above. In order to conquer this resistance you need much higher power such as offered by gas turbine machines. We list a table to compare the various thrust-over area ratio of different turbofans as shown in Table 9.3.

The thrust is indeed proportional to the area of the turbofan, except that the electric-driven fans exhibit only 1/10 thrust/area ratio compared to the gas turbine machines (0.04 vs. 0.4). One key reason is that the turbine power is much stronger than the battery. For example, the DA's electric duct fan only delivers ~ 200 kg thrust with just 70 kW power. If we use the comparable 700 kW power from the gas turbine, the DA's 80 cm duct fan shall deliver 2000 kg thrust. Likewise, our 50 cm RDF jet could achieve our target thrust value of 900 kg with enough power of over 450 kW.

Additionally, from Table 9.3, we also see that 35 has a higher thrust/area ratio of 0.63 than ordinary gas turbine machines, which implies that a more dedicated approach on designing the fan may help to enhance the thrust of the RDF jet. For example, the use of a rim driven with fat fan blade may allow a larger air intake rate compared to the conventional shaft-driven fan, and the shrunk jet outlet enhances the jet speed for a larger thrust. In fact, by observing the outlook of the ordinary electric

Table 9.3 Thrust versus the fan size of various fans

Vendor	Fan diameter (cm)	Fan thrust (kg)	Ratio (thrust /area) (kg/cm^2)	Power	Notes
DARPA	80	195	0.04	70 KW	*Ordinary turbofan: electric power, shaft driven fan*
Taobao1	9	3.6	0.06	N/A	
Taobao2	7	1.2	0.03	N/A	
Taobao3	5	0.85	0.04	N/A	
RDF Jet*	**50**	**900**	**0.46**	**450 KW**	*New propulsion: rim-driven Taichi fan + 3D HK SC e-power*
CFM56-5C1	183.6	10,875	0.41	10–50 MW	*Ordinary gas turbine fan. The thrust from turbofan account for 70–80% of the total thrust depending the BPR from 6 to 9*
CFM56-3C1	152.4	7000	0.38		
CFM56-2	173.5	8560	0.36		
JT-3D	135	5600	0.39		
GEnX	270	22,720	0.40		
F35	127	8000	***0.63***	1 MW	*Special-designed elevate-fan driven by turbine power*

Note the fan diameter and thrust data are from
1. https://www.amazon.com/s?k=duct+fan, https://s.taobao.com/search?q=duct+fan
2. https://www.cfmaero-engines.com/engines/cfm56/
3. https://defense-update.com/20160303_vtol_x_plane.html

duct fan in the Amazon market, one can see that blades and architecture of the fan are not optimized as compared to the most advanced commercial gas turbine fans used in GEnX (Fig. 9.21), which implies that there still a lot of room to improve.

Fig. 9.21 Comparison of the gas turbine fan (left) and the Amazon duct fan (right)

9.7 The Building Material

An electric aero-engine must be built as small and lightweight as possible. Lightweight and firm materials should be used to build e-engines. In comparison to the gas turbine machines, the electric engine does not involve the tough requirement on high-temperature building materials such as heavy Ni–Cr alloys, which makes it fairly easy to choose its building material. Electric engines should consider only functional materials possessing electric and magnetic properties together with sufficient mechanical strength. The working temperature ranges between − 40 °C (1000 m height ambient) and + 200 °C (high-speed rotating rim in RDM).

9.7.1 Weight Estimation

There are two RDF jet engine options with different sizes of 30x30 cm and 50x50 cm cylindrical shapes, with expected maximum of 600 kg and 900 kg thrust, respectively. Table 9.4 lists a rough estimate of the weight. 3D nested FRC (fiber-reinforced composite [2]) and CMC (ceramic matrix composite [3]) lightweight materials are used to build the RDF jet (more details in Chapter 11). Roughly the ϕ30 RDF jet weighs 20 kg and Φ50 RDF jet weighs 60 kg. We use ϕ30 engines for small VTOLers and Φ50 engines for van-airplane with ~ 20 pax or to aid the Boeing 787 for STOL.

Table 9.4 Weight estimation φ30 and Φ50 RDF jet

rim weight	diameter D (m)	outer rim (m3)	inner rim (m3)	outer rim(kg)	inner rim(kg)	outer+inner(kg)
	0.3	0.008	0.008	5.935	5.546	11.480
	0.5	0.024	0.023	16.485	15.832	32.317
		the thickness of the rim is 3cm, 700kg/m3				

RDF Jet weight (kg)

30L x 30D	50L x 50D
19.193	57.842

=rim+magnet+blade

magnet weight	D	width (m)	dimensions	PM size /weight (m/kg)	EM size /weight (kg)	PM+EM (kg)
		3.14*D/12	width	0.07	0.07	
	0.3	0.079	length	0.1	0.1	
			thickness	0.005	0.005	
			numbers	12	12	
			weight	3.15	3.15	6.3
	0.5	0.131	width	0.12	0.12	
			length	0.2	0.2	
			thickness	0.005	0.005	
			numbers	12	12	
			weight	10.8	10.8	21.6
	Notes:	Nd-Fe-B permenant magnet (PM) density:7.5g/cm3				
		Electro-magnets (EM) has the same weight as PM				

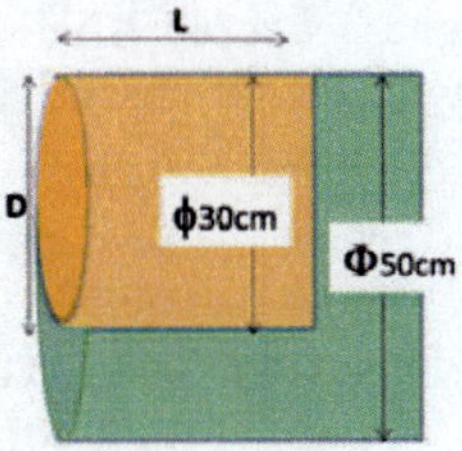

magnets+blades weight(kg)

30Lx30D	7.713
50Lx50D	25.525

fan blade weight	D	volume (m3)	weight (kg)
	0.3	0.001	1.413
	0.5	0.004	3.925
		blade is 2cm thick, density 1g/cm3,CMC built, assume a flat plate	

9.7.2 Permanent Magnet

Unlike the Tesla's EV cars, which use purely electromagnetic coils instead of permanent magnets as rotors, PM is an indispensable material to build rim-driven aviation engines due to its distinct weight advantage. Neodymium Nd-Fe-B is the key building material for permanent magnets processing the highest magnetic intensity. China is a major rare earth country accounting for 36 ~ 43% of the world Nd element. Baotou, Jiangxi, and seven other southern provinces have abundant rare earth PM material to build high-intensity neodymium Nd-Fe-B magnets. The maximum magnetic energy product is as high as 512 kJ/m^3, which is known as the "magnetic king". Permanent magnet materials often need to work in the temperature range of − 20 ~ 400 °C. Temperature is usually not a concern because the rotational heat may compensate for the low temperature at a height over 1000 m. The preparation of neodymium Nd-Fe-B magnet includes ingot and powder for sintering, calibration of surface orientation for magnetization to form the PM materials, and pulse magnetization to generate a strong magnetic field for the magnetizer/magnetizer [17]. The drawback of the Nd-Fe-B PM is that the density is still too high (7.5 g/cm^3). Research on developing lightweight $Nd_2Fe_{14}B$ permanent magnet material is necessary considering its use in aviation scenarios, which is often missed for the car industry and others.

Brushless Motor Versus Conventional Motor

Table 9.5 compares the old and modern electric engines, PM/EM (Permanent magnet rotor plus the electromagnetic stator used in BYD's EV car) engines vs. the EM/EM (electromagnetic rotor plus electromagnetic stator used in Tesla's EV car) engines. Key features related to the aviation e-engine are marked in red and in boldface including power density, efficiency, and reliability. PM/EM electric motor plus the IGBT high-power electric control form the bases of the future aviation electric propulsion technology.

In summary, the RDF jet engine is composed of a rim-driven motor, the Tai Chi fan blade, and the duct jet. Among them, rim driven motor is good at larger torque, Tai Chi fan blade is good at more air intake, and duct and shrunken outlet is for higher jet speed. The RDF jet is a small, light, and flexible rotatable e-propulsion engine for various aircrafts with VTOL / STOL features.

In addition:

(1) There are two sizes of RDF jets: one is a small cylinder with a diameter of 30 cm and length of 30 cm, with the highest thrust of 600 kg, and the estimated weight of 20 kg. The other is a 50 cm diameter and 50 cm long cylinder, with a maximum thrust of 900 kg and an estimated weight of 60 kg. Smaller engines are used for smaller e-airplanes, such as car-type and delivery e-airplanes. Bigger engines can be used for high-power thrusts, such as the STOL for Boeing 787.
(2) There are two types of tail nozzle designs for RDF jets. In the 1st kind, the diameter of the tail nozzle gradually decreases and shrinks to the center to form a focused jet beam. The second is the conical jet ring. The first way of tail nozzle is simple in structure with less weight. The advantage of the 2nd kind

Table 9.5 Electric motor comparison

Classification	Permanent magnet brush DC motor	PM brushless motor	AC induction motor
Time, user	Old motor	BYD, aero-engine, UAV	Tesla
Stator	Permanent magnet	Electromagnet, multiwinding	Electromagnet, multiwinding
Rotor	Electromagnet, multiwinding	Permanent magnet	Electromagnet, squirrel cage winding
Weight	Heavy	Lighter	Light
Electronic controller	Not necessary except to adjust the speed	Must have, IGBT + SiC	Not necessary except to adjust the speed
Operating voltage	Low	High	High
Efficiency	LOWER	High, rotor has no loss	Low, has rotor loss
Power density	Lower, rotor is not easy to dissipate heat	High, stator winding is easy to dissipate heat	Lower, rotor has loss and heating
Reliability	Low because of brush	High, mainly determined by bearing	High
Maintenance	Regular cleaning and maintenance needed	Less frequently maintained	Less frequently maintained
Cost	High, permanent magnet materials	High, permanent magnet materials/controllers	Low cost

is that the engine is more conformal cylindrical shape, and the rear spare in the center can be used as a compartment to store the battery/supercapacitor as independent mobile electric power supply. The whole tail nozzle or tail jet ring structure is formed by 3D printing or casting at one time without a welding process. According to the same concept, the manufacturing method of rim and Tai Chi fans can be 3D printed or molding/casting as well.

An electric power supply with both high energy and power densities, such as a lightweight gas turbine generator and a 3D HK supercapacitor, is needed to achieve a higher thrust, although conventional lithium and fuel cell batteries can still support RDF jets for smaller thrust in various AAM scenarios.

The RDF jet is a joint effort of the AIAA and IEEE fellows—the world's two largest engineering families. The integration of AIAA and IEEE is the interdisciplinary interaction between the two groups of smart people—a teamwork uniting both minds for both needs by both people. Combining the efforts of IEEE and AIAA, the next generation of hybrid/full-electric aero-engines will come true—just as the 1st Toyota's Prius hybrid car in 1997 to be updated with the Elon Musk's Tesla electric car. From the combustion engine car to electric car we have used 20 years efforts with the help of advancement of brushless electric motor and more powerful electric generation / storage technology. By taking advantage of the advancement of electric cars and other relevant technologies such as AI, lightweight materials, etc., the new generation of electrified propulsion and VTOL e-airplane will come into being very soon.

References

1. Biser, S., Atanasov, G., Hepperle, M., et al.: Design space exploration study and optimization of a distributed turbo-electric propulsion system for a regional passenger aircraft. In: AIAA Propulsion and Energy 2020 Forum (2020)
2. Chen, H., Zhang, L.M., Jia, G.Y., et al.: The Preparation and characterization of 3D-Silica fiber reinforced silica composites. Key Eng. Mater. **249**, 159–162 (2003)
3. Jamra, F.R.M.: A brief review on ceramic matrix composites, it is attributes and it is utility in future generation gas turbine. IJIRST Int. J. Innovative Res. Sci. Technol. **1**(6), (2014)
4. Hsieh, M.F., Chen, J.H., Yeh, Y.H., et al.: Integrated design and realization of a hubless rim driven thruster. In: IECON 2007—33rd Annual Conference of the IEEE Industrial Electronics Society, pp. 3033–3038 (2007)
5. Hinrichs, R., Dirks, K., Urone, P.P., et al.: Coll. Phys. (2013)
6. Dickinson, H.B.: Propeller-Design problems of high-speed airplanes, special report. Techn. Rep. Arch. Image Library (1941)
7. Webster, W.C., Shacham, I.: A water-turbine driven propeller for high performance ship propulsion. J. Hydronaut. **11**(1), 11–17 (2015)
8. Hsieh, M., et al.: Integrated design and realization of a hubless rim driven thruster. In: IECON 2007—33rd Annual Conference of the IEEE Industrial Electronics Society, Taipei, pp. 3033–3038. https://doi.org/10.1109/IECON.2007.4460298
9. Tan, W., Yan, X., Liu, Z.L., et al.: Technology development and prospect of shaftless rim driven propulsion system. J. Wuhan Univ. Technol. (Transp. Sci. Eng.) **2015**(39), 601–605 (2015)
10. Yakovlev, A.Y., Sokolov, M.A., Marinich, N.V.: Numerical design and experimental verification of a rim driven thruster. In: Proceedings of Second International Symposium on Marine Propulsors, pp. 396–403 (2011)
11. Dubas, A.J., Bressloff, N.W., Sharkh, S.M.: Numerical modeling of rotor–stator interaction in rim driven thrusters J. Ocean Eng. **106**, 281–288 (2015)
12. Song, B., Wang, Y., Tian, W.: Open water performance comparison between hub-type and hubless rim driven thrusters based on CFD method. J. Ocean Eng. **103**, 55–63 (2015)
13. Lai, S.H.: Design optimization of a slotless brushless permanent magnet DC motor with helically wound laminations for underwater rim driven thrusters. University of Southampton (2006)
14. Bolam, R.C., Vagapov, Y., Day, R.J., Anuchin, A.: Aerodynamic analysis and design of a rim driven fan for fast flight. AIAA J. Propul. Power **37**(2), 179–191 (2021)

Chapter 10
The Electric Power—Energy and Weight

The outline of this chapter is:

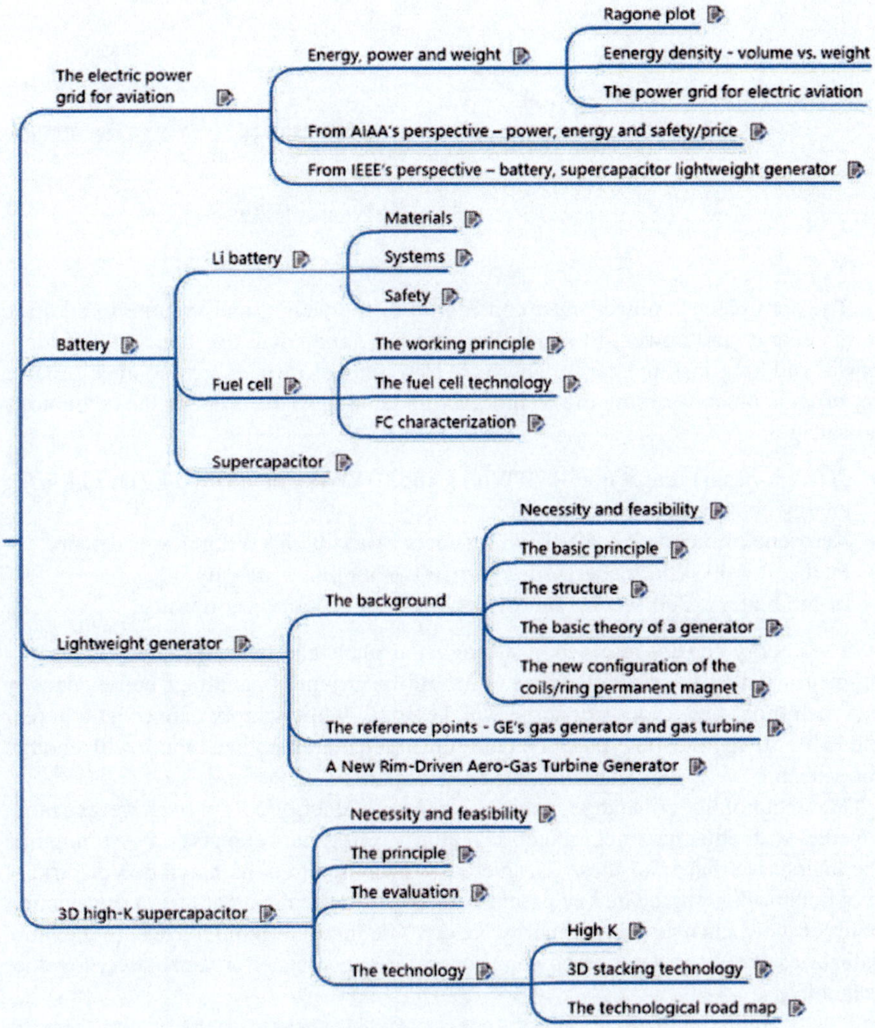

The electric power for aviation: energy and weight

© The Author(s), under exclusive license to Springer Nature Singapore Pte Ltd. 2023

F. L. Duan, *When AIAA Meets IEEE*,
https://doi.org/10.1007/978-981-19-8394-8_10

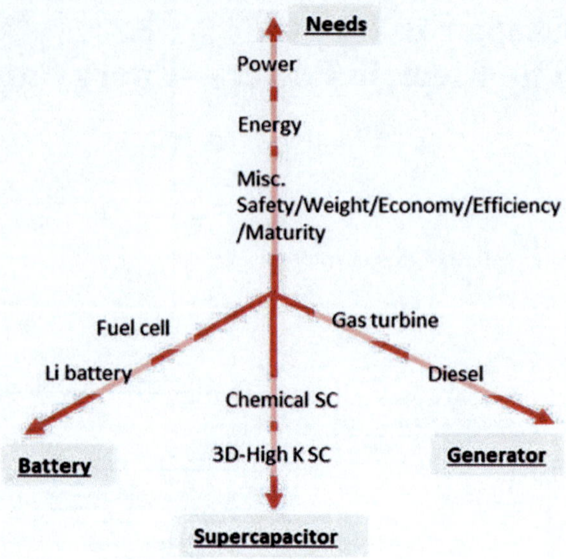

Fig. 10.1 The needs versus technologies for aviation electric power

The aero-electric power must consider both the weight and volume in addition to the energy and power—it should be both light and small together with its high-power and long-lasting energy. There are four main electric energy sources suitable for propulsion considering the weight factors ranking in the order of the technology advantages:

- 3D high-K supercapacitor: ~ 70 Wh/kg and 10 kW/kg or 700 Wh/kg and 1 kW/kg energy density/power density;
- Kerosene or bio-fuel: ~ 4000 Wh/kg energy density, 5 kW/kg power density;
- Fuel cell: 500 Wh/kg energy density, 200 W/kg power density;
- Li-ion battery: 200 Wh/kg energy density, 500 W/kg power density.

The energy density refers to how long an airplane can fly, and the power density refers to whether the aircraft is able to lift off the ground. In analogy, energy density is a marathon, power density is the weight lifting—you simply cannot let a 5-year kid to lift 50 kg load, but you can let him finish the marathon in a month with enough food/sleep.

The technology advantages mainly refer to its energy/power over weight ratio, together with other parameters such as maturity, safety, and expenses. We summarize the interactive matrix of these parameters versus the three main aviation electricity suppliers in Fig. 10.1. The key parameters on the vertical axis are from the aviation point of view, such as power, energy, safety. The three horizontal axes are the three categories of electric aviation powers: generator, supercapacitor, and battery together technologies.

The calculation/estimation of energy density/power density, or specific energy/specific power, is pretty comprehensive which involves many factors. For example, the weight factor includes both the system and fuel. For example, the evaluation and energy density of gas generator is different from that of the battery. In gas turbine machine, the electric generator and the amount of kerosene are two separate units and the weights of the fuel is not fixed number. In battery, the weight of Li battery package is unchanged. Therefore, the energy density evaluation of the gas generator is a variable and the energy density of a battery is a fixed number.

The price comparison is also comprehensive. When we are comparing the price, we must clarify whether we are comparing the fuels or we are comparing the systems or we are comparing both. For example, the Tesla Li battery car is more expensive than most of the fossil fuel cars in America, mainly due to its high battery cost. However, its electric charging fee is cheaper than gasoline. This phenomenon is even more obvious in China—the charging fee of EV car with lithium batteries is only ~ ¥0.09/km. For gasoline, the cost per kilometer is approximately ¥0.72. For fuel cell EV car, the cost of hydrogen filling is ¥0.33/km. Therefore, the price ranking of the fuels are batteries < fuel cell < gasoline. However, the ranking of the car price is: Fuel cell EV car > Li battery EV car > Gasoline car. Overall, the cost of travel is 0.10, 0.07, and 0.04 USD/km for the Toyota Mirai FC car, gasoline car, and Tesla Li battery car, respectively (in US). Unless being environmentalists, if one just considers the price when buying a new car, he just needs to balance the car and fuels [1]. We also need to consider the government policies. For example, in China, the government encourages new energy vehicles by subsidizing its industry. Such a policy already redirects the whole automotive industry to electric vehicles. In the next 5 or 10 years, almost 80–90% of cars will become electric.

In the vertical axis in Fig. 10.1, some parameters are even contradictory to each other. For example, in the Li battery, safety may contradict to its performance (the NCM battery has higher performance yet less safety, and vice versa for the LFP Li battery). Gas turbines generators possess much higher power density yet much more expensive than diesel machines. Boeing 747-400 costs 200 million dollars with four PW4000 gas turbine engines. Each engine cost 12 million dollars, while the diesel generators only cost less than a few thousand dollars.

In this chapter, we review the current electric power for aviation in terms of the energy/power over weight versus battery, generator, and supercapacitor.

10.1 The Electric Power Grid for Aviation

First, we separate three distinct types of objects which offer the energies for aviation:

① Generate energies,
② Store energies,
③ Both generate and store the energies.

To generate and to store the energies are two different capabilities. Sometimes these two capabilities are in one, such as in Li batteries, the electricity just sits in the system. In many cases, they are two separate units. For example, the gas turbine

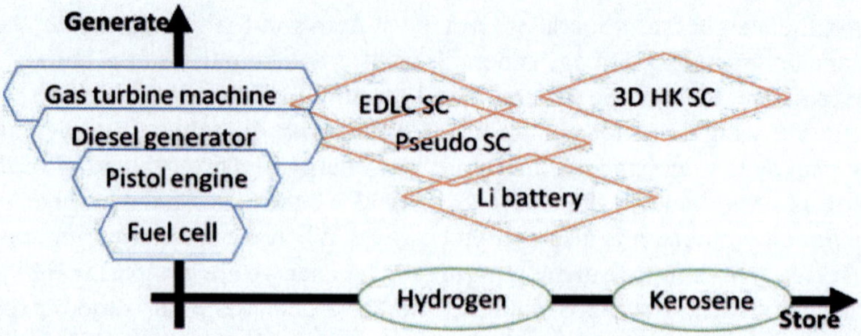

Fig. 10.2 There are two categories to offer the energies: to generate or to solve, and there are various medias to achieve these two goals

generator belongs to the first category which generates energies, but the kerosene belongs to the second category which stores the energies. To store and to generate refer to two different qualities. In the aviation, for example, to store the energy refers to the energy density, how much the energy stored, how much fuel the airplane carries. The quality is like the endurance in the marathon. To generate electricity refers to the power density, how fast the energy can be created per second. It is like the burst power in weight lifting. The capability of the storage refers to the range of an airplane, how long and how far it can fly. The capability of a generator refers to the power and thrust, whether the airplane can takeoff and fly fast. It is very important to clarify these concepts clearly when we are developing the VTOL machine, which must be capable of both horizontal and vertical capabilities on the same machine with a rotatable wing, which has not yet fulfilled so far in any aircrafts, and which is our goal illustrated in this book. We therefore draw their connections in Fig. 10.2.

Below, we classify the three main compositions of aviation electricity power grid: generator, battery, and supercapacitor with their weight factors and other aviation market concerns, such as technology maturity, economy, and efficiency. Then, we examine the needs from the AIAA's perspective. Afterwards, we adapt these needs from IEEE's perspective.

10.1.1 Energy, Power, and Weight

Figure 10.3 classifies a multi-dimensional matrix containing 5 families of parameters. Each interactive connection of items on these 5 axes may establish a distinct story. For example, the mature GE's LM6000 gas turbine generator using kerosene as fuel can generate 4000 Wh energy per weight of 1 kg, such a feature can be made use of when developing the LTG lightweight turbine generator. This single sentence links "Gas turbine → Maturity → Kerosene → Power → Weight" in the 5 axes together, involving a comprehensive story associated with many technical data, technology details, and engineering endeavors. The other example is the Li battery connecting the

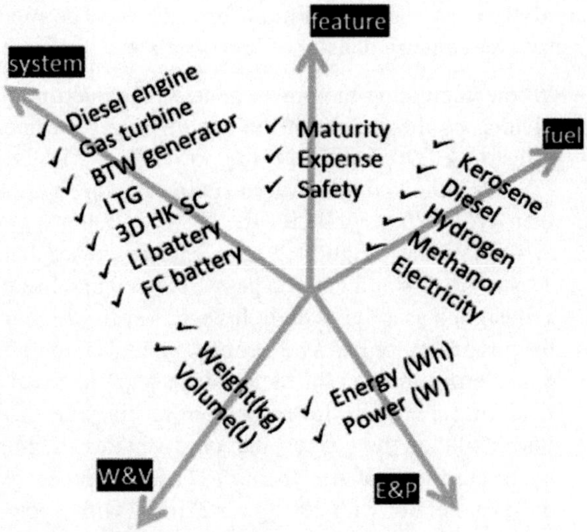

Fig. 10.3 Matrix of the five categories relating to the aviation electric powers

"System, Feature, Fuel, E&P, W&V" together: the Li battery, which uses electricity as a fuel (recharging) and has an *E* density of 200 Wh/kg and *P* density of 450 W/kg, is a mature yet less safe technology but can be utilized as a power source in the early development of e-airplane. Below, we try to analyze these interactions in details, intending to draw some meaningful conclusions for future research.

Ragone Plot

Ragone plot [2] is used to compare and characterize the energy and power densities of various power sources, such as batteries, super capacitors, and generators. The positions of a power source located on the Ragone plot is the indication of the strength and endurance of its energy. There are two ways to plot this chart:

(1) Considering the weight of the machine and fuel or
(2) Just considering the fuel.

There are two types of energy media: charging type (lithium battery and supercapacitor) and fuel type (fuel cell, diesel generator, and gas turbine). For the first kind, the energy and power densities are dependable on the weight of the equipment. For the latter, the power and energy density are variable (for example, depending on how much fuel the airplane carries).

Ragone Plot Evaluation

(1) Ragone plot considering both equipment and fuels

There are three kinds of energy sources: battery, capacitor, and generator. There are two main characteristics: how strong and how far. In academic terms, the power is that you can provide enough to take off an airplane, and the energy is the range of an airplane with the maximum fuel carried. The method of calculating the power density and energy density of kerosene are as follows:

- When calculating the power density, the maximum thrust or maximum power is divided by the weight of the engine or gas turbine. For example, the maximum thrust of GE90 is 380 kN, the power $P = TV$, the speed at takeoff is 270 km/h (75 m/s), the weight of the aviation engine is 6 tons, and the power density is then 4750 W/kg, while for the GE LM6000 gas generator, the generating power is 40 MW, the weight is 7 tons, and the power density is 5700 W/kg. It is fairly easy to understand that the gas turbine engine has a higher power density since it is designed as a lightweight high-power machine to drive the airplane. Although the gas generator has a higher energy conversion efficiency, it is designed without considering the weight factor since most of them are operating on the ground, rendering a slightly lower power density. The diesel generator is certainly less powerful than the gas turbine, but its mature technology can be quickly adopted by the current airplane. Boeing 787 used the most advanced weight-efficient VSVF diesel generator with 200 kg for 220 kW with a power density level of 1100 W/kg.
- When calculating the energy density, Specific Fuel Consumption (SFC) from official website [3] data are used. For example, the GE's LM6000 gas generator has a SFC of 0.210 kg/kWh. For the aero-engine, the SFC is 55.44 kg/kNh. Take 1 N = ~ 0.25 kW at navigation speed of 900 kmh. The equivalent SFC is 0.2218 kg/kWh, slightly higher than LM6000. The SFC of the diesel generator is ~ 0.2 kg/kWh [4].

Considering both the machine and fuel, the energy density evaluation of Ragone plot is shown in Fig. 10.4, which is a variable depending on how much fuel the system carries. Such situation applies for kerosene generators and fuel cells only. Obviously, the more fuels, the higher the energy density. That is why in the Ragone plot, the energy density for the fossil fuels is in a range instead of just a point. We use a fuel weight of 1000 kg as the starting point in the range, the final point in the range is its SFC^{-1}. For example, Boeing 787 carries 100 tons of fuel for the whole journey, and the two aero-engine only weighs 11 tons. In this case, the energy density is very close to the reciprocal of its SFC. For the gas generator, its energy density is slightly higher since it is designed to minimize the energy loss, but it is much heavier than the gas turbine engine. The weight of the generator is from 7 to 100 tons, rendering a very low energy density of 433 Wh/kg as a starting point with just 100 kg kerosene. However, in the long run, the energy density is also close to its SFC^{-1}. Considering the weight factor, diesel generator has a greater advantage since it can be built very lightweight. For example, Boeing 787 uses 6 diesel generators [5, 6] and each weighs 200 kg to generate 220 kW power. With 1000 kg fuel, and its starting energy density is close to 4000 Wh/kg. Diesel engine is less powerful than gas turbines, rendering a lower specific power.

The technical data of kerosene in the above estimation are as follows:

- Energy density and power density of fuel during gas turbine propulsion ~ 4000 Wh/kg and 12,000 W/kg, respectively (e.g., CFM56-3C1)

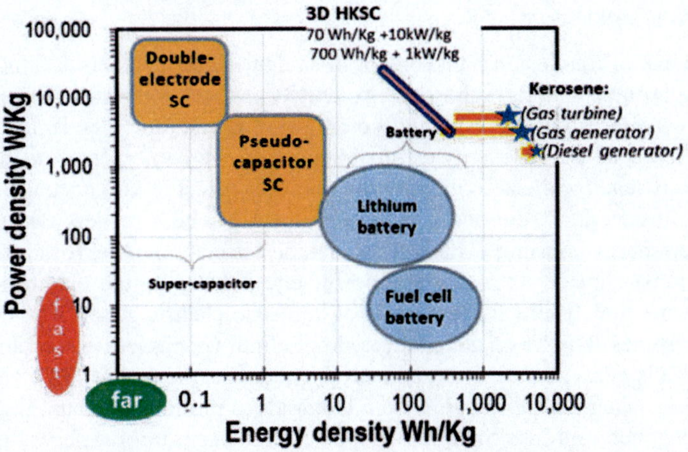

Fig. 10.4 Ragone plot showing the energy versus power of various power suppliers

- Energy density and power density of fuel oil in gas turbine power generation ~ 4600 Wh/kg, 6000 W/kg (e.g., GE's LM6000)
- Energy density and power density of diesel generator ~ 5000 Wh/kg, 1100 W/kg (e.g., Boeing 787, Generator Source, LLC, Colorado).

The Ragone plot shown in Fig. 10.4 basically tells us what kinds of electric energy can be used for aviation power and what their capabilities are. In general, the energy density levels of the battery/SC are still far from comparable to kerosene fuel. Although the current battery's energy density can meet small e-aircraft, it is still not sufficient to meet the need for larger aero-vehicles. The coexistence of the hybrid power of the gas turbine generator and the battery/SC will still dominate the future electrified aviation industry for a while.

(2) Ragone plot considering equipment fuel separately

The second type of Ragone plot only uses the reciprocal of SFC as the energy density. According to previous SFC data from official websites, the specific energies in weight are 3571 Wh/kg for gas turbine aero-engines [7], 4761 Wh/kg for gas turbine generators [8], and 5000 Wh/kg for diesel generators. They are just the maximum of the energy density when the fuel weight is much greater than the equipment, i.e., the GE's LM6000 ground electric generator house feeding constantly with the non-stopping kerosene feeding. The power density is the indication of the ability of the machine to generate power and is only depending the weight of the machine and indifferent to the amount of kerosene. Therefore, the second kind of Ragone plot is shown in the "STARs" signs in Fig. 10.4.

It is fairer to use the first type of Ragone plot if the weight of the machines is comparable to the fuel carried. The second Ragone plot is only applicable for a very long journey plane carrying many fuels or the ground electric generator power plant, which is constantly running with the streaming kerosene fuel.

Ragone Plot Explanation

(1) The general conclusion is that the fossil- and bio-kerosene fuels are still the main energy source with its highest energy density and power density. However, there are two interesting things to mention in this Ragone plot. One is that there are two kerosene power/energy estimations due to the two types of operations: to use kerosene for the aero-engine (the output is thrust in kN) or to use it for the electric generator (the output is power in kW), which renders two estimated power/energy densities. The other is that the estimation of the fossil fuel generator must consider the weight and efficiency of the engine together with the kerosene fuel. There are two kinds of kerosene electric generators: diesel and gas turbines. The diesel generator is simple/light yet has less power, and the gas turbine is more powerful yet too heavy. Therefore, the estimation incurs a long span of power density ranging from hundreds to millions of watts. Lightweight diesel generators are simple yet less powerful than gas turbine generators. Some company [9] claims an electric generator with 1000 Wh/kg and 500 kW/kg equivalent energy density and power densities, yet its accurate data should be calculated based on the total weight of fuel and engines, and the total weight of the kerosene carried on an aircraft depends on the capacity and the need. The simple diesel generator shall be used as a transition at this moment before the lightweight gas turbine generator is ready.

(2) Three-dimensional supercapacitor technology has the distinct technical superiority of high-power density, pure electric, no chemical process, and no combustion process. It can help the aircraft leave the airport and ground quickly and quietly, either takeoff vertically or make the takeoff distance shorter, which cannot be replaced by any other electric power such as battery or gas turbine. A 3D high-K capacitor is a physical-type supercapacitor that can provide burst-mode power to drive the 900 kg-thrust RDF jet. In general, there are two types of SCs: chemical and physical. The working principle of a chemical SC is similar to a battery but with a larger power density. The weakness of chemical SC is the shortage of energy density, i.e., the burst power cannot last long, even for a few seconds. The 3D high-K capacitor is able to meet both the power and the energy need simultaneously. The estimated energy density can achieve 700 Wh/kg better than current Li ion or FC batteries. 100 kg SC can provide 70 kWh energy which already matches the Tesla's Model 3 EV car (top model in 2021) with the mileage of 700 km. 70 kWh electric energy releasing can offer at least 1-h flight for small or medium-size airplanes. Optionally, the SC can be built as high-power type which provides 10 kW/kg maximum power density (higher current output) with energy density of 70 Wh/kg. With 100 kg SC the total power 1000 kW which is sufficient to drive the RDF jet to achieve the 900 kg thrust. In this scenario, the total energy of 100 kg SC is 7000 Wh which can provide the burst power of 1000 kW for nearly 30 s (7000 Wh/1,000,000 W = 25.2 s), sufficient to aid 50-ton airliner for STOL by leveraging its takeoff weight and shorten the takeoff distance by 1/7–1/5, saving a lot of airport resources. If

we use the current supercapacitor in Fig. 10.4, this high-power output can only last 1 s, not enough to support the VTOL or the STOL.

(3) Fuel cells and lithium batteries have high-energy density, but the power level is just too low. The battery is not able to meet the high-power need to support large airplanes, although it is powerful enough to meet small-scale lightweight AAMs such as 2-passenger aircraft. For example, a 100 kg Li battery with an energy density of 200 Wh/kg can provide 20 kWh for a 2-passenger airplane, which is sufficient to navigate for 3 h. However, its maximum power is just 100 W/kg * 100 kg = 10 kW, which is too low to drive RDF rendering enough thrust.

Table 10.1 is a list of the current popular aviation electric power supplies together with their pros/cons in their performance such as energy and power densities, energy conversion efficiency, environmental friendliness, technology maturity, and manufacturing expense.

In summary, the kerosene fuel in general has a much higher energy density level, which is still the only option for long journey aviation power sources. The advantage of the 3D HK SC becomes more obvious from its power density point of view since the supercapacitor does not involve any heavy mechanical parts. This will greatly enhance the takeoff performance by providing just one minute burst power. Once taken off, this supercapacitor can be recharged by the fossil fuel electric generator. The 3D HK SC does not involve any mechanical and combustion processes, hence reducing the airport noise and shortening the takeoff distance. Such a distinct feature

Table 10.1 Electricity suppliers' pros and cons

Type (device or fuel)	Energy density	Power density	Eta	Green	Maturity	Cheaper
Gas turbine generator	++	++	++	−−	++	+
LTG	++	++	++	−	−−	−
Diesel generator	++	−	−	−−	++	++
Li ion battery	+	−	+	−	++	+
Fuel cell battery	+	−	+	++	−	+
Chemical supercapacitor	−	+	++	++	−	−−
Physical supercapacitor	−−	++	+	+	−	−−
3D HK supercapacitor	++	++	++	++	−−	−
Kerosene	++	++	+	−−	++	+
Methanol	+	+	++	++	−−	++
Hydrogen	−	−	+	+	−	−

Notes (1) Bold denotes hot spots for promising technology. (2) "+" is pro, "−" is con. (3) *LTG* lightweight turbine generator is not compatible by other electric resources in the Ragone plot. Its disadvantage is its insufficient energy for very long journey flight.

Energy Density–Volume Versus Weight

There are two types of energy densities or specific energies: the amount of energy per volume (Volumetric) or the amount of energy per weight (Gravimetric). The preference for specific energy in volume from aviation's perspective lets the gas turbine choose kerosene instead of gasoline as fuel. In most cases, volumetric energy density is similar as gravimetric energy density for most of the liquids fuels. For hydrogen gas tank the situation is very different. That is why the fuel cell prefers liquid methanol as a fuel instead of the compressed hydrogen air tank.

The Kerosene Versus Gasoline

The automobiles vehicles use gasoline, heavy trucks use diesel, why do airplane use the kerosene? The reasons lie in the optimized choice of the energy density. The following concerns are listed in Table 10.2. The energy density in volume is higher for kerosene (35 MJ/L compared to 31 MJ/L of gasoline), while the energy density in weight is 44 MJ/kg for gasoline and 43.28 mW/kg for kerosene, which are slightly higher. Obviously, although the aircraft is sensitive to weight, it has higher requirements for smaller volume. In addition, kerosene is much more stable than gasoline. On the other hand, diesel is too viscous. Therefore, considering the overall performance, the best choice for aviation fuel is kerosene.

Liquid Ammonia

Ammonia has been produced and utilized for the past 100 years as a fertilizer, chemical raw material, and refrigerant. It was recently reported a few successful applications of ammonia as fuel in gas turbines and in industrial furnaces [10]. This suggests that ammonia could be an alternative to replace fossil fuel, although several challenges in ammonia combustion, such as low flammability, high NO_x

Table 10.2 Comparison of gasoline, kerosene, and diesel

Fuel	Gasoline	Kerosene	Diesel	
Calorific value (MJ/kg)	43.71	43.28	42.65	
Density (kg/m^3)	720	810	855	
Flash point (°C)	– 45	38	55	
Viscosity (mm^2/s @30–40 °C)	0.55–0.69	1.0–2.0	3.9	

emission, and low radiation intensity needs to be overcome. NH_3 has an energy density of 3.5 kWh/L, similar to gasoline. The specific gravity of liquid ammonia is 0.771 kg/M^3, also similar to that of gasoline. The biggest advantage of synthetic ammonia is that it is much easier to store than hydrogen [11]. Liquid NH_3 can be stored at − 33 °C under normal pressure or 9 atmospheres under normal temperature, and the requirements for storage containers are not high [12]. The production of ammonia can only rely on water, air and electricity, which means that theoretically, it is as green as hydrogen.

Despite the above advantages, the ammonia production process is mature yet, with problems such as poor efficiency, unstable reaction process and unclear economy. The reaction mechanism for ammonia combustion is not completely clear, and the key parameters such as combustion speed, flame structure, ignition delay, and pollutant formation are still imperfect [13, 14]. Therefore, the manufacturing technology of ammonia as well as its combustion mechanism need further scientific research. Once it has made great progress, it will play a great role in aviation industry.

Fuel Cell: The Volume Versus Weight

Although hydrogen has a very high-energy density per weight, it has a very low specific energy in volume. The hydrogen gas is very light (0.89 g per L), its gravimetric specific energy (Wh per kg) is 14 times higher than its volumetric density (Wh per L). To carry more hydrogen fuels, it is usually compressed to 35 MPa and 70 MPa, corresponding to ~ 21 g/L and ~ 42 g/L, respectively. With 70 MPa, the volumetric energy density improves 43 times, rendering an ~ 1000 Wh/L energy density with a 30 L tank (gasoline tank) holding 1 kg of hydrogen gas. The weight of 26 L (1092 g of hydrogen) type IV 70 MPa quantum tank weighs 24 kg [15], rendering an equivalent gravimetric energy density of 1454 Wh/kg[1] (the energy density is 33,427 Wh/kg for pure hydrogen gas). Regarding the mileage, 5 kg hydrogen is necessary to be compatible to gasoline cars. Quantum's 118 L tank of 73 kg holding 5 kg hydrogen with an energy density of 1183 Wh/L will satisfy the need. However, the tank size is almost four times larger than that of ordinary gasoline automobiles. On the other hand, enlarging the tank does not help the energy density too much. A 26-L 24 kg tank weight holding ~ 1 kg hydrogen has an energy density of 1142 Wh/L. With the increased tank size to 5 times (118 L), the energy density becomes 1183 Wh/L, just a slight improvement.

The above volumetric energy density of compressed hydrogen may satisfy the need for electric cars but may not be suitable for airplanes. The energy per volume with 70 MPa compressed hydrogen gas is only 1/10 of kerosene—far from being comparable to the current aviation need. The physical storage of hydrogen is just too bulky.

The Power Grid for Electric Aviation

---

[1] The energy density of H_2 is 33,427 Wh/kg. The weight of the type IV 70 MPa quantum tank weighs 24 kg. The tank holds 26 L of H_2. The 26 L of 70 MPa H_2 weighs: 26 L * 42 g/L = 1.092 kg, the energy carried is: 1.092 kg * 33,427 Wh/kg = 36,502 Wh. The total weight (tank + H_2) is 24 + 1.092 = 25.1 kg. The equivalent energy density is 36,502/25.1 = 1454 Wh/kg.

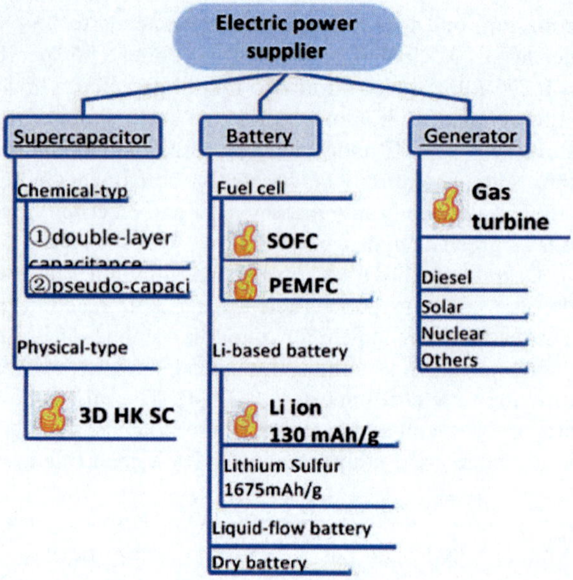

Fig. 10.5 The three families of mobile electric resources and five promising power sources (thumb)

Figure 10.5 outlines the three key electric sources for the aviation power grid. There are few hot spots (thumbs in the chart) suitable for aviation need among the three sources.

- Lithium-ion battery technology is the only mature technology that can be used for electric aero-engines, since the Li batteries have been fully adopted in the current EV market such as Tesla and BYD. The mileage of electric vehicles is already comparable to ordinary fossil fuel automobiles. From an aviation perspective, Li-ion batteries have two disadvantages: one is safety and the other is insufficient power density. The latter shortcoming is attributed to its first principle of chemical batteries; a liquids chemical reaction process is slower than a combustion thermal chemical process from fossil fuels. The safety issue is an intrinsic weakness of lithium-ion batteries, but we can still use it to verify the new technology of electric aviation, such as our RDF jet concept. In terms of sustainability, fuel cells (FCs) is the best green energy with no environmental contaminations, but their technology is far from being matured.
- At present, diesel and gas turbine fossil fuel electric generators still possess the highest energy and power density. To ensure the enough range, the hybrid gas generator must be used as a navigation fuel, only the battery is not enough. Developing lightweight gas turbine generator (LTG) is the incoming target to provide hybrid electric power for aviation propulsion.
- It is also highly necessary to develop a "quiet" pure electric power—the 3D HK supercapacitor. It has both great power and energy density. It's pure electricity with no chemical process involved and has no combustion process. It aviation

advantage is obvious: either the VTOL or the STOL relies on the sufficient long-lasting (> 30 s) burst power for quiet and 0 pollution takeoff, which cannot be replaced by battery or generator.

The "coupling" of the LTG and 3D HK SC provides the best configuration for the next generation electric power grid to ensure both the mileage and takeoff performance. Hence, there are two distinct approaches in parallel to achieve the best electricity for aviation in the Ragone plot: to develop a lightweight gas generator together with the 3D high dielectric supercapacitor.

(1) LTG is a joint effort from two groups of engineers: the gas turbine engine and gas turbine generator. The main focus of aero-engines is thrust and lightweight, while the target of turbine generators is mainly the maximum power and efficiency but ignoring the weight factor. The key is to balance the power and the weight. The balancing point could be the 100–2000 kg generator generating 1–10 MW electric power. Both the scale and configuration of the machine must be re-engineered. For example, one can use the GE's LM6000 heavy electric generator of 35.3 MW with 97 ton weight [16] and GE's CF700's lightweight gas turbine engine of 2 ton thrust with 333 kg weight [17] as prototypes to develop a dedicated gas turbine electrical generator. This can be achieved by skipping the front big fans, reducing the levels of turbines, and shrinking the overall size and weight.
(2) Different from LTG which greatly involves AIAA people, it is a purely IEEE's work to develop 3D supercapacitor. Although 3D HK SC has its distinct market place in IEEE such as the mobile power for laptop computer and other mobile devices, when we are using it in AIAA, the main focus is the high-power output. For this reason, there are two distinct options for the 3D HK SC: large energy density mainly for longer navigation range, and high-power density (larger electric current) mainly used for the burst thrust of vertical takeoff. Technically, the larger current density is ensured by thick metal layer which allows larger current to flow through, but of course, this will limit the numbers of layers of the capacitor which renders a lower energy density (the slant line in the Ragone plot in Fig. 10.4). The dedicated effort to develop the 3D HK SC will be elaborated in detail in a later chapter.

10.1.2 From AIAA's Perspective—Power, Energy, and Safety

In general, what the AIAA fellows care is the electric powers to match these three key criteria:

- Energy—how far and how long can the airplane go;
- Power—whether able to lift the plane vertically;
- Misc.—including (1) the weight and size; (2) the safety, reliability, endurance, etc.

To fulfill the need for:

- Car-type: 2-passenger airplane
- Van-type: 19-passenger airplane
- Bus-type: 50–400 passenger airplane.

Regarding the criteria, power refers to the rate to generate the energy that is crucial to accelerate the aircraft for horizontal takeoff or just to lift it up (VTOL). The energy refers to the distance and duration of the aircraft's traveling. The thrust is the key related to the power, the kerosene fuel and the amount of electricity stored in the battery is the energy.

As to the needs, car-type airplanes refer to the category of 1–2 seater small airplanes, including some catering airplanes, rescue aircrafts, and various AAM scenarios. Van-type aircraft should be the most market-efficient air flight and can be flexibly adapted to best fit of air flow for local/domestic traveling. Multiple mid-size van-flights is more effective than a big 400-pax flight even with less occupancy. Bus-type commercial airliners are still necessary for long journey international flight and yet electric propulsion can still help for STOL in takeoff or thrust reversal propulsion for landing (elaborated in Chap. 7; Sects. 7.3.3 and 7.3.4).

The above need and criteria are matched in Table 10.3. There are three types of aero-vehicles: car, van, and bus type airplanes, corresponding to two passengers, 20 passengers, and over 50 passengers' aircrafts. For the first two types of airplanes, the VTOL can be achieved by multiple electric engines. For the big commercial aircraft, it is impractical to achieve the VTOL, but we can do the STOL, i.e., to use the electric engines to leverage the MGTOW in order to shorten the takeoff distance. For the first two kind of aircrafts we can use batteries and super capacitors, while for the bus-type airplane gas turbine machines and kerosene fuels must be used. Although the small airplanes can be purely electric driven with the batteries or super capacitors but its range is limited. The total energy of a commercial aircraft (A320) depends on how much fuel it carries, and its power depends on its engines such as CFM56' LEAP-X1C. Its kerosene fuel can also be made use of by the car- and van-airplanes to increase the range by the aid of lightweight gas generator. Small planes can be built with VTOL fitting in many AAM scenarios such as a helicopter type of 3 × 3 rescue aircraft, despite the fact that the only existing in field test models for the car-planes are RR's Spirit of Innovation or Airbus's E-fan, both are still using the traditional propellers driven by electric motors and the Li-ion batteries without VTOL capability. The van-type aircraft with long mileage by hybrid power and driven by RDF jets with VTOL features should have a bigger marketplace.

Considering both the feasibility and necessity, the feasible near-future electrified aviation market is hybrid, which may last the next 20 years. The comparative analysis is listed in Table 10.4 for the few options of the hybrid aviation scenarios. The three future promising markets of the hybrid electrified propulsions (with three smile faces) are (1) STOL, (2) VTOL-type RDF Jet propulsion for van-type 20-passenger aircraft, and (3) drone-type AAM aircrafts.

For the 1st configuration, the STOL can be achieved by the aid of a 3 × 3 RDF jet configuration attached to the commercial airliner. During takeoff, the RDF jets are

Table 10.3 Electric power for 3 types of aircraft

Aircraft	Criteria						
	Energy	Power	Thrust	Weight	E.g.	Market	References
Car: 2-passengers private aircraft	70 kWh	40 kW	100 kg	500 kg	SoI E-Fan	**AAM**	[18, 19]
Van: 19-passengers urban aircraft	10 MWh	2.5 MW	10 ton	9 ton	Sweden ES-19	**Hybrid**	[20]
Bus: 50-passenger commercial aircraft	< 1000 MWh	< 30 MW	16 ton	77 ton	A320	**STOL**	[21, 22]

Table 10.4 A few markets for electrified propulsion

Propulsion \ Features	Energy	Power	Efficiency	Maturity	Sustainability
RDF jet: STOL + BTW + 3D HK SC (A320)	++	++	+	– –	+
RDF jet: VTOL + LTG + 3D HK SC (20-seater)	++	+	++	– –	++
RDF jet: Drone + LTG + 3D HK SC (AAM)					
Popeller + Diesel Generator	– –	– –	–	++	– –
Popeller + FC Battery	–	—	+	—	++
Popeller + Li Battery	– –	– –	+	++	+

powered by a 3D HK SC. After taking off this supercapacitor, it can be recharged by the by-the-way gas turbine generator during navigation. These distributed RDF jets can manipulate the airplane during flight in addition to main stream gas turbine propulsion. For example, an RDF jet can provide the jet to the front to help the airplane decelerate during landing and achieve a short distance landing—saving airport resources.

For the 2nd configuration, the 20-seater mid-size aircraft can be vertically lifted and then redirect the RDF jets gradually moving forward. The 3D HK SC can help this VTOL and then start the LTG for longer navigation range and recharging the supercapacitor. This greatly reduces the ground pollution and airport takeoff noise. The 20-passenger van-type airliner allows more flexible and economic controls of passenger flow compared to the big liner of 400 passenger with just half occupied seats in the flight.

For the 3rd kind, delivery aero-vehicles play an indispensable role for the Internet of Things. The IoT greatly enhances the national flow of merchandise and people, making the country more energetic and prosperous. The drone delivery aircraft only carry the goods during the whole journey—making the aircraft loading more efficient and lessening the safety cost. The VTOL and navigation principles are similar to the 2nd configuration except for the more advanced UAS control of the drone aircraft operated from the ground station. Drone delivery is a huge market place for the domestic and local merchandise communication.

For the rest of the propulsion alternatives in Table 10.4, they are either short of technology maturity (e.g., fuel cell) or technology advantage (e.g., propellers). These alternatives can be used as references when developing the new aircrafts.

Safety

When talking about the safety issues, one has to separate the R&D endeavor with industry marketing—the latter has a much more stringent requirement on safety, especially for the mid-size and large-size commercial airliners. One also needs to separate the UAS and PAX-airplanes. For R&D development work, the main focus is on the verification of the new concept/principles/technologies. which has a bigger allowances on trial-and-errors.

Aviation requires more stringent quality control on the safety than those used in EV cars for obvious reasons. From this perspective, the lithium battery is less advantageous [23] due to its intrinsic chemical process during charging/discharging. For the aviation purpose, the power requirement is higher than the automobile esp. for VTOL purpose, which needs high current output. The high-power output for aero-propulsion exacerbates this safety concern since most of the safety issues related to the high-power use of the Li batteries. At present, scientists and engineers have not been able to fully understand the failure mechanism and effectively evaluate or predict its failure rate and reliability [24]. In other words, one cannot detect whether the safety issue is random failure or due to the poor quality control that exhibits the systematic fall-out. The latter is controllable with tight manufacturing control, while the former involves uncertainty (e.g., local overheating) for unknown reasons. Fuel cells may be more suitable to provide electric power for aero-engines until the failure mechanisms of Li-based rechargeable batteries are fully understood so that prevention measures can be implemented. Nevertheless, the more competitive fuel cell of higher energy and power densities with the competitive price still needs more work from IEEE people. More safety issues will be further addressed in the "Battery section" in this chapter.

10.1.3 From IEEE's Perspective—Battery, SC, LTG

When working on an electrified aviation project with AIAA fellows, the IEEE people should always keep in mind of the above three crucial criteria. The pros and cons, together with the relevant references in Fig. 10.5 are compared in Table 10.5, the "+" signs show the advantage, "–" is the weakness, and bold indicates its current maturity.

Table 10.5 Comparison of battery, super capacitor, and generators suitable for airplanes

Type	Criteria							References
	Power	Capacity	Economy	Efficiency	Wt/Vol	Safety	Maturity	
Battery	+	+	++	+	++	++	++	[25]
Fuel cell (PEMFC) Fuel cell (SOFC) Li-battery (Li-ion) Li-battery (Li-Sur) VRB flow battery	++	+	−	+	++	+	−−	[26]
	+	++	++	+	+	−	++	[27, 28]
	++	++	+	+	+	−−	−−	[29]
	+	+	++	++	−−	+	+	[30–32]
Super capacitor	++	−−	−−	−	−−	++	++	[33]
SC (Chemical) SC (Physical)	++	++	−−	−	+	++	−	[34]
Fossil fuel	++	++	+	++	−	++	++	[35–37]
Gas turbine hybrid Diesel generator	+	+	+	−	−−	++	++	[38]

Battery

At the current stage, the battery is the best matured electric energy, taking advantage of the concurrent advancement in Li-based batteries used EV cars. The basic principle of battery is the "quiet" chemical reaction that separates electrons and ions via an oxidization/deoxidization process. This chemical process has an intrinsic limitation on power comparing to the chemical combustion, i.e., the amount of energy released per second is much less. The low power can satisfy the need to drive some AAM applications, but it is difficult to satisfy the VTOL/STOL for larger aircraft.

Mainly, there are two types of batteries which can be used for the aviation purpose. Li-based batteries, also called "rocking chair batteries", where Li ions flow back and forth between two electrodes during the charging and discharging process. The fuel cell battery uses hydrogen or methane as a fuel to flow to a catalyst plate where the separation of electrons/ions occurs. The electrons flow outward to provide electric current, and oxygen flow provides the re-oxidation process to neutralize hydrogen ions into H_2O. Logically speaking, the fuel cell is the most ideal energy source, which provides environmentally friendly clean energy and creates little waste. Increasing the efficiency of the hydrogen flow and the use of catalysts in the electrolyte membrane are two keys to improving the performance of proton exchange membrane fuel cells (PEMFCs).

In addition to these two major types of batteries, NASA proposed the concept of an nanoelectrofuel (NEF) liquid flow battery [39], claiming its blueprint of achieving a higher energy density of 530 Wh/kg, 3 times that of the current Li-ion battery (125 Wh/kg), together with its integration with RDPs. However, no proven cases of RDP propulsion and the powerfulness of lightweight NEF have been reported since 2018.

Supercapacitor

The second mobile power is the supercapacitor (SC). SC is much more advantageous for its rapid charge/discharge and burst-mode power delivery—very suitable to provide an immediate high power for aircraft take-off. Ordinary SC is still the chemical type that operates similar to a battery—the exchange of electrons/ion via the oxidized/deoxidized process. The main difference lies in the fact that SC has a similar outer performance as a capacitor. Unlike the operation of a battery, the electron/ion separation process of SC is more reversible, and the charging/recharging time is much faster (the adsorbed ions do not react with the atoms on the electrode rather than the charge transfer). However, the energy density of ordinary SC is too low compared to the batteries. With its current energy density, the SC can only provide the burst power for less than one second, which is far more sufficient for the VTOL purpose.

In this chapter, we propose a new type of SC—3D multi-layer super dielectric supercapacitor, which operates based on the first principle of physics: $C = \varepsilon\frac{S}{d}$. We use the direct method to make a capacitance "super" by just increasing ε (from ~ 10 to ~ 10,000 [34]), reducing d (from μm to nm by using the magnetron sputtering PVD process) and enlarging S by 3D stacking the layered MIM structure.

The manufacturing of this 3D-packed high-K capacitor is a standard MEMS microfabrication process involving a combination of thin-film depositions plus simple photolithography operations.

Lightweight Gas Turbine Generator

The third option is the turbine gas generator specifically built in lightweight designs. A lightweight aero-generator (LTG) is re-engineered from a conventional gas turbine engine such as: GE (GE's LM2500 [40]), PW (TPM's FT8 [37]), SIEMENS's SGT-800 [35], and Rolls Royce's RB211 [41], where the experiences can be borrowed to fit in lightweight/medium power output aviation scenarios. Developing a LTG is a new project that is different from either the traditional gas turbine engine or the ground/marine electric generators. The main focus of engines is thrust and light weight, while turbine generators are mainly seeking for high efficiency in generating electricity but ignore the weight factor. The feature of an aviation gas generator is to generate electricity with light weight. Both the scale (no large fan) and configuration (e.g., fewer levels of turbines) of these engines must be re-engineered by balancing the electric power and the weight.

For example, one can use the CFM56-3C series commercial aero-engines as a prototype to develop a dedicated gas turbine electrical generator by skipping the front large fans and shrinking the size and weight. It is possible to create a dedicated lightweight 10 MW 1 ton turbine generator to provide the 10^4–10^5 W/kg power density and over 10^3 Wh/kg energy density for 9 RDF Jets to provide 8.1 ton thrust.

Electrified aviation technology is disruptive, disruption needs transition. Hybrid use of turbine engines together with batteries/super capacitors will coexist for a while. Looking back on the evolution of the automobile industry, starting from the first hybrid car of Toyota's Prius as early as 1997, it took over 20 years to transition from fossil fuel to the fully electric automobile. Currently, many countries are starting to adopt pure electric cars, while hybrid and fossil fuel vehicles will soon become history. Likewise, the hybrid power period using both electricity/fossil fuel is an inevitable process, although the transition may be faster in the aero-engine electrification roadmap.

From the IEEE's point of view, there are three basic questions need to be asked: where is the electricity from, how can they be stored, and how can they be transferred? It must be mentioned here that the electric vehicle (EV cars) in English is not translated as it is in Chinese. In Chinese, the proper translation should be NEV, the new energy vehicle. The reason is that the EV car is actually taking the advantage of the green new energy (solar, wind, water) as electricity instead of using the fossil fuels which is not sustainable. Electricity is just a media to create, store and transfer this new energy, fossil, coal, sun light, tidal and wind are the sources of the energies.

It is more suitable to replace the word EV (Electric Vehicle) with NEV (New Energy Vehicle)

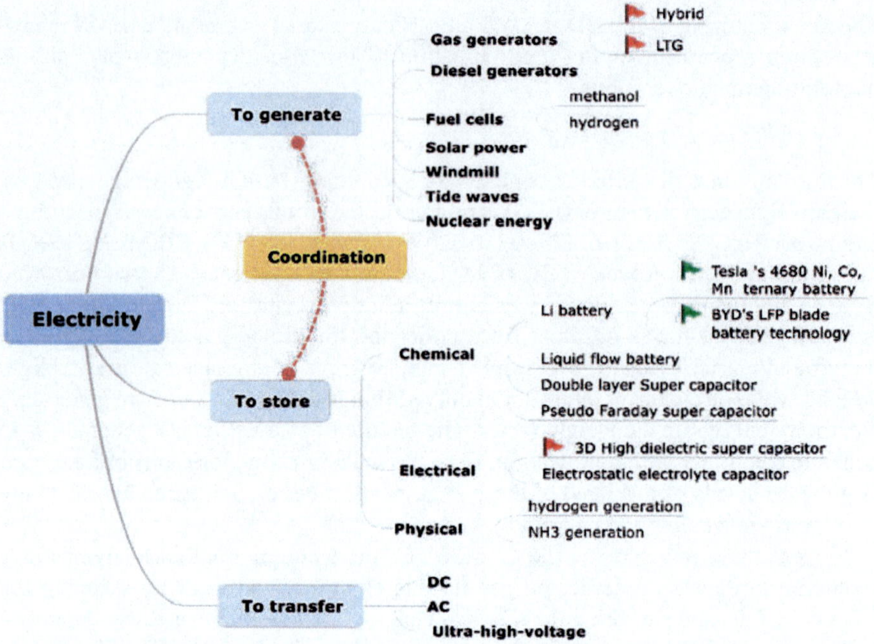

Fig. 10.6 The electricity network from various facilities regarding the creation, storage, and transfer. The three red flags are the hot spots of the concurrent aviation electricity, and the two green flags are the hot spots of electricity of the current electric cars, which may be adopted for AIAA

The power grid configuration is shown in Fig. 10.6. Electricity as an energy power source includes:

- Power generation: converting other forms of energy into electrical energy. This is similar to converting a seed into a vegetable.
  Fossil fuel is still the main source of the global electricity. Yet the three "free" energies (light, wind and water) are the future mankind sustainable energies, i.e., to use the solar cells, windmills, and hydraulic/tidal energies to replace the thermal powers from oils and coals.
- Power storage: store energy in some form and release it when needed. This is similar to the refrigerator.
  Refrigerator is needed to stop the vegetables to save our times for shopping. Likewise the amount of electricity generated by the power plant may not be always the same as need of users. Battery is needed as a buffer to modulate the vendors and providers. The liquid battery is a very important technology to store this large amount of the electricity.
- Power transmission: transfer the electric energy from far away power plant to the destination—less loss and stabilization are two keys for this transmission between the power supply and the load.

In this field China did a great contribution by its advanced super high voltage (> 5000 V) electricity transmission technology and it's already been in use to transplant the abundant solar, windy, and hydraulic energy from its far West to its highly populated East.

Some technical details of these three electricity families are also illustrated in Fig. 10.6.

Although there are many ways to generate electricity, there are only a few options that are suitable for aviation purpose as shown in the red flags and green flags:

(1) The first red flag is the by-the-way gas turbine electric generator. It is a BTW generator that makes use of the rotating shaft to drive permanent magnet to rotate inside the electric coils for electricity for the supplement RDF jets on the current commercial airliners. This BTW generator does not cost too much by some modified configuration just behind the big turbo fans on modern gas turbine engines such as GEnX with added small electric generator supplement. To transfer the transmission energies from the turbine to the front fan via the shelf was already a clever idea to make full use of the turbine energy. To make use of this energy to create BTW electricity is another clever idea. The front fan contributes nearly 80% of the total thrust with a larger BPR ratio but consumes most of the shaft energy by conquering the air resistance on the fan blades. In comparison, the rotation of the magnets to create electricity adds up only a small amount of mechanical power on the shaft when it drives the large fan to create duct jet power. There have been some previous efforts on integrated generators with general aircraft engines [42]. This project started in 1995 with a preliminary result of 350 kW electric power by integration design of an electrical starter/generator without major modifications. It did not continue further with the possible reason for the lack of sufficient necessary. It is just an R&D research topic other than a project which is aiming at the pragmatic engineering need (the electric power is too small [43] compared to 30 MW for gas turbine machines). More work is worthwhile to continue considering our new need for all RDF jets and the above BTW electrical generator of our architecture would still have its value for further engineering endeavors.
(2) The second red flag in Fig. 10.5 is the LTG—lightweight turbine generator, which is purely dedicated to creating electricity and using this electricity to drive the RDF jets. The key here is a lightweight—both small and light. Its technical details will be discussed in Sect. 10.3.
(3) The two green flags refer to the concurrent Li-ion battery achievements, which could be made use of by AIAA's R&D work. Very recently, the two big EV companies, Tesla and BYD have been developing their own advanced Li-ion batteries to improve safety and to enhance efficiency. The analogy of Tesla's approach versus BYD is like to connect 10 resistors in series or in parallel; the latter connection renders a faster and wider delivery of electrons from point A to point B. The BYD's idea is cleverer which uses different blade battery architecture to resolve the safety issues and enhance the energy density of the

Li-ion battery system. The details will be elaborated in the following battery section (Sect. 10.2.1).

(4) The third red flag is our 3D HK super capacitor, which is discussed in Sect. 10.4. This is a purely electricity device—no machine parts, no thermal process, no chemical reaction—quiet, clean, and safe.

The electricity storage is indispensable to balance the electric power plant (grid power) and the electricity users (factory, city …). There are three ways to store electricity: chemically, electrically, and physically.

- Batteries operated electrochemically have the advantage of higher energy storage density but usually have safety issues. The prices are relatively higher, and charging/discharging cycles are usually limited. Liquid batteries are safer and more economical but usually take considerable space and are very heavy.
- A supercapacitor can also be used to store the electric energy. The advantage of the supercapacitor in comparison with the battery is its very fast charging discharging ability. The recharging cycles are also much higher than those of the battery. It is just too expensive and energy density is too low.
- One can also use the redundant electricity to generate the hydrogen, NH_3 as the energy fuel for later use, such as fuel cells. The technology to generate hydrogen and energy is not matured yet. However, this alternative is much more advantageous than the previous two options.

In addition to electric creation and storage, the coordination of generation and storage is also very important for hybrid electric systems. For example, the electric power management to incorporate the 3D HK SC and LTG is highly necessary to ensure the more efficient operation of electric aircraft. For instance, one should use a battery and supercapacitor to take off the airplane. Once the airplane is in the air, the hybrid generator starts to operate to provide long-term navigation. In the meantime, it recharges the energy loss of the battery/SC during takeoff. At landing, the 3D HK SC empowers the RDF jets to decelerate the airplane to shorten the landing distance. In this way, airport takeoff noise and takeoff/landing distance can be greatly reduced and the journey of the 20-passenger aircraft light is more optimized.

10.2 Battery

The battery is an energy storage device, but not necessarily a mobile power supply. There are some slight differences between the two. Some batteries are used as energy regulating devices of electricity power grids in power plants. For example, the liquid flow battery is similar to a reservoir. It is basically an immovable energy storage device. The mobile power supply does not connect with the power grid by wires and exists as an independent device. This power supply is irreplaceable in many situations, such as new energy cars and electric aviation. As far as electric aviation

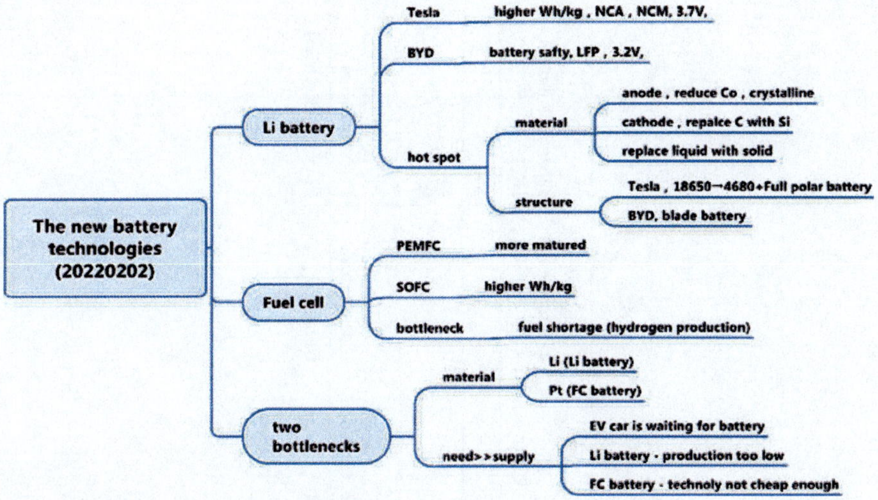

Fig. 10.7 Comparison and development of the latest battery technologies

is concerned, what we need is a mobile + electric power supply. The following map lists the mobile battery families together with their frontier developments (Fig. 10.7).

Batteries are chemical reactions in comparison to "violent" thermal chemical reactions in combustion engines. This first principle basically limits the reaction speed in batteries which is far slower than thermal chemical combustion—batteries are less powerful than fossil fuel electric generators. However, batteries have a higher energy conversion efficiency because there is no thermal process involved—avoid energy loss during heat dissipation, and there is no mechanical process involved—avoiding mechanical energy loss. Conventional supercapacitor has a comparable high-power density as fossil combustion generators but this high-power output lasts too short a time (less than 1 s, check Fig. 10.4). A 3D purely electrostatic supercapacitor is invented to eliminate the shortcoming of the conventional chemical-based SC. Such physical-operated SC possesses both high-power density as well as the energy density. Table 10.6 list the comparison of the three types of electricity media.

The battery chemical reaction involves three steps, "flow, contact, and react", i.e., a medium flows inside a system and perform the chemical reactions on the surface of the anode and cathode. The flowing media are either Li ions or the H^+ protons in the Li battery or in the fuel cell. After the chemical reactions, the electrons and ions are separated. The electrons flow outside from the anode to the circuit to provide the electric current and then flow back to the cathode to recombine with the positive ions there. The separation and recommendation process are called oxidation and re-oxidation. The loss of electrons at the anode is an oxidation reaction, the acquisition of electrons at the cathode is a reduction reaction. Oxidation refers to the separation of electrons from a substance and the release of electrons to produce current. Reduction

Table 10.6 Comparison of popular electricity resources

	Diesel generator	Gas generator	Li ion battery	Fuelcell battery	Chemical supercapacitor	Physical supercapacitor (3D HK)
Burst output power	+	++	–	–	+	++
Amount of energy stored	+	++	+	+	–	++
Energy loss	High	High	Low	Low	Low	Low
Energy storage media	Diesel, biofuel	Kerosene, NH_3	Battery	Hydrogen, methanol	Electric device	Electric device
Mechanism/process	Mechanical, thermal, chemical	Mechanical, thermal, chemical	Chemical	Chemical	Chemical	Physical, electrical

means that electrons flow from the external circuit to the negative electrode and are absorbed by the cathode to complete a redox cycle.

The battery has three main components: electrodes, diaphragm, and electrolyte.

- Positive and negative electrodes provide the media for chemical reactions on their interface.
- Electrolytes function as fluids allowing ions and electrons to flow freely between positive and negative electrodes. Traditionally, electrolytes are in liquid form. The broad meaning of the electrolyte is not necessarily a liquid. It can be a solvent electrolyte solution, solid electrolyte, and molten salt electrolyte.
- The function of the diaphragm is to prevent short circuits caused by direct contact of positive and negative electrodes while allowing ions to pass through freely. Most of the Li battery accidents occur when this thin layer got penetrated-through which shorten the positive and negative electrodes, causing high current, the burning and explosion. Proper structure design should be implemented to avoid this damage in diaphragm such as the blade battery design in BYD.

Figure 10.8 shows an example of the chemical reaction in a fuel cell. The hydrogen lost one electron to become a proton and then flowed via the electrolyte toward the negative electrode. In the meantime, the electron flows outside to the circus to provide electricity. On the other hand, oxygen reacts on the negative electrode to form oxygen ions. Hydrogen ions and oxygen ions combine to form water. This is a whole loop of the FC battery operation. The only by-product is water. Therefore, the FC is a clean battery.

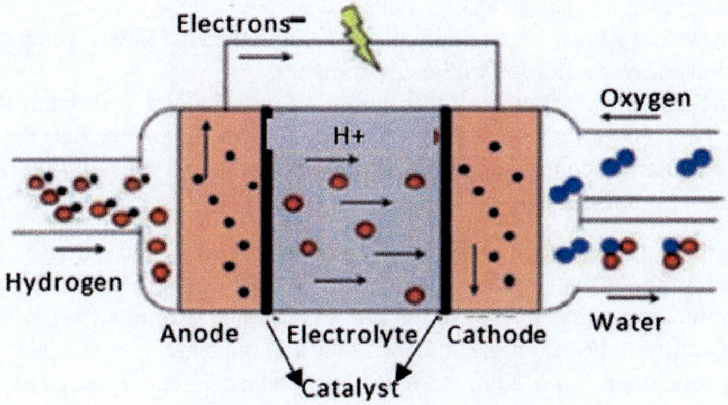

Anode : $H_2 \rightarrow 2H^+ + 2e$

Cathode: $2H^+ + 2e + \frac{1}{2}O_2 \rightarrow H_2O$

Total : $2H_2 + O_2 = 2H_2O$

Fig. 10.8 The basic physical and chemical operation of a FC battery

The chemical reaction is much slower than the thermochemical reaction produced by gas combustion, which is mainly reflected in:

Step 1. The supply of reactants is not fast enough. For example, to generate electricity, fuel cell battery must continuously feed in the fuel for oxidation. If the supply flow is not fast enough, the battery will "starve". The flow field plate combined with the porous electrode structure can most effectively help the efficient transport of reactants. But the maximum flow rate is still limited.

Step 2. The electrochemical reaction speed is not fast enough. Once the reactants are transported to the electrode, they must undergo electrochemical reactions. The faster the electrochemical reaction is, the greater the current generated. Obviously, the electrochemical reaction rate of a battery cannot be compared with the combustion process of fuel.

Step 3. Ions or electrons do not flow fast enough. Ions or electrons must be transferred from the region they produce to the region they consume. There must be an electrolyte to provide a path for the flow. In many electrolytes, ions move through the "jump" mechanism. Compared with electronic transmission, this process is very inefficient. Therefore, significant resistance loss may occur in ion transmission. This reduces the speed again.

Compared with thermal electric power generation, the advantages of chemical batteries are as follows:

(1) Efficiency. The power generation efficiency is as high as 50 ~ 60%. If it can be combined to form a circulating power generation system, its power generation efficiency can be as high as more than 70%.
(2) Green. For example, compared with traditional thermal power generation, fuel cells have lower pollution to the environment.
(3) Quiet. The electrochemical reaction does not involve mechanical/thermal machines, rendering a very quiet operation. For example, because the fuel cell has few internal components, it will not produce large noise during operation. The general noise is 50 ~ 70 dB.

The performance of a battery is usually characterized by the output voltage (V) versus energy (mA/h) at different release currents [44]. As shown in Fig. 10.9, the battery can be discharged with different output currents. C represents the rate of the discharging current. For example, 1 C indicates that the battery discharges the whole capacity in one hour. For a 2200 mAh 18650 standard Li battery, it takes one hour to discharge 2200 mAh with a discharging current of 2200 mA. If discharged at 15 C, i.e., 16.5 A, it takes only 4 min with a higher releasing power output, but with a slightly lower output voltage. There is a range of power densities in the Ragone plot for battery performance. In this case, the output powers of 1 C and 15 C are approximately $3.8 * 2.2\ \text{W} = 8.4\ \text{W}$ and $3.3 * 16 = 54.5\ \text{W}$, respectively, ~ 7 times output power difference.

Lithium battery can discharge at a high rate (max rate up to 20 C) with a power density up to 2.6 kW/kg. It has to be mentioned hear that the power density of battery

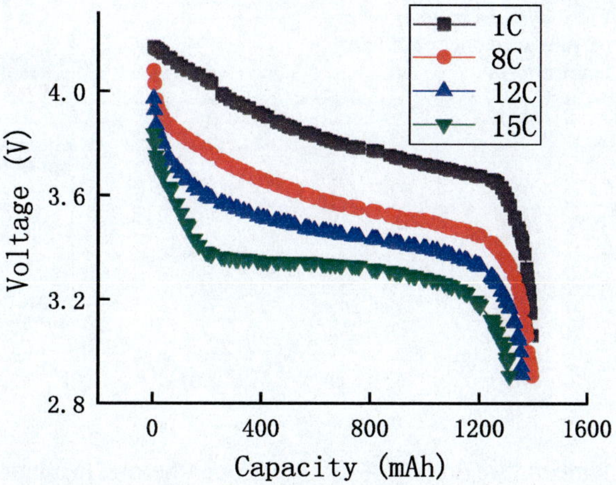

Fig. 10.9 Li battery performance at different discharging rates

is not fully studied because of no need to take the risk to output its power in such a high rate. Generally, the safer discharged rate is 4 C, that is, 2200 mAh (the weight of 18650 lithium battery of 2200 mAh is about 43–45 g). Its discharge voltage is 3 V, the discharge power is 24–120 W. With the weight 0.044 kg and the energy density of 200 Wh/kg. One 18650 Li battery of 44 g can offer 9.2 Wh. If the above 4–20 C is used to discharge, the max power density is 545–2727 W/kg. If there is a 100 kg lithium battery, it can reach the maximum output power of 54.5–272.7 kW. Of course, under the condition of high rate discharge, the temperature will increase sharply causing safety issues of Li battery, and this topic need further research before the Li becomes a feasible electric power supplier.

On the other hand, the higher of the discharging current, the lower the output voltage becomes. In the whole discharge process, the voltage curve of the lithium-ion battery can be divided into three stages:

- In the initial stage, the terminal voltage of the battery drops rapidly. The greater the discharge rate is, the faster the voltage drops;
- The battery voltage enters a flat, which is called the platform area. The smaller the discharge rate is, the longer the duration of the platform area, the higher the platform voltage and the slower the voltage drop. In the actual use of lithium-ion batteries, it is hoped that the battery will work in the platform area as much as possible;
- When the battery is almost discharged, the battery load voltage begins to drop sharply until it reaches the discharge cutoff voltage. It should be noted here that the Li battery should be protected to avoid a voltage drop below this level; otherwise, the battery can no longer be recharged.

The influence of the temperature on the battery performance is also important, especially for aviation purposes [45], because the temperature decreases 6 °C when the altitude increases 1000 m. The battery operation decreases sharply when the

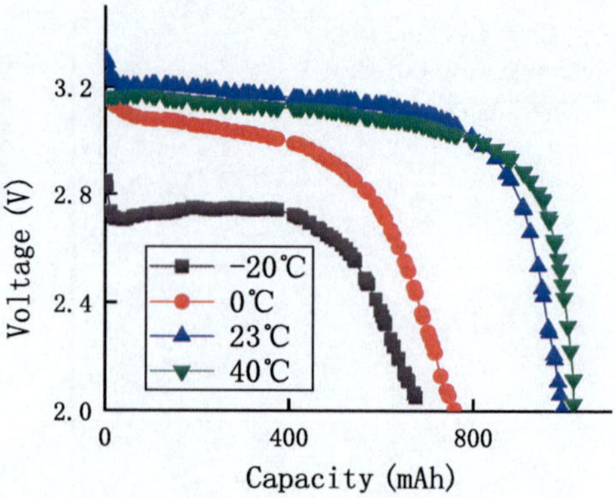

Fig. 10.10 Li battery performance under different temperatures

temperature drops to − 20 °C. Proper thermal insulation is needed to maintain the Li battery operation in the air. The best operating temperature of lithium-ion batteries is 0 ~ 35 °C, not more than 60 °C. Take the 18650 battery as an example in Fig. 10.10, the battery capacity is greatly reduced at the lower temperature of − 20 °C.

10.2.1 Li Battery

There are two kinds of Li batteries: Li-ion batteries and LiS batteries. LiS is still in the R&D stage, and in the following, the term "Li battery" refers to the Li-ion battery, which is a fairly mature technology with a comfortable price.

The Li battery is also called the "rocking chair" battery: Li^+ ions migrate back and forth between the positive and negative electrodes in the charging/discharging cycle. When the battery is charged, the lithium ions are de-embedded from the positive electrode and embedded into the negative electrode through the electrolyte. The carbon material of the negative electrode has many micro-pores to trap in these ions in charging process. During discharging, electrons are released from the negative electrodes to the external circuit as electric current, and the lithium ions embedded in the carbon material fallout from the negative electrode and migrate back to the positive electrode through the electrolyte inside the battery.

Compared with traditional NiCd and NiMH batteries, lithium battery is a revolutionary event in the battery industry. Goodenough, Whittingham, and Yoshino's contribution in the field of lithium-ion batteries led them win the Nobel Prize in chemistry in 2019. There are quite a few advantages of the Lithium battery compared to the previous generation. Number one, the lithium battery has a 3.8 V voltage rendering a higher energy density. Number two, Li battery is highly rechargeable. At the level of medium/high-power applications such as EV cars, lithium batteries basically replace all traditional chemical batteries.

The Li-ion battery is a mature technology that has already been adopted by the new energy vehicle (NEV), such as Tesla and BYD. The Li battery can be improved

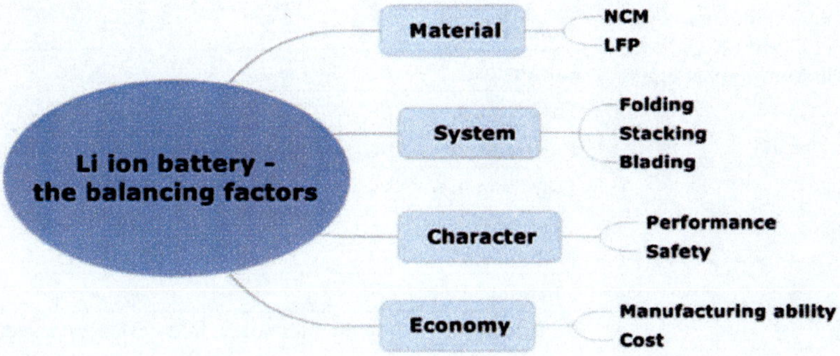

Fig. 10.11 The four factors to improve the Li battery

either from the material or from the architecture. However, these two large companies have their own approaches to improve their Li batteries. Figure 10.11 describes the balancing factors of enhancing the battery performance, including material, system, character, and economic concerns. Most of them conflict with each other. For example, the NCM battery has a performance advantage but has more safety issues. Blading battery is the best architecture yet you have to consider its manufacturing ability. Therefore, each company should make its own choice based upon its overall judgment of the pros and cons of these balancing factors.

We also need to distinguish the differences between the three current main Li battery producers: CATL, Tesla, and BYD. CATL is a dedicated battery vendor from Ningde Era. Tesla and BYD are just by-the-way battery vendors. Their main products are EV cars and batteries are just the key component to build the cars. Tesla used to manage hundreds of cylindrical 18650 NCM battery together with more advanced battery management systems, but this architecture is very cumbersome and is not optimized. BYD uses an integrated package of blade batteries to improve its energy level and safety. BYD also uses cheaper and safer LFP battery, which has a lower energy density in general than NCM. The final outcome of battery efficiency depends on both the material type and its architecture. The overall performance of BYD's and Tesla's batteries are compared in Table 10.7 together with the CATL battery. As a dedicated battery manufacturer, the CATL battery has an overall high Li battery performance.

Table 10.7 Energy density: battery pack compared

	Volumetric	Gravimetric	LFP/NCM
	Wh/L	Wh/kg	
CATL pack	450	250	NCM
CATL pack	290	160	LFP
Tesla Model 3 pack	238	163	NCM
BYD blade pack	280	140	LFP

Table 10.8 Comparison between LFP and NiCoMn ternary lithium-ion batteries

Criteria	NiCoMn	LFP
Lower cost	–	+
Higher safety	–	+
Energy density	+	–
Low temp. limit	+	–
Charging performance	+	–
Charging cycles	+	–
Technology features	4680, individual, fold	Blade, packed, insert
Car	Tesla	BYD

Materials

The Li battery is divided into two distinct families based upon the different cathode materials: lithium iron phosphate ($LiFePO_4$, LFP) and NCM ternary system (Li and Ni, Co, Mn). Both are currently used in new energy vehicles by Tesla and BYD. Neither approach is perfect. Each battery family has its own pros and cons listed in Table 10.8.

The two largest new energy car companies, Tesla and BYD, have successfully implemented these two types of Li batteries by making use of their pros and trying to avoid their cons. Balancing these pros and cons, they both have their success in their marketing shares. Generally, LFP batteries are cheaper and safer with slightly inferior performance compared to NCM batteries. Geographically, China is the main user of the LFP battery, and Japan/Korea and the US are the main users of the NCM battery.

- *LFP battery*. China is the leading market for LFP batteries and occupies the vast majority of the market share of electrochemical energy storage batteries because of its safety and the rapidly decreasing cost in recent years. In 2019, the three companies with the largest shipments of LFP batteries in China were BYD, Ningde Times and Sunshine Power, with a total shipment of 3.8 GWh, a year-on-year increase of 26.7%. In terms of R&D capacity, the technical level of LFP batteries represented by the BYD and Ningde Era has reached the world top level.
- *NCM battery*. The market for NCM ternary batteries is also in the field of electric vehicles but is dominated mainly by Japan and South Korea. Panasonic, Toshiba and LG Chemical occupied the main market share in the early years. At present, the cost of NCM battery cells is ¥1000 ~ 1500/kWh, which is expected to fall below ¥700/kWh by 2030.

Systems

From the architectural perspective, there are three typical configurations to compose the Li-ion battery, as shown in Fig. 10.12: folding (a), stacking (b), and inserting (c). Lithium batteries generally contain three-layer structures, a positive electrode, a diaphragm, and a negative electrode. The whole lithium battery is formed by superimposing these three layers together in different ways. There are three methods: fold, stack, and insert. The blading architecture (c) is more advantages among the three. From its technological easiness, the ranking is (a) > (b) > (c) [46, 47]. If ranking from the general performance, the order is (c) > (b) > (a) [48, 49]. The performance criteria are: the space usage efficiency (volumetric energy density), thermal dissipation ability and batter mechanical strength (safety), internal resistance reduction (efficiency and power).

(a) Fold. The three layers can be rolled into a cylinder or a square. Obviously, square folding saves space at the corner and increases the volumetric density.

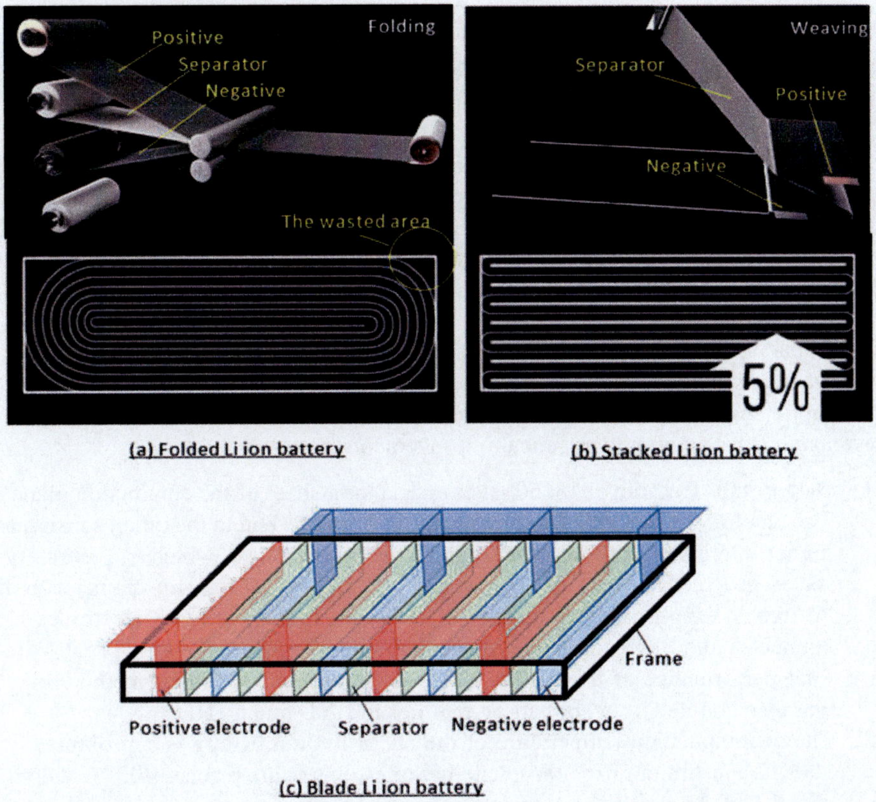

Fig. 10.12 The three ways to form the Li-ion battery. The Tesla cars use the battery structure in (**a**), and the BYD is using (**c**)

(b) Stack. The manufacturing process is like weaving a cloth back and forth with a shuttle, and adding the triple layers in turns. Obviously, the laminated battery is square and flat, saving considerable space at corners. However, the mechanical strength and safety factors are low if the plane area is too large.
(c) Insert. Blade battery technology uses long narrow positive/negative blades inserting between the diaphragms membrane. Obviously, these architecture is more advantageous. Compared with the stack battery, the direction of the battery membrane is perpendicular to the flat direction, and the cross-sectional area is narrow. The blade is inserted vertically on the battery rack, with a stable structure and small stress surface. Therefore, blade batteries are a more optimized design of Li battery architecture. It is more suitable for the use of new energy vehicles, rather than the arrangement of hundreds of discrete 18650 (18 mm diameter 65 mm length and 0 represents cylindrical shape) batteries in old Tesla cars, which wastes considerable space. The old method is very simple and easy to use to move the car quickly to the market, but it is not optimized in the long run.

Safety

The Li-ion battery has an intrinsic safety issue based upon the first principle of its chemical operation. The burning event of Li battery happens when the chemical decomposition occurs driven by the high heat caused by the high electric current. The chemical decomposition process is almost like an explosion in which the chemical reaction multiplies once it happens. Such a chemical reaction is dangerous because the high temperature encourages the chemical reaction, and it is very difficult to control this avalanche process, just as one cannot stop the bomb once it explodes. This decomposition disadvantage only occurs in Li batteries. From this perspective FC is relatively safer, even if hydrogen is a flammable gas. A physically operated supercapacitor that is purely electric does not involve any chemical/thermal process and is a much safer energy source than a Li battery.

Despite this intrinsic safety shortcoming, both Tesla and BYD are trying to leverage this issue through technical improvements:

(1) Battery tab. Traditional 18650battery has a long sheet of the conduction plane. The electrons take a long journey traveling from one end to the other, causing a higher internal resistor and generating more internal heat, which causes safety issues and reduces battery efficiency. Tesla's 4680 (46 mm in diameter and 80 mm in length) battery technology adds a series of parallel electrodes to connect A and B, greatly reducing the internal resistance and internal heat—the total performance of the battery is greatly enhanced. A similar methodology was also adopted by other battery families in BYD and CTRL.
(2) The decomposition temperature of the NCM lithium battery is approximately 200 °C. For lithium iron phosphate batteries, this value reaches 400 °C. Obviously, LFP batteries have an intrinsic advantage in terms of safety and are also cheaper to manufacture. That is why BYD prefers such LFP instead of

NCM. BYD uses an alternative way to improve the battery quality by the blade architecture instead of shifting to the NCM cathode material.

In summary, the traditional formation method of the Li battery is to fold the three layers of anode/separator/cathode together. This usually causes the waste of the areas in the four corners. The method of stacking the three layers together successfully resolved this issue and improved the volumetric density. However, the open area of the anode/separator/cathode sheet is too large, which is not favorable to ensure strong mechanical strength. Obviously, the architecture design of the blade battery has a distinct advantage in overcoming the above shortcomings for the most efficient use of the space and improved mechanical strength. As soon as its manufacturing maturity/expenses become satisfactory in the competing marketplace, such a blade battery architecture should be very well adopted.

10.2.2 Fuel Cell

Unlike the conventional concept of fuels such as kerosene used in the gas turbine, no thermal chemical reaction is involved in the fuel cell battery. A fuel cell (FC) is an electrochemical system that generates electricity via an electrochemical reaction with hydrogen and oxygen gas as fuels. FC is a green battery and has a higher safety factor than Li batteries. Fuel cells are silent and clean, with no vibrations and no NO_x emissions.

Although FC is more green and safer than the Li battery, its technology at the current stage is not as matured as Li-ion battery, especially on its marketing advantage in the EV cars. Reducing the hydrogen generation expense is the key. DFMC (direct methanol fuel cell) is another alternative with cheaper and liquids fuel rendering a better energy density. However, the technology is farther less mature than the hydrogen FC.

Regarding to the price, hydrogen FCs and methanol FCs may have different perspectives. The price of H_2 FC mainly depends on the cost of hydrogen production and storage, while the main concern for DMFC is to improve its technology for higher efficiency at an affordable price. As a rough estimate for commercial automotive hydrogen FC systems, the price is \$45/kW in 2017, \$43/kW in 2020, and \$36/kW in 2025 [50], not including fuels.

Considering the maturity and price, using Li batteries is a more pragmatic choice in the early R&D stage for electric propulsion and airplanes. Although FC also use "fuels" to generate electricity, its efficiency and process are far less superior to kerosene fuels. Besides, FC uses the electrical chemical reaction instead of thermo-chemical reaction and therefore is less powerful than the combustion process using kerosene.

The Working Principle

Unlike the Li battery, the chemical reactants do not come from the battery itself in FC. The ox- and redox-chemical reaction on positive and negative electrodes occur through the continuous delivery of "fuel"; in contrast, Li battery is a rechargeable-type closed system which has no external physical fuel exchange. In other words, fuel cells need extra fuel to maintain battery operation, while Li batteries maintain operation by itself. Their commonality is that the separation and movement of electrons and ions on positive and negative electrodes via a membrane. The feeding fuel of FC is usually methanol, ethanol, and pure hydrogen, together with oxygen. In FC (Fig. 10.13), hydrogen on the anode produces a chemical reaction to release electrons. The external circuit conducts electrons to the cathode on which redox with oxygen. The anode and cathode contain catalysts to accelerate the electrochemical reaction. The ions pass through the electrolyte through the diaphragm and then react with hydrogen protons to form water. The two poles are separated by the electrolyte diaphragm. Obviously, the by-product of the FC is just water, which is a very environmentally friendly operation compared to the Li battery.

Among the various types of fuel cells, proton exchange membrane fuel cells (PEMFCs) and solid oxide fuel cells (SOFCs) were mostly considered for aviation applications. The SOFC operates at high temperature and uses a dense ceramic layer as an electrolyte, while the PEMFC operates at low temperature and employs a proton conducting membrane as the electrolyte. SOFC has higher efficiency but needs a much higher operating temperature (800–1000 °C) which is not as practical for electric aviation. In 2020 the ratio of PEMFC 78.08%, SOFC 11.19%. The working principles are shown in Fig. 10.14.

The Fuel Cell Technology

Fuel cell technology includes the system and fuel. There are few good articles of fuel cell technology [25, 26, 51] as references. Below is just a quick review.

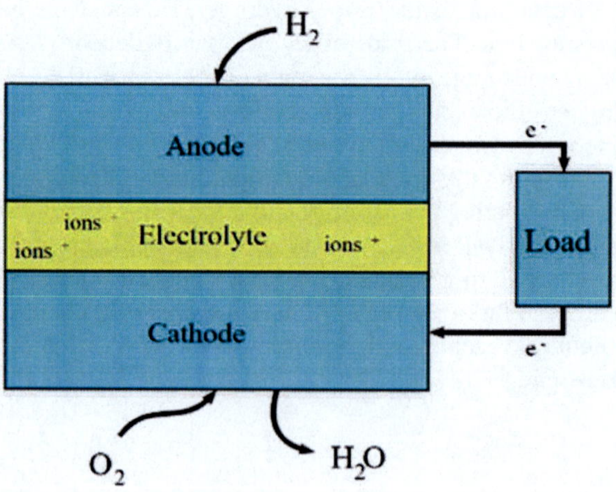

Fig. 10.13 Schematic of the fuel cell structure

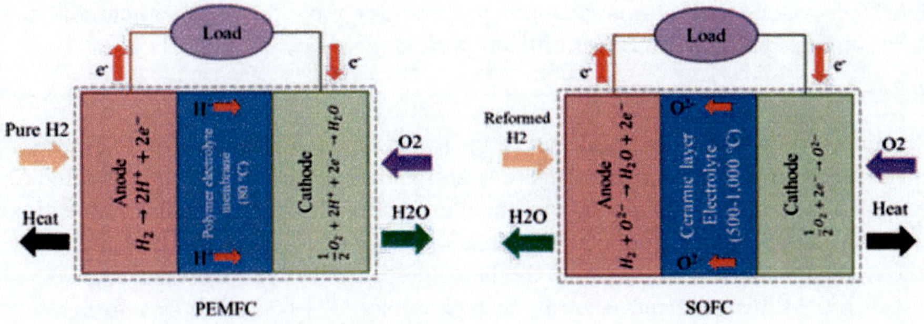

Fig. 10.14 Summary of PEM and SOFC fuel cell

Fuel Cell System

The FC system is basically the 3D stacking of the 3 basic components: anode/membrane/cathode and these three components form a repeating unit. Shown in Fig. 10.15, this repeating unit stacks together alternatively with the space where hydrogen and oxygen air flow in between. There are many 3D channels built on the surface of the anode and cathode for the best hydrogen/oxygen gas flow and more contact area for the chemical reaction. Catalysts are also applied on these two surfaces to enhance oxidation and re-oxidation.

DMFCs use methanol as an alternative fuel. The difference lies in that methanol is in a liquid form, saving a lot of space as compared to hydrogen gas. Additionally, its volumetric energy density is also higher than that of the gaseous hydrogen FCs, and its storage is much easier. Figure 10.16 illustrates the working principle of the fuel cell using methanol. There are certain level of CO_2 emission but not as serious.

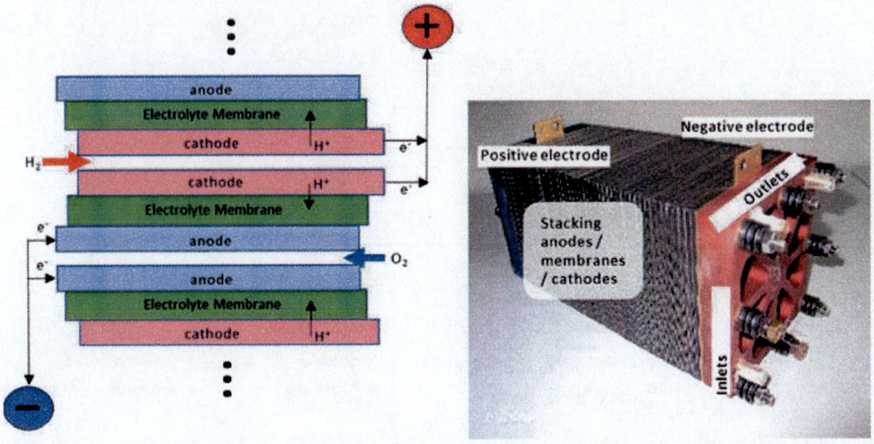

Fig. 10.15 3D stacking FC system: left: schematic, right: product

DMFC is much less matured technology. Enlarging the effective chemical reaction area is a key approach for better efficiency and higher power output.

Fuels—Creation and Storage

Obtaining hydrogen from fossil fuel is a matured but less green approach. Approximately 90% of the global hydrogen is currently produced from fossil fuels, including natural gas and coal [52]. This principle to generate hydrogen is against our original will to use hydrogen as green energy. Ideally, from the first principle of chemistry, extracting hydrogen from water is the best approach. It is just its technology is not yet ready. At this moment, electrolysis technology [53] is used to transform water into hydrogen and oxygen [54], with thermochemical water splitting by solar power or wasted heat of nuclear power reactions. It's still in the R&D stage.

Compared to hydrogen production, hydrogen storage technology is fairly mature. Figure 10.17 shows the various hydrogen storage methods. There are two ways to store hydrogen gas as fuel: chemically or physically. Hydrogen storage using chemical media is more efficient but it's still in a hot R&D stage [55], while the physical method is straightforward and its product has already been commercialized.

Figure 10.18 shows the tank used to store the gaseous pressurized hydrogen. Gas tanks evolve from the first generation of metal tank Type I to the current generation

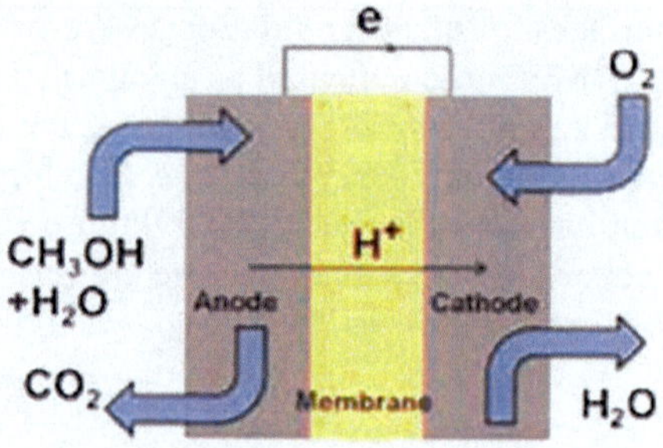

Fig. 10.16 Schematics of a DMFC

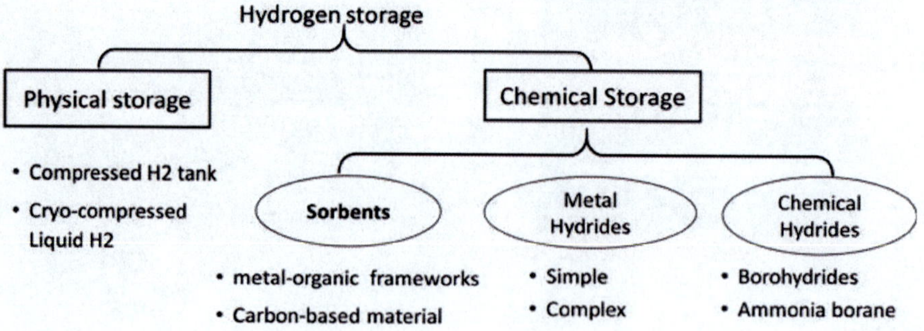

Fig. 10.17 Hydrogen storage technologies

of Type IV using carbon fiber composite with a polymer liner (thermoplastic) [56] from the commercial vendor Quantum Fuel Systems [57].

There is an optimized compression rate for hydrogen compression depending on the size of the container. For example, as shown in Fig. 10.19, 35 MPa is most appropriate for the 100 L container for the best mass fraction rate. For the 30 L tank, 70 MPa is the best.

Table 10.9 compares the various hydrogen storage methods [26, 58]. It is seen that the 70 MPa compressed hydrogen has a similar volumetric energy density to the liquid hydrogen, which is the ultimate limit of hydrogen storage. Even for liquid hydrogen, its energy density is just a quarter of that of kerosene. Therefore, hydrogen fuel has intrinsic shortcomings compared to kerosene in its energy density.

It should be noted here that the whole cooling system must be taken into account when we are considering the liquid hydrogen (LH) storage. The total volumes of the vessel are much larger. The liquid hydrogen supposed to be only 0.04 L with 2.8 kg

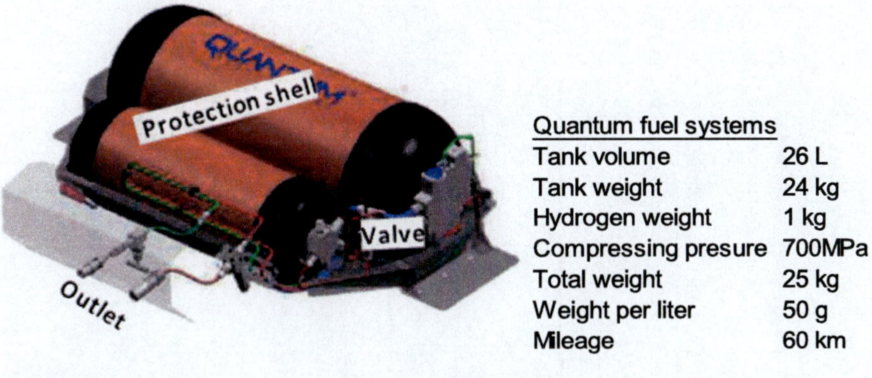

Fig. 10.18 Structure of one typical Quantum hydrogen tank

Fig. 10.19 Efficiency comparison of a few Type IV tanks for storing hydrogen

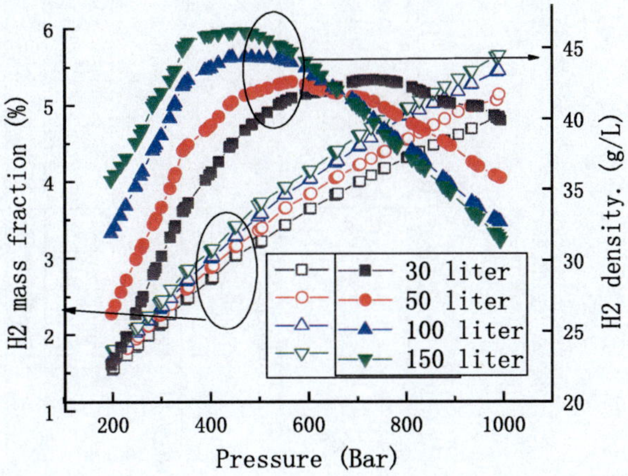

Table 10.9 Comparison of various hydrogen storage methods with their key parameters

Liter		Container (L)	Hydrogen weight (kg)	Total weight (kg) (container + hydrogen)	Wt ratio (%)	Energy (Wh)	Effective energy density (Wh/kg)	Effective energy density (Wh/L)	Density (g/cm^3)	References
	70 MPa compressed H_2	100	3	50	6	1E+05	2002	1001	0.030	[58]
0.039437	Cryogenic liquid H_2	90	2.8	40	7	93,408	2335	1038	0.031	
	Pure gaseous H_2 (1 atm)						33,360	3	8.99E−05	
	Pure liquid H_2						33,360	2369	0.071	[26]
	Kerosene						12,010	9702	0.808	
	Liquid methane						13,900	5894	0.404	
	Bio diesel						10,814	9424	0.870	

weight (density 0.071 g/cm^3), and the LH container is 90 L—just 2.8 kg of LH need 90 L-cryogenic system to hold it! Therefore, when considering the containers, the equivalent volumetric and gravimetric density of hydrogen is greatly reduced due to the comprehensive system either by the compressing tanks or the cryogenic cooling.

FC Characterization

V–I and P–I curves are used to characterize an FC, with units of W/cm^2 to quantify the power density instead of Wh/kg. The reason lies in that the chemical reaction happens on the surface of the anode/membrane/cathode. Such multi-layers can be 3D packed. In this way, multiple layers of FCs can be formed to save space and generate a higher volumetric density—the total area is multiplied by the number of layers in the 3D stack. Therefore, using W/cm^2 is more meaningful. The detailed evaluation of its energy/power density will be elaborated in a later section. The commercialization level of the FC battery for electric cars on the current market is also shown on the same chart. Basically, the operating voltage is close to 1 V, the power density is approximately 0.25 W/cm^2. The data as shown represent the current commercialized level of FC batteries for electric cars [59].

Volume Versus Weight—Hydrogen or Methanol

Although hydrogen has a very high-energy density per weight, it has a very low specific energy in volume. In contrast, methanol has a much better volumetric energy density. Many researchers have shown that the direct methanol fuel cell (DMFC) is an appropriate alternative, although many issues need be resolved before its use. Table 10.10 compares the energy densities in volume and in weight—methanol fuel has much advanced privilege for aviation purpose.

Fuel Cell—Power Density Versus Energy Density

Unlike the Li battery, the quantification of the power and energy densities for FC are different. The power density is used to characterize the capability to generate power of the FC system, while the energy density depends on how much fuel the FC system carries and how long it can last. Therefore, the power density depends on the weight of the FC systems and the energy density is mainly depends on the weight of the tank.

(1) The power density

Table 10.10 Comparison of hydrogen and methanol as the fuel in FC

FC fuel	Weight		Volume	
	(MJ/kg)	(Wh/kg)	(MJ/kg)	(Wh/L)
Methanol	50	13,900	21	5894
Hydrogen (1 Pa)	120	33,427	0.1068	30
Hydrogen (70 MPa)	120	33,427	4.1077	1142

Table 10.11 Energy density of hydrogen with other fuels

Fuel	Gravimetric energy density		Volumetric energy density	
	(MJ/kg)	(Wh/kg)	(MJ/L)	(Wh/L)
Kerosene	43	12,065	35	9647
Methane	50	13,900	21	5894
Ethanol	27	7570	22	6010
Hydrogen 1	120	33,427	0.1068	30
Hydrogen 2	120	33,427	4.1077	1142
Hydrogen 3	7	2014	4.2539	1183
Li battery		150		450

Hydrogen 1: 1 Pa
Hydrogen 2: 35 MPa
Hydrogen 3: 70 MPa

The specific power of FC is usually defined as W/cm^2. At the current stage of 2020, the typical power density is approximately 0.25 W/cm^2. To convert W/cm^2 into W/kg, we need to estimate the weight of this system. With 300 μ anode/membrane/cathode [60] stacking with 300 μ separation, 150 W/kg power density is expected without considering the weight of the frame, valve and other supporting accessories. In general, the power density is below 100 W/kg for FC, which is less than Li battery and far less than fossil fuel engines. Therefore FC is not a preferred electricity for aviation purpose esp. for the larger aircraft.

(2) The energy density

Regarding the energy density estimation, one cannot ignore the weight of the vessel to hold hydrogen (we evaluate the energy density for the gasoline purely based on its weight because they are in the liquid form and the weight of the tank is usually ignored). Table 10.11 compares the energy density in weight and in volume of the hydrogen stored in different forms together with their fossil fuel counterparts. Clearly, the energy density of the FC is not comparable with that of fossil fuel but better than that of the battery.

10.2.3 Supercapacitor

The term "super" or "ultra" in supercapacitor (SC) shall be more understandable if compared with conventional capacitor and battery. Compared to the physical electrolytic capacitor, SC is chemical-based and stores up to 100 times more energy per unit volume or mass. Unlike ordinary capacitors, this supercapacitor, either double-layer capacitance or pseudo capacitance, is chemical-based instead of physically based [61, 62], which operates similarly to a battery—the exchange of electrons/ion via the oxidized/deoxidized process. The main difference lies in the SC having a

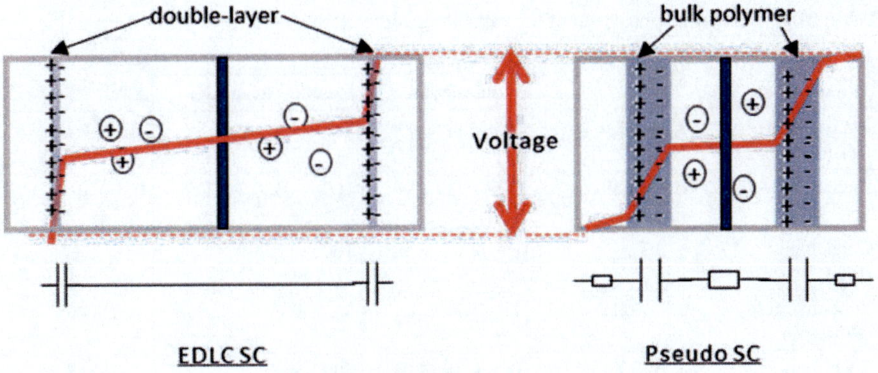

Fig. 10.20 Illustration of the operation of the EDLC and Faraday capacitor

similar performance of charging/discharging as a capacitor. Its charging/recharging time is much faster—suitable to provide an immediate burst of power. Compared to battery, SC releases electrons much faster although SC's operation is also an OX/re-OX chemical process. SC's power density is much higher, and the rechargeable cycle is much more than that of rechargeable batteries. The drawback, however, is that its energy density is too low, which means that it can only provide the burst power for a few seconds which is far from being sufficient for the 30–60 s takeoff for VTOL.

There are two categories of SCs: double-layer capacitors (EDLCs) and Faraday pseudo capacitors.

- An electric double-layer capacitor (EDLC SC) generates charge separation on the solid–liquid *interface* between the carbon electrode and electrolyte. The positive and negative ions accumulate in two double layers connected to the positive and negative electrodes, respectively. In the discharging process, electrons can be promptly released from the double layers (Fig. 10.20).
- Faraday capacitor (also called pseudo supercapacitor) uses a metal oxide or conductive polymer as an electrode and produces charge separation by a redox reaction in a shallow layer of *bulk* polymer. Compare with EDLC, its capacitance is several times higher in the same area. Pseudo capacitors therefore store many more charges than EDLCs. However, the internal resistance of the pseudo capacitor is higher due to a comprehensive ox-redox process—the power density is lower than that of the EDLC.

Table 10.12 summarizes the various parameters of the SCs together with the conventional electrostatic capacitors and battery.

Comparing these two SCs, the chemical reaction media only occurs on the surface for the ELDC, while the chemical reaction occurs in the bulk for pseudo SC, allowing more chemical reactions and a higher energy density. However, its chemical reaction speed is slower, rendering a slightly lower power density. SC has the advantages of high-density power, high response speed and high service life, but its energy density is just too low, and the product consistency is poor. For example, it takes

Table 10.12 General comparison of the capacitors and batteries

Parameter	Electrolytic capacitors	Supercapacitors		Lithium-ion batteries
		Double-layer	Pseudo-capacitors	
Maximum volts (V)	4 ~ 630	1.2 ~ 3.3	2.2 ~ 3.3	2.5 ~ 4.2
Recharge cycles (k)	< Unlimited	100 ~ 1000	100 ~ 1000	0.5 ~ 10
Capacitance (F/kg)	≤ 2.7	10 ~ 470	100 ~ 12,000	–
Specific energy (Wh/kg)	0.01 ~ 0.3	1.5 ~ 3.9	4 ~ 10	100 ~ 265
Specific power (W/kg)	> 10,000	3000 ~ 10,000	500 ~ 4000	300 ~ 1500
Working life (y)	> 20	5 ~ 10	5 ~ 10	3 ~ 5

approximately one minute for an airplane to takeoff, but the SC can only provide this instant power for just a few seconds due to its very low energy density. In addition, SCs are not a mature technology compared to batteries, and their price is still high. In short, the energy stored from the chemical SC is not comparable with Li batteries. Therefore, it is not at all a good candidate for future aviation power. In comparison, our new generation of 3D HK SC is a future candidate for an electrical propulsion energy source. This will be explained in Sect. 10.4.

10.3 Lightweight Generator

Lightweight turbine generator (LTG) is a combination and moderation of the current gas generator and gas turbine industry, focusing on the light weight and high-power output for electrified aviation, because the current gas generator is high efficient but too heavy, the current gas turbine engine is lightweight but cannot output electricity.

10.3.1 The Background

Necessity and Feasibility

The Necessity

At present, fossil fuel still has the highest specific power and energy compared to battery technology. In general, there are two types of fossil fuel electric generators, the diesel generator and gas turbine generator. Table 10.13 compares the various performances of these two types of generators. Clearly, the diesel generator is simple and lightweight but less efficient, gas turbines are much more efficient but need a fairly complicated system. At the current stage, the gas turbine generators are large and heavy. This is not quite feasible as aviation electric power. A gas turbine machine right now can be built either as a powerful electric generator or as a lightweight aero-engine but not yet as a lightweight electric generator.

Feasibility

GE's LM6000 big gas generator and J85 small gas turbine engine are two existing mature prototypes that provide a good reference point to combine into an LTG. It is an integrated innovation for the first time—the project of lightweight small gas turbine generator used to generate high-power mobile electricity for electrified aviation. LTG can be used as the hybrid electric power supplier of electric planes for long-range navigation, taking advantage of the largest energy density of kerosene which can also be substituted by other bio-fuels in future. For VTOL purpose, the purely electric 3D supercapacitor is preferred for cleaner, greener, and peaceful takeoff and LTG can be used to recharge it once taking off.

The Basic Principle

The gas turbine compresses air and mixes it with fuel to heat up to a very high temperature. The combustion air flow makes the gas turbine rotate at high speed and the fast-spinning turbine drives a generator to generate electricity by using a shaft to deliver this mechanical energy to spin the magnet to rotate inside the coils and generate electricity. A gas turbine generator converts thermal chemical energy into mechanical energy and then converts mechanical energy into electrical energy. The difference from the aircraft engine is that it only converts thermal chemical energy into mechanical energy with the minimum weight. In principle, the energy conversion efficiency of the generator should be better since the loss of the exit hot jet flow energy can be minimized. In addition, the heat flow can be reused for the ground gas turbine generator to enhance the overall efficiency more. However, the cost of this high efficiency is that the gas turbine machine is too heavy. The lightweight strategy

Table 10.13 Comparison of diesel, gas turbine, and lightweight generator

	Weight (kg)	Power (kW)	*P*/*W* ratio (kW/kg)	SFC (kg/kWh)	Source
10 kg LTG	10	50	5.00		Under development
7-ton, GE LM6000 gas generator	7000	40,000	5.71	0.20	*
220 kg, Boeing 787 diesel VFSG generator	220	250	1.14	0.20	#

* GE LM6000: https://www.geaviation.com/marine/engines/military/lm6000-engine

Generator Source: https://www.generatorsource.com/

of gas turbine engine shall be integrated with the current gas turbine generator to form a new product—the LTG.

The Basic Theory of a Generator

Faraday's law of magnetic induction is the basic theory of the electric generator. According to Lenz's law, current is induced in the coil if the line is perpendicular to the magnetic flux moving direction (Fig. 10.21). The basic equation is that the electric current:

$$I = BLV/r \tag{10.1}$$

where

B = magnetic flux density
L = length
V = velocity
r = resistance of the movable arm PQ.

From Eq. (10.1), one can easily see that in order to increase the electricity, the length L, the rotating speed V, and the magnetic intensity shall be increased accordingly. Multiple winding numbers per coil (n) can increase the value "L". In addition, when the moving speed is higher and the permanent magnet is strong (NdFeB), the total electromotive and current shall be tremendously improved.

The New Configuration of the Coils/Ring Permanent Magnet

We propose a new design of the rim driven electric generator with the special coil structure. In this new electric generators, the electric generator coils do not move and function as stator. The permanent magnet ring rotating by rim driven turbine machine. This rotating magnetic field cuts the electric line in the coil to generate electric current in a series of rectangular electric coils according to Lenz's Law. A strong magnetic field using high-spinning NdFeB permanent magnets built in a ring (inside is the North Pole, outside is the South Pole) plus the multiple coils situated on the outside ring (fixed) renders a large electromotive force (emf) and a strong electric current output.

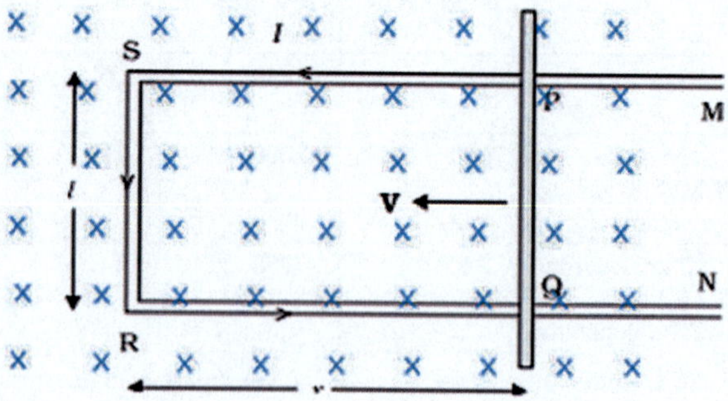

Fig. 10.21 The illustration of the electric generation principle

10.3.2 The Reference Technologies

GE's LM6000 gas generator and J85 small gas turbine engine are two existing mature technologies that provide a good reference point to combine into an LTG. Table 10.14 compares the key parameters of weight and power concerning the LTG based on the existing datasheet from General Electric. As a reference, a simple light 3 kg weight 20 kg thrust jet plane can be used as a prototype for a simple 10 kg 50 kW LTG development.

GE LM6000 Marine Gas Turbine

The LM6000 is a simple-cycle, two-shaft, high-performance gas turbine that is derived from GE's CF6-80C2 high bypass turbofan aircraft engine. LM6000 is a compact LM6000 with a 42% high thermal efficiency and is an ideal gas turbine for consideration in ship propulsion system design when high performance and high power are needed. LM6000 weighs 7411 kg, and the entire unit is 4.9 m long, 2.16 m wide, and 2.05 m high, providing ~ 50 MW [63] power with SFC ~ 203 g/kWh [64].

GE J85 Lightweight Aero-engine

Although GE's CF700 replaced the J85 old model as a lightweight gas turbine engine for supersonic propulsion serving XB1, J85 is still the prototype to start with when developing the LTG. The General Electric J85 is a small single-shaft turbojet engine [65]. It is very small, length 286 cm and diameter 53 cm; very light, dry weight 310 kg, thrust 16 kN (~2 MW), SFC 0.84 kg/kWh. J85 is a jet engine, and we are aiming to convert it into a lightweight electric generator.

To develop an LTG is an integrated innovation—the project of a lightweight small gas turbine generator used to generate high-power mobile electricity for electrified aviation. To achieve this goal, the GE's LM6000 gas generator and GE's J85 gas turbine can be re-configured as follows:

Table 10.14 The prototypes and targets for the LTG

Gas turbine/gas generator	Weight (kg)	Power (kW)	Power/weight (kW/kg)	SFC (kg/kWh)	Source
GE LM6000	7000	40,000	5.71	0.20	
GE J85 *	310	2672	8.62	0.75	¥
Our temptative 10 kg LTG	10	50	5		$ / Under development

¥ GE LM6000: https://www.geaviation.com/marine/engines/military/lm6000-engine
$ GE J85: https://en.wikipedia.org/wiki/General_Electric_J85
* Converting the thrust to power: 1 N = 0.167 kW based on $T = P/V$, i.e. 1 kN is equivalent to 167 kW at a speed of 600 kmh

- Reduce the weight of LM6000: reduce the number of compressor and turbine stages, reduce the overall diameter, restructure the gas turbine;
- Convert the J85 engine into a generator: transfer the turbine shaft power into the rotating Nd–Fe–B permanent magnet, reconfigure the shaft-driven to rim driven.

Our target parameters for the LTG are:

- Size: diameter < 0.5 m, length < 1.5 m
- Weight: 1 t–2 t
- Power: 1–10 MW.

Below, we propose a new configuration of this LTG by making use of the rim driven principle inside the gas turbine machine for a higher efficiency as a lightweight electric generator.

10.3.3 A New Rim Driven Generator

Here, we propose a rim driven gas turbine generator. The key feature is that all the rotating parts are rim driven instead of axis driven. The rim driven rotating parts include the rim driven turbines and the rim driven permanent magnet, together with the front fan and compressors. The electric generator is located on the rear part with a high-speed spinning permanent ring (rotor) inside the electric coils (stator). Unlike the RDF jet, there is a center shaft to hold the combustion chamber and the other fixed parts, such as the outer shell and electric coils. The process of the operation is:

- The combustion chamber creates hot air flow with burst thermal power.
- The directed hot gas flow drives the rim driven turbine to rotate.
- The rotating rim ring drives the magnets to spin to generate electricity.
- At the same time, this rotating energy is transferred to the front fan and compressor via the rim driven ring to provide compressed air for combustion.
- The whole system is optimally designed as a lightweight electric generator instead of a jet, although the exhausted air can also be used as a propulsion jet.
- Several advanced achievements in the gas turbine industry can be incorporated, such as CMC and ceramic bearings in material engineering.

10.4 3D High-K Supercapacitor

Traditionally, the so-called supercapacitors are based on electric-chemical reactions but not on the basic electrostatic concept (two electrical conductors separated by a dielectric medium), simply because the traditional electrostatic capacitor is just one-layer of MIM sandwich with an ordinary dielectric material with maximum ε value of just 30. Comparing to the chemical-based supercapacitor, this physical or electrostatic capacitor has a very high-power density but its energy density is just too low. Although the battery-like SC has higher energy density, its power density got degraded due to its chemical operation process. Here, we propose an electrostatic capacitor whose working principle is purely physical instead of either chemical or thermal, allowing a noiseless and very fast charging/discharging operation to provide a high-power density as well as a good energy density. This physical supercapacitor achieves its "super" by adopting a 3D-packed architecture with a super high-K dielectric thin film in order to meet both needs on power and energy simultaneously. Since the electrostatic SC is purely electronic, its system is much lighter than the gas generators, which may render a similar power and energy density as the gas turbine generator.

10.4.1 Necessity and Feasibility

The aircraft needs an immediate burst power to take off, that is, the "power density" should be large. "Large power" is needed either to accelerate or vertically lift off the aircraft—both need enough strength or thrust. The second factor is "density": the lighter the aircraft system, the easier to fly. On the other hand, the duration of this burst power output must be long enough for the airplane to take off, usually > 1 min. Neither the conventional Li battery nor the chemical supercapacitor can fulfill this goal. This is the unique advantage of 3D HK SC, possessing both high-energy and high-power density.

- Its discharging process does not involve ion–electron exchange chemical process, rendering a very fast discharging speed and can produce high burst power.
- Its system is simple and light because it does not involve any mechanical/thermal machines like the gas turbines.
- It is a purely green and peaceful power for takeoff causing neither airport pollutions nor noises.

This multi-layer supercapacitor can maintain the high-power output for at least one minute to aid the airplane to take-off.

3D HK SC uses three distinct techniques to ensure that the capacitor is "super", i.e., the very high dielectric constant material as the dielectric media, 3D stacking increases the contact area, and thin film technology to make the dielectric layer as thin as possible, in the formula $C = \varepsilon * S/d$. What we need to do is to increase the value of C by improving ε, increasing S and decreasing d. High dielectric constant materials (40,000 as compared to 30) have been revealed recently as a successful precedent to build super high-K capacitors. The 3D micro-manufacturing technology can be borrowed from the existing MEMS process. Traditionally, micro-manufacturing is just to build a one-layer structure for MEMS devices such as sensors. Here, we use it to build a 3D structure. 3D configuration save a lot of planar areas. America has more houses and China has more buildings, the US has 400 million people and China has 1.5 billion people. The Chinese 3D buildings can hold more people. The multi-layer 3D building accumulates more space in the Z direction so that more people can reside

Table 10.15 Structure of 3D HK SC and the principle to enhance its capacitance values

Principle	Structure
In $C = \varepsilon \frac{S}{d}$ ε: from 5 to 50,000–10,000 times S: from 1 to 200 layers, 200 times d: from 200 μm to 200 nm, 1000 times In total, C is 2,000,000,000 times larger	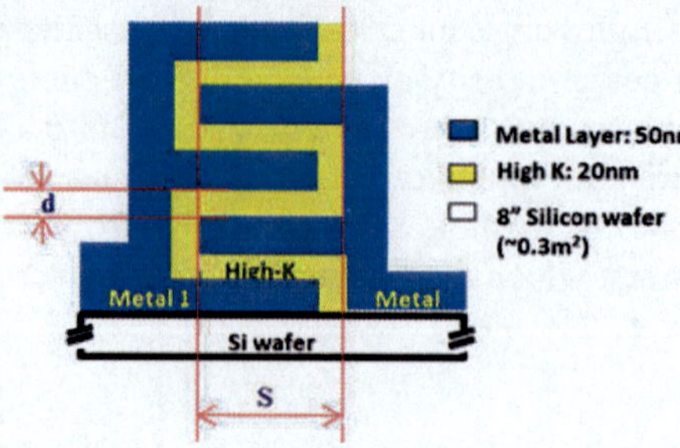

in the same surface area. Our 3D HK SC follows this principle to superimpose the metal/HK/metal structure for more useable areas to multiply capacitance in the same silicon chip area.

10.4.2 The Principle

The 3D HK SC follows the most basic physical formula of the capacitance:

$$C = \varepsilon \frac{S}{d}$$

where ε, S, and d are the dielectric constant, area, and thickness, respectively. The technical details to enhance ε, enlarge S, and reduce d are revealed in Table 10.15.

10.4.3 The Evaluation

The approximate estimation of the energy and power density of the 3D HK SC packed on a 12″-silicon wafer is as follows:

(1) The capacitance

Assuming the dielectric constant is 50,000 and the thickness d is 200 nm, then the capacitance density is 22 F/m^2. The effective area of a 12″-wafer is 7.1×10^{-2} m^2, and the capacitance is 1.56 F per wafer. If we pack this thin film capacitor in 3D by 64 times, the total capacitor is 100 F, which is equivalent to a 0.6 Ah. One laptop computer needs 4.4 Ah energy then only seven wafers of SC can fulfill this need (weighs ~ 120 g compared to 500 g of Li battery).

(2) The weight

- The silicon wafer of 12 in., 100 μm thick weighs 16 g.
- The 3D stacked supercapacitor of 64 layers of MIMs weighs approximately 2.9 g, assuming the metal layer thickness is 500 nm and the high-K dielectric layer is 200 nm.
- In total, one 12″ silicon wafer with 3D SC weighs 18.9 g. Approximately 53 wafers make a 1 kg SC with 5300 F of total capacitance.

(3) The density

This includes both the power and energy density. The limiting factor for the power density is the highest current flowing through the cross section which is mainly limited by the max current density and the cross section. Regarding to the energy density, basically it is related to the numbers of layers and the capacitance of each layer in 3D HK SC.

The Power Density

The maximum power density refers to the ability to provide enough electric power to support the RDF jet engine for enough thrust to lift up the airplane for takeoff—the VTOL. When the aircraft weighs 7 tons, 9 RDF jet providing 8.1 ton thrust can lift up this plane. Then at least 710 kW is needed for each RDF jet. This high-power output mainly depends on the discharging speed and the maximum electric current. Since the 3D HK SC discharges physically rather than chemically, the discharging time is much faster. The main limiting factor is the max electric current output, which is mainly limited by the maximum current density of the metal thin film together with the its cross section area. For example, assuming the thin film metal layer is 1 μm, if the metal line width is 300 mm (the max width of the 12 in. silicon wafer) then the cross section is 300×10^{-9} m^2. The maximum electric current density of metals is ~ 3 A/mm^2. Then the maximum allowed current to flow through this cross section is about one ampere. With 10 layers of supercapacitor, the max output current is 10 A. One wafer of supercapacitor weighs 19 g, then 53 silicon wafers weigh 1 kg, providing the maximum output current of $53 * 10 = 530$ A. If the output voltage is 22 V, the max power density is then $22 * 530 = 11{,}660$ W/kg. Therefore the rough estimate of maximum power density of the 3D HK SC is ~ 10 kW/kg. Since the film is relatively thick (1 μm), we only build 10 layers of supercapacitor on one wafer for larger current output level. This reduce its capacitor by ~ 1/10, rendering a lower energy density to 71 Wh/kg. The 100 kg of such supercapacitor can provide $71 * 100 = 7.1$ kWh energy. If this amount of electric energy is discharged in 710 kW power rate then the lasting time is $7.1/710 = 36$ s. In other words, 100 kg 3D HK SC can supply 710 kW power output for RDF jet to thrust at 9 kN level for 36 s—sufficient for VTOL operation of the aircraft. After takeoff the SC can be recharged by the hybrid electric generator.

It should be noted that the maximum power density can be higher than 10 kW/kg with proper topology design of 3D HK SC. Although the maximum allowed current density is fixed at 3–5 A/mm^2 for a certain metal material, the cross section area can be enlarged by fabricating the wider metal electrode through the manipulation of the device design. For example, if we build a 3×3 SC blocks on the 300 mm wafer

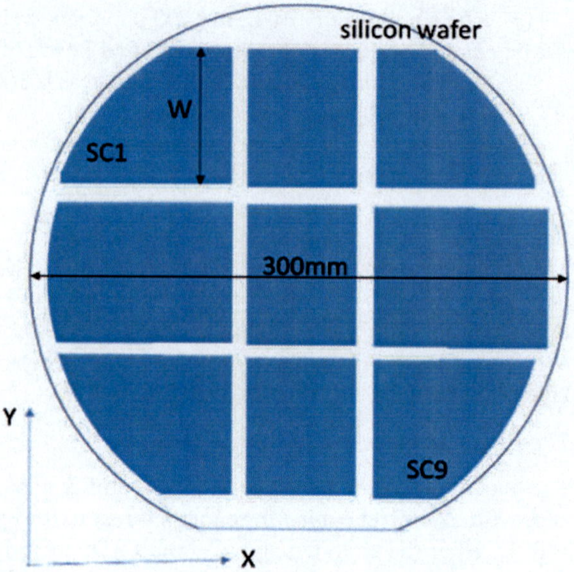

Fig. 10.22 The mask design which can enlarge the total width of the electrodes

one can multiply the width by a factor of 3 (Fig. 10.22). Linking the 9 blocks of SCs in parallel afterwards will enlarge the output cross section and render a high-power output. In this layout design, the total cross section area is enlarged by 3 times. Of course, more power output level can be achieved by splitting the more SC blocks in *X* directions depending on the need. It should be noted here that the manipulation of the layout can only the increase the power density level, but cannot change the energy density. Although the power density is increased in this way, the energy density keeps the same. Besides, the high output rate, the less lasting time.

The Energy Density

The energy density is mainly determined by $C = \varepsilon \frac{nS}{d}$, the ε is dielectric constant, n is the number of the layers, S is the area of the silicon wafer and d is the thickness of the high-K dielectric layer. The 3D HK SC can be built with moderate power output level for larger energy density. In this option the metal layer is 1/20 times thinner, allowing 100 layers of SC built on one wafer. One wafer of 100 layers supercapacitor (with super high dielectric constant of 40,000) can provide the total capacitance of 100 F, which has an equivalent of 710 Wh/kg energy density, better than the current Li battery technology.

The energy density is the product of the electric charges and the output voltage ($E = Q * V$). The amount of charges is the product of capacitance and voltage. In other words, with the same capacitance value of 100 F, the higher voltage can make the supercapacitor store more energy as long as the voltage is within its highest tolerance. On the other hand, when charging the supercapacitor using a higher voltage but operating it at a lower voltage, the SC lasts longer. This is the distinct advantage of the supercapacitor which provides a flexible power output level comparing to the fixed voltage such as Li battery (3.8 V).

10.4.4 The Technology

The key technologies involved in the 3D HK SC consist of two parts: the super high-K dielectric material and the 3D micro-fabrication technique.

High K

The first key technology is to develop super high dielectric constant (ε) dielectric material as the target for magnetron sputtering. Generally, ε is below 10 for in semiconductor industry, while recently more evidence appears that some materials exhibit four orders of magnitude larger ε. Through our intense research and investigation, it was found that materials with high-K values are the composites of titanium zirconium barium, barium titanate, cobalt oxide, niobium oxide, and rare earth oxide ceramics with ultrahigh dielectric constants of 400–40,000 [66]. One family is based on $BaCO_3$, TiO_2 composite (BT) [66–68]. The dielectric constant can reach 10,000. The other family is the SrO_3 TiO_2 composite [69]. Tecdia company in Japan already has supercapacitor products with a dielectric constant of 30,000 and a resistivity above 10 MΩ [34], but its products are mainly for mobile phone circuits. However, this fact shows that it is feasible to develop high dielectric constant thin film sputtering targets. We just need the "first person to try tomato".

The chip ultra-capacitor product from Tecdia is shown in Fig. 10.23. We have verified the value of the capacitor using the testing and characterization apparatus. In the experiment, the sample is fixed by a TRL calibration clamp and connected to a vector network analyzer for measurement.

The measurement is carried out in the frequency range from 0.2 to 8 GHz. The measured value of the capacitances is very close to data given by the Tecdia. Our SEM view (Fig. 10.24) of the supercapacitor structure is also in agreement with the product data sheet. Combining the measure capacitance and SEM dimensions we have verified that the dielectric constant ε is indeed 40,000.

From the EDS analysis, we identified the ingredient is C, O, Ti, Sr by evaluating the amount in Fig. 10.24, we identify the building material is $SrCO_3$: SrO: TiO: TiO_2 = 7:11:6:10. In the follow-up research work, we can prepare the sputtering target for PVD from this ratio. Super high-K dielectric thin films can be made by physical vapor deposition, chemical vapor deposition or other methods based on the proportion to form our 3D HK SC.

3D Stacking Technology

The 3D structure multiplies the single layer capacitance n times by stacking the n layers of MIM with the help of standard IC/MEMS technology [70] to enlarge the total area to nS in the formula $C = \varepsilon \frac{nS}{d}$.

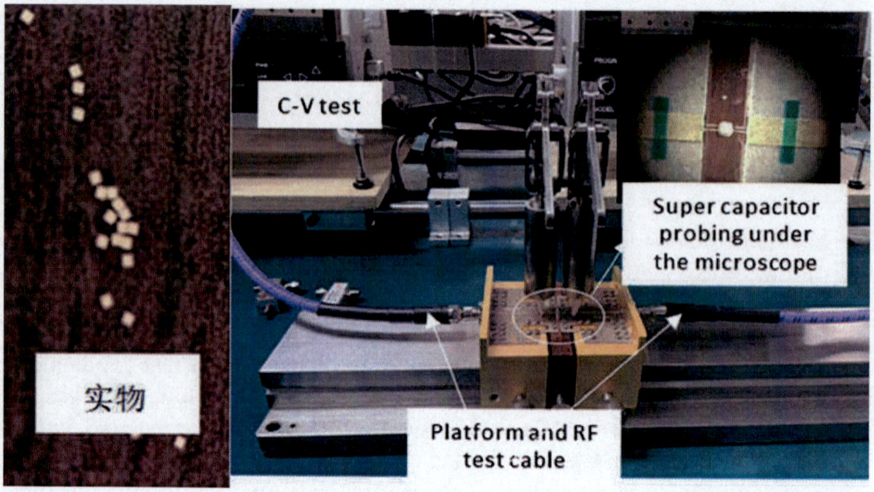

Fig. 10.23 The physical structure of the chip ultra-capacitor product and measuring system

MIM Capacitor

Each layer of the SC is composed by three layers of metal/insulator/metal (MIM) layers as shown in Fig. 10.25.

Quite a few teams of people have been working on the process to build this MIM capacitor, but still using the ordinary dielectric material. In 2015, Srivastava et al. used the PLD method to prepare HfO_2-La_2O_3 composite dielectric layer-based capacitors, and the specific capacitance reached 208 nf/mm^2 [71]. In 2016, Chaker et al. prepared metal dielectric metal (MIM) capacitors based on a TiO_2 dielectric layer by the ALD method [72], and the specific capacitance reached 185 nf/mm^2. On the same year, Neve et al. prepared silicon-based capacitors based on a HfOx dielectric layer by the CVD method. The specific capacitance reaches 96 nf/mm^2 [73]. Although these initial efforts on MIM may provide some references when developing our SC, these MIMs are mainly for RF electronics rather than high-energy aviation. Its manufacturing process should be adjusted to fit in our new application scenario. The supercapacitor used in an aviation power source must tolerate high-voltage and high-power output rather than its high-frequency character. We are going to use the PVD method to deposit the thin films of metal and the high-K layers. More uniform and high-quality thin films can be achieved by physical vapor deposition method. The drawback of the PVD method is lower productivity as compared to CVD method. At the current R&D stage we'll start with PVD method to establish the 3D HK SC first and then we consider its productivity later.

After building the MIM capacitor, there are two 3D construction modes: horizontal and vertical, to stack the MIM layers. The horizontal method is analogous to dividing a large room into small rooms, and the vertical method is equivalent to building a skyscraper. The first 3D structure uses the trench to build sidewall area to multiply

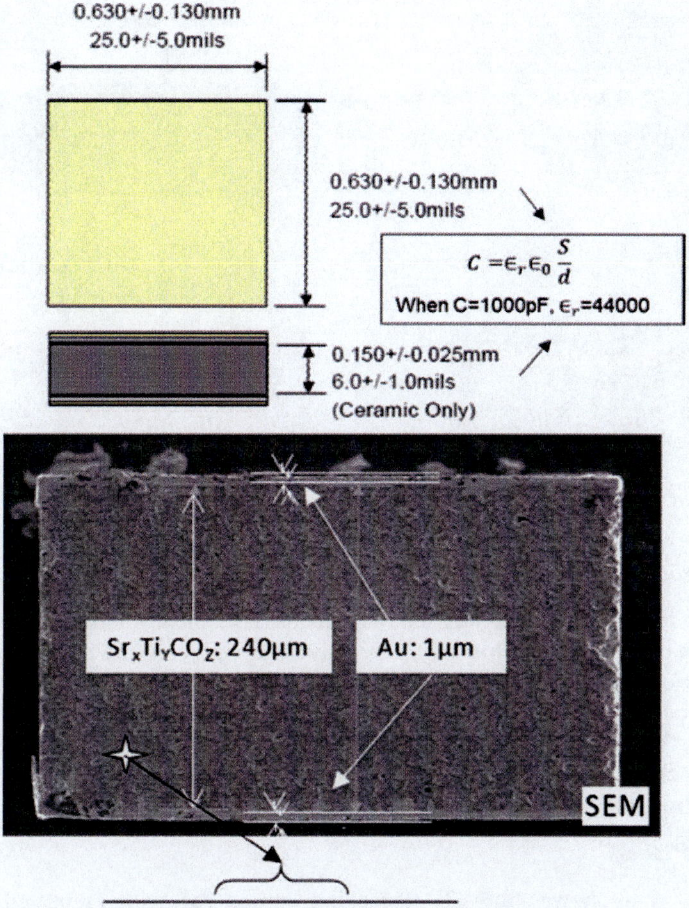

	C	O	Ti	Sr
#1	7.23	58.68	15.66	18.12
#2	7.09	59.26	15.55	17.89
#3	6.96	58.24	15.91	18.68

EDS

Fig. 10.24 The combined info for supercapacitor product

the effective surface area. The second approach uses the pile-up stacking to build the multi-layer MIMs vertically. Both methods are based on the IC/MEMS process. The first kind is to use deep trench etching and filling the trench layer by layer to construct 3D SC in the horizontal direction. The second is to use repeated thin film sputtering and stripping processes to form vertical 3D structures.

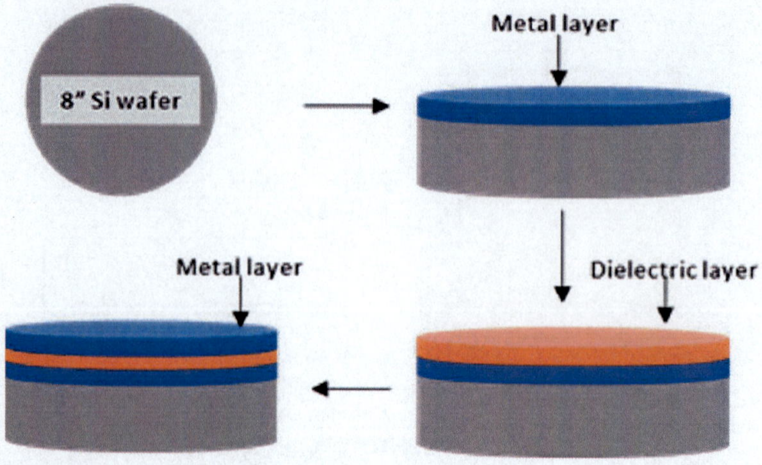

Fig. 10.25 The basic MIM process

Trench Process

As shown in Fig. 10.26, the deep sidewall inside the trench can increase the surface area of the capacitor by multiplying alternatively more layers of MIMs. The connections of two MIMs can be established on the two sides of the wall. The advantage of this method is that trench technique is a matured process in IC industry. The disadvantage is that the number of layers in the horizontal direction is limited by the width of the trench. Nevertheless, a supercapacitor can be also made with this architecture by replacing ordinary silicon dioxide with $SrO + TiO_2 + TiC$ ultra-high-K material.

Stacking Process

Figure 10.27 shows the 3D stacking structure with repeated dielectric/metal/dielectric/metal layers alternatively from button to top. The connection

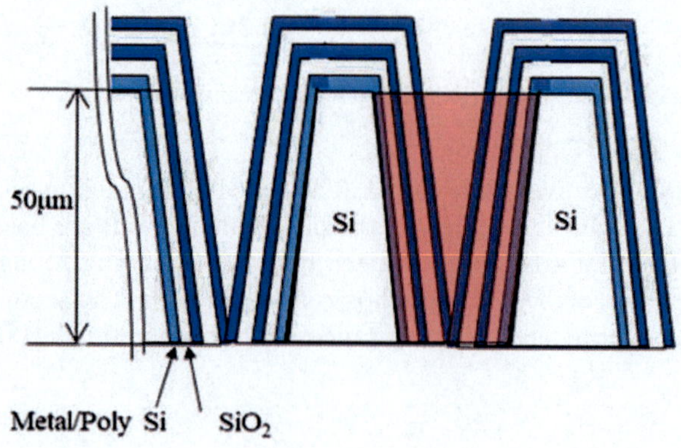

Fig. 10.26 Schematic diagram of the trench SC structure

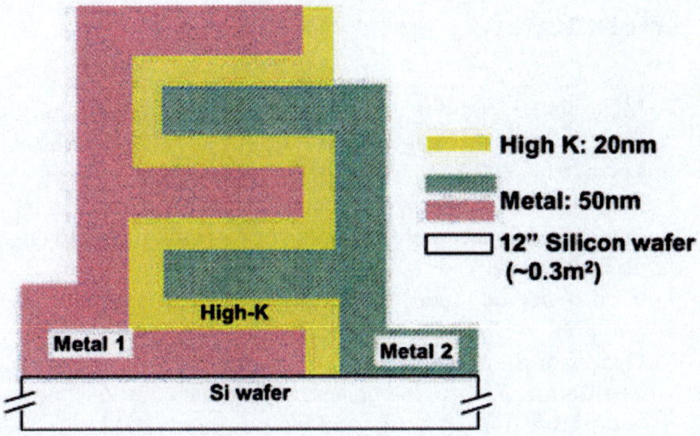

Fig. 10.27 The vertical stacking of the supercapacitor

between the adjacent MIM is arranged on the two sides of the stack. In this embodiment, the left metal layer, the dielectric layer, and the right metal layer remain synchronized and detour upward to form a meandered structure for the most efficient use of the silicon wafer area.

10.5 Summary

The best future candidate for future aviation electricity is the hybrid power combining 3D HK SC and LTG. The 3D HK SC provides the power for VTOL and aids a clean takeoff, and the LTG ensures long-range aviation (Fig. 10.28).

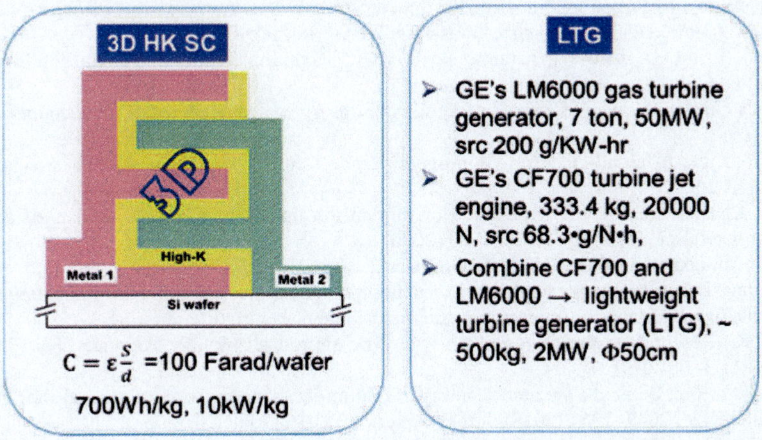

Fig. 10.28 The promising combination for future electric aviation power supply for both VTOL and long navigation range

References

1. Mekhilef, S., Saidur, R., Safari, A.: Comparative study of different fuel cell technologies. Renew. Sustain. Energy Rev. **16**(1), 981–989 (2012). ISSN 1364-0321. https://doi.org/10.1016/j.rser.2011.09.020
2. Fan, et al.: A high-performance supercapacitor-battery hybrid energy storage device based on graphene-enhanced electrode materials with ultrahigh energy density. Energy Environ. Sci. **6**, 1623–1632 (2013)
3. Thrust-specific fuel consumption. https://en.wikipedia.org/wiki/Thrust-specific_fuel_consumption
4. Generator Source, LLC, Brighton, Colorado. https://www.generatorsource.com/
5. The Boeing 787 engine generators produce 230VAC power. https://www.energy.gov/sites/default/files/2014/03/f9/sofc_for_aircraft_pnnl_2012.pdf
6. The Boeing 787 is equipped with six VSVF diesel generators for startup. https://www.youtube.com/watch?v=Sf6H8kSunRA
7. GE's CFM56 gas turbine fan aero-engine. https://www.vennershipley.co.uk/wp-content/uploads/2020/07/Europes_new_aviation_vision_is_electric_the_future1.pdf
8. GE's LM6000 gas turbine generator. https://www.geaviation.com/marine/engines/military/lm6000-engine
9. VerdeGo Aero™. https://www.verdegoaero.com/
10. Kobayashi, H., Hayakawa, A., Kunkuma, K.D., Somarathne, A., Okafor, E.C.: Science and technology of ammonia combustion. Proc. Combust. Inst. **37**(1), 109–133 (2019). ISSN 1540-7489. https://doi.org/10.1016/j.proci.2018.09.029
11. Salmon, N., Bañares-Alcántara, R.: Green ammonia as a spatial energy vector: a review. Sustain. Energy Fuels **5**(11), 2814–2839 (2021)
12. Ammonia as a renewable energy transportation media. ACS Sustain. Chem. Eng. **5**, 10231–10239 (2017)
13. MacFarlane, D.R., Cherepanov, P.V., Choi, J., Suryanto, B.H.R., Hodgetts, R.Y., Bakker, J.M., Simonov, A.N.: A roadmap to the ammonia economy. Joule **4**(6), 1186–1205 (2020)
14. Chen, D., Li, J., Huang, H., Chen, Y., He, Z., Deng, L.: Research progress on ammonia combustion and reaction mechanism. Chem. Bull. **83**(6), 508–515 (2020)
15. Quantum's hydrogen cylinder general specifications. https://www.qtww.com/wp-content/uploads/2019/06/H2-Tank-Specifications-June-2019-All-Tanks-1.pdf
16. GE's LM6000 electric generator. https://www.geaviation.com/marine/engines/military/lm6000-engine
17. GE's CF700 lightweight turbofan jet engine. https://www.geaviation.com/bga/engines/cf700-engine
18. Airbus, the E-fan all-electric twin propeller aircraft. https://www.airbus.com/search.html?q=e+fan+2.0
19. Li, K.: Research on the application of lithium-ion batteries in electric unmanned aircraft. Aeronaut. Sci. Technol. **31**(05), 1–10 (2020)
20. Sweden heart aerospace, ES-19. https://heartaerospace.com/
21. Airbus, E-Fan X, complex hybrid-electric flight demonstrator disruptive technologies project. https://www.airbus.com/innovation/zero-emission/electric-flight/e-fan-x.html
22. Li, K.: Research on the development of electric aircraft technology. Aeronaut. Sci. Technol. **30**(01), 1–7 (2019)
23. Liu, X., Ren, D., et al.: Thermal runaway of lithium-ion batteries without internal short circuit. Joule **2**(10), 2047–2064 (2018). https://doi.org/10.1016/j.joule.2018.06.015
24. Wang, L., et al.: Safety accidents of Li-ion batteries: reliability issues or safety issues. Energy Stor. Sci. Technol. **10**(1) (2021)
25. Sharaf, O.Z., Orhan, M.F.: An overview of fuel cell technology: fundamentals and applications. Renew. Sustain. Energy Rev. **32**, 810–853 (2014). ISSN 1364-0321. https://doi.org/10.1016/j.rser.2014.01.012

26. Baroutaji, A., Wilberforce, T., Ramadan, M., et al.: Comprehensive investigation on hydrogen and fuel cell technology in the aviation and aerospace sectors. Renew. Sustain. Energy Rev. **106**, 31–40 (2019). https://doi.org/10.1016/j.rser.2019.02.022
27. Airbus E-fan. https://www.airbus.com/innovation/zero-emission/electric-flight.html
28. Rolls-Royce ACCEL, 2019, Li-ion battery of 367.5 kW power for 320 km flight, weight 1200 kg. https://www.rolls-royce.com/innovation/accel.aspx
29. Bruce, P.G., Freunberger, S.A., Hardwick, L.J., et al.: Li-O_2 and Li-S batteries with high energy storage. Nat. Mater. **11**(1), 19–29 (2011)
30. Maleki, M., Tichter, T., ElNagar, G.A., et al.: Hybrid electrospun nanofibers as electrocatalyst for vanadium redox flow batteries: theory and experiment. ChemElectroChem **8**(1), 218–226 (2021). https://doi.org/10.1002/celc.202001380
31. Trovò, A.: Battery management system for industrial-scale vanadium redox flow batteries: features and operation. J. Power Sources **465**, 228229 (2020)
32. Papathakis, K.: Review of AQUIFER technology feasibility. In: 2020 AIAA Aviation Forum, Oral Report, 15–19 June 2020, Virtual Meeting
33. Skeleton Technologies, a Germany ultracapacitors company. https://www.skeletontech.com/
34. Tecdia, a Japanese high-K ceramic capacitor. http://www.tecdia.com/
35. Lörstad, D., Lindholm, A., Pettersson, J., et al.: Siemens SGT-800 industrial gas turbine enhanced to 50 MW: combustor design modifications, validation and operation experience. In: Turbo Expo: Power for Land, Sea, and Air, p. 55119: V01BT04A038. American Society of Mechanical Engineers (2013)
36. El-Suleiman, A., Samuel, O.D., Amosun, S.T., et al.: Gas turbine performance forecast and assessment: GE LM2500 in outlook. IOP Conf. Ser. Mater. Sci. Eng. **1107**(1), 012025 (2021)
37. Day, W.H.: FT8: a high performance industrial and marine gas turbine derived from the JT8D aircraft engine. In: Turbo Expo: Power for Land, Sea, and Air, p. 79245: V002T03A005. American Society of Mechanical Engineers (1987)
38. Vignesh, P., et al.: Biodiesel and green diesel generation: an overview. Oil Gas Sci. Technol. **76**(1), 6 (2021)
39. Lechniak, J.A., Salazar, M., Abbigail, W., Morello, J., Papathakis, K.: Nano-electro fuel energy economy and powered aircraft operations. In: AIAA Scitech 2020 Forum. AIAA 2020-0117, Jan 2020
40. GE's aeroderivative and heavy duty gas turbine electric generators. https://www.ge.com/power/gas/gas-turbines
41. Walsh, P.P.: Gas Turbine Performance, 2nd edn., Chap. 1.4 (2008)
42. Richter, E., Anstead, D., Bartos, J., Watson, T.: Preliminary Design of an Internal Starter/Generator for Application in the F110-129 Engine. SAE Technical Paper 951406 (1995). https://doi.org/10.4271/951406
43. Andersson, F.: Integrated generator for use in aircraft engines (2018)
44. Ceder, G.: Opportunities and challenges for first-principles materials design and applications to Li battery materials. MRS Bull. **35**(9), 693–701 (2010)
45. Tarascon, J.M.: Key challenges in future Li-battery research. Philos. Trans. R. Soc. A Math. Phys. Eng. Sci. **368**(1923), 3227–3241 (2010)
46. Kim, D.H., Lee, J.H., Hwang, H.Y.: Aerodynamic analysis, required power and weight estimation of a compound (tilt rotor + lift + cruise) type eVTOL for urban air mobility using reverse engineering techniques. J. Adv. Navig. Technol. **25**(1), 17–28 (2021)
47. Liu, Y., Zhang, R., Wang, J., et al.: Current and future lithium-ion battery manufacturing. iScience **24**(4), 102332 (2021)
48. Tranter, T.G., Timms, R., Shearing, P.R., et al.: Communication—prediction of thermal issues for larger format 4680 cylindrical cells and their mitigation with enhanced current collection. J. Electrochem. Soc. **167**(16), 160544 (2020)
49. Schwunk, S., Armbruster, N., Straub, S., et al.: Particle filter for state of charge and state of health estimation for lithium–iron phosphate batteries. J. Power Sources **239**, 705–710 (2013)
50. Fuel cell vehicle cost analysis. http://www.hydrogen.energy.gov/pdfs/progress17/v_e_5_james_2017.pdf

51. Winter, M., Brodd, R.J.: What are batteries, fuel cells, and supercapacitors? Chem. Rev. **104**(10), 4245–4270 (2004)
52. Cecere, D., Giacomazzi, E., Ingenito, A.: A review on hydrogen industrial aerospace applications. Int. J. Hydrogen Energy **39**(20), 10731–10747 (2014). ISSN 0360-3199. https://doi.org/10.1016/j.ijhydene.2014.04.126
53. Bičáková, O., Straka, P.: Production of hydrogen from renewable resources and its effectiveness. Int. J. Hydrogen Energy **37**(16), 11563–11578 (2012). ISSN 0360-3199. https://doi.org/10.1016/j.ijhydene.2012.05.047
54. Hydrogen production: thermochemical water splitting. Department of Energy. [Online]. Available: https://www.energy.gov/eere/fuelcells/hydrogenproduction-thermochemical-water-splitting
55. Hwang, H.T., Varma, A.: Hydrogen storage for fuel cell vehicles. Curr. Opin. Chem. Eng. **5**, 42–48 (2014)
56. Dornheim, M., Doppiu, S., Barkhordarian, G., Boesenberg, U., Klassen, T., Gutfleisch, O., Bormann, R.: Hydrogen storage in magnesium-based hydrides and hydride composites. Viewpoint set no. 42 Nanoscale materials for hydrogen storage. Scr. Mater. **56**(10), 841–846 (2007). ISSN 1359-6462. https://doi.org/10.1016/j.scriptamat.2007.01.003
57. Quantum fuel systems for hydrogen tanks. http://www.qtww.com/
58. Schlapbach, L., Züttel, A.: Hydrogen-storage materials for mobile applications. In: Materials for Sustainable Energy: A Collection of Peer-Reviewed Research and Review Articles from Nature Publishing Group, pp. 265–270 (2011)
59. Kadyk, T., Winnefeld, C., Hanke-Rauschenbach, R., et al.: Analysis and design of fuel cell systems for aviation. Energies **11**(2), 375 (2018)
60. Muthukumar, M., Rengarajan, N., Velliyangiri, B., et al.: The development of fuel cell electric vehicles—a review. Mater. Today Proc. **45**, 1181–1187 (2021)
61. Tehrani, Z., et al.: Large-area printed super capacitor technology for low-cost domestic green energy storage. Energy **118**, 1313–1321 (2017)
62. Conway, B.E., Pell, W.G.: Double-layer and pseudo capacitance types of electrochemical capacitors and their applications to the development of hybrid devices. J. Solid State Electrochem. **7**(9), 637–644 (2003)
63. GE's LM6000 gas turbine datasheet. https://www.geaviation.com/sites/default/files/datasheet-lm6000.pdf
64. GE's LM6000 engine. https://www.geaviation.com/marine/engines/military/lm6000-engine
65. GE's J85 small single-shaft turbojet engine. https://en.wikipedia.org/wiki/General_Electric_J85
66. Wang, X., et al.: Ultra high dielectric constant, temperature stable multilayer ceramic capacitor material and its preparation method. Patent CN1397957A, Tshinghua University, 2003
67. Hennings, D., Klee, M., Waser, R.: Advanced dielectrics: bulk ceramics and thin films. Adv. Mater. **3**(7–8), 334–340 (1991)
68. Xia, W., Liu, Y., Wang, G., Li, J., Cao, C., Hu, Q., Chen, Y., Lu, Z., Wang, D.: Frequency and temperature independent $(Nb_{0.5}Ga_{0.5})_x(Ti_{0.9}Zr_{0.1})_{1-x}O_2$ ceramics with giant dielectric permittivity and low loss. Ceram. Int. **46**(3), 2954–2959 (2020). ISSN 0272-8842. https://doi.org/10.1016/j.ceramint.2019.09.292. https://www.sciencedirect.com/science/article/pii/S0272884219328287
69. Tan, F., Zhao, H., Zhang, Q., et al.: Dielectric performance of $(Pb_{1-x}Sr_x)Nb_2O_6$-$NaNbO_3$ thin film materials system: substitution effects. Mater. Sci. Forum **898**(pt.3), 1699–1704 (2017)
70. Pavlidis, V.F., Friedman, E.G.: 3-D Integrated Circuit Fabrication Technologies, pp. 37–63. Morgan Kaufmann (2009). ISBN: 9780123743435. https://doi.org/10.1016/B978-0-12-374343-5.00003-4
71. Srivastava, A., Mangla, O., Gupta, V.: Study of La-incorporated HfO_2 MIM structure fabricated using PLD system for analog/mixed signal applications. IEEE Trans. Nanotechnol. **14**(4), 612–618 (2015)

72. Chaker, A., Szkutnik, P.D., Pointet, J., et al.: Understanding the mechanisms of interfacial reactions during TiO_2 layer growth on RuO_2 by atomic layer deposition with O_2 plasma or H_2O as oxygen source. J. Appl. Phys. **120**(8), 1–3 (2016)
73. Neve, C.R., Detalle, M., Nolmans, P., et al.: High-density and low-leakage novel embedded 3D MIM capacitor on Si interposer. In: 3D Systems Integration Conference. IEEE (2016)

Chapter 11
The Integrated Design of Aero-engines and Aero-craft

The outline of this chapter is:

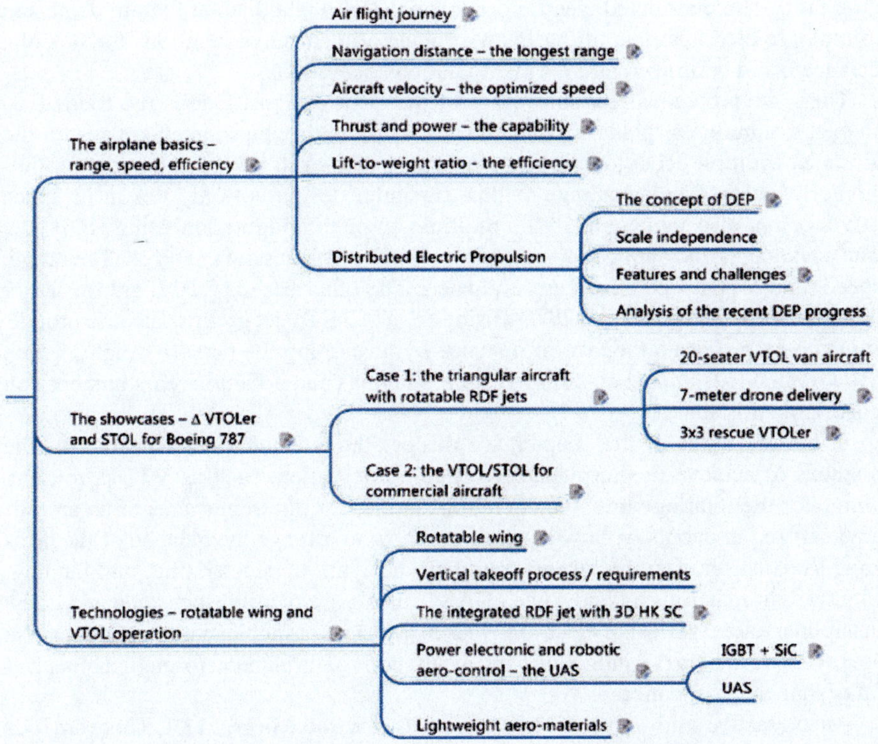

The outline of the e-plane regarding to its design and technology

Traditionally, the airplane and aero-engine are two separate units and are not closely related. In electric aircraft, since multiple small engines can be equipped

© The Author(s), under exclusive license to Springer Nature Singapore Pte Ltd. 2023

F. L. Duan, *When AIAA Meets IEEE*,
https://doi.org/10.1007/978-981-19-8394-8_11

on one airplane, the integral design of aero-engines with the aero-vehicle becomes feasible together with its obvious advantages. The integrated design of an airplane with multiple engines requires three key elements:

(1) Multiple and flexible propulsion technology, i.e., a small and lightweight aero-engine with a compact structure that is easy to operate in multiple orientations.
(2) Aviation power technology, such as using 3D HK SC to provide enough electric power for vertical takeoff and kerosene LTG for long navigation ranges.
(3) Rotatable wings allowing free-scale engine orientation, including strategies and methodologies.

The topics discussed in this chapter seem similar to those in Chap. 7. The difference lies in their different focuses: This Chapter addresses the technical details, while Chap. 7 addresses the general goals and management issues, the benefits, and basic concepts. Thie Chapter contains three aspects: the overall integrated aircraft design concept, the showcase, and enabling technology. In the first section, some general aviation concepts are re-visited in the context of electric aviation and eVTOL, such as the range, speed, power and thrust, efficiency versus thrust-weight ratio or lift-to-drag ratio. The distributed electric propulsion (DEP) is a distinct feature dedicated to multiple electrical jets fitting in an airplane at distributed locations for flexible, balanced, and optimized aircraft manipulations/operations.

Then, we propose a few showcases of the DEP designs. One is the triangular-shaped streamline airplane with three rows of rotating wings attached beneath the airplane. Multiple RDF jet engines are arranged on each wing for flexible propulsion/orientation. The advantage of this triangular design not only lies in its space privilege but also its best fit in the multiple engine configuration using RDF jets. The advantages and application scenarios of such airplanes are analyzed. The range, speed, and VTOL operations are evaluated. The other case is STOL aid for larger commercial aircraft such as B787 by using the 3×3 RDF jet groups. Electric propulsion can help shorten the takeoff distance by leveraging the takeoff weight, which has a tremendous impact on airport resource savings and pollution reduction both on noise and emissions.

In the third part of the chapter, we discuss the technological aspects, i.e., the methods to achieve this airplane/aero-engine integration, such as VTOL, rotating wing. For the rotating wing, the operations at the five different stages of an aircraft are analyzed in details on how to rotate and how to arrange the balance of the horizontal/vertical thrusts: the takeoff, ramping, navigation, decelerating, and landing. Lightweight materials and advanced UAS controls are two indispensable and interdisciplinary accessories for electric airplanes. Most recent achievements and progress are reviewed to help our judgments and evaluations of airplane/aero-engine blueprint on design and fabrication.

An interactive study of an electric airplane is illustrated in Fig. 11.1. There are four basic elements and four basic parameters fitting the three main aviation scenarios, corresponding to the relationship between the Z-axis versus the X and Y axes—the application versus the methods and features:

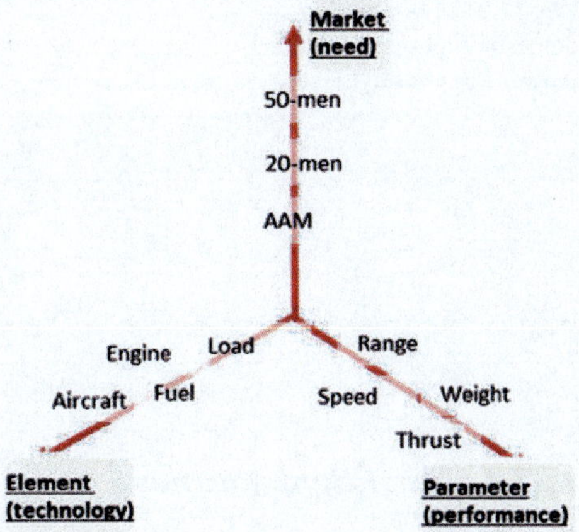

Fig. 11.1 Interactive chart between the aviation marketplace versus technology and performance

- *X* axis: an aircraft consists of four basic elements, especially under the context of weight factors: airplane, engine, load, and fuel. For example, the load includes the payload, aircraft (body, wing, tail) dry weight, aero-engine, and generator/fuel. The fuels include kerosene, battery/SC/FC with the energy and power density of Wh/kg, and W/kg.
- *Y* axis: an aircraft has four key parameters: the mileage and speed, thrust and weight, etc.
- Z axis: there are three main market places: the AAM market, which covers a vast range of applications, such as the two popular cases we proposed: 500 kg drone and 3-men rescue VTOLer. The other two popular markets are hybrid 20-seater van flyer and hybrid STOL commercial airlines such as C919 and Boeing787/Airbus380.

11.1 The Airplane Basics—Range, Speed, Efficiency

The aircraft/engine integrated design includes basic aircraft parameters such as range and speed versus technology abilities such as the power/thrust and lift-to-drag ratio. Let us first analyze the whole airplane journey.

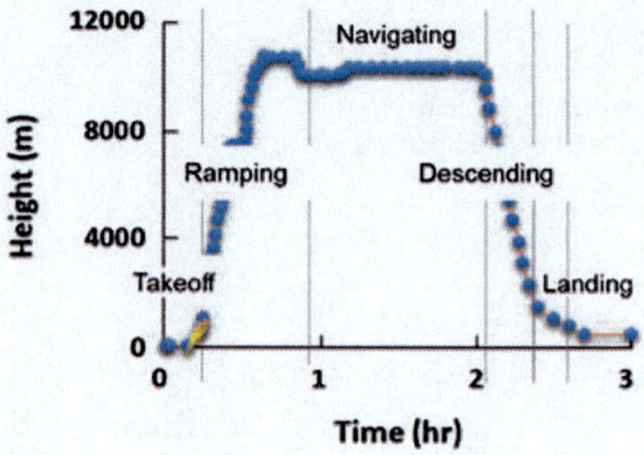

Fig. 11.2 Five stages of a domestic flight profile of an Airbus 320 aircraft

11.1.1 Air Flight Journey

There are five basic stages for a flight.

(1) Take off: leave the ground to an altitude of 460 m.
(2) Climb: from 460 m to cruise altitude.
(3) Cruise: maintain a certain speed and altitude until approaching the destination.
(4) Descend: decrease the altitude to approach the runway.
(5) Land: from the height of 15 m to the complete stop on the ground.

The profile of a flight is illustrated in Fig. 11.2. Most of the power consumption rate with the maximum thrust is during the takeoff, although it is very short (~ just 5 min). Most of the fuel consumption is during navigation although the thrust is just 1/5 of its peak. The engine is not completely shut off during landing. In fact, reverse thrust is needed to decelerate the airplane when landing for shorter distance. It is noted that the airport contamination during landing is comparable to the takeoff considering its time span and its subcombustion state (see detailed analysis in Chap. 7).

11.1.2 Navigation Distance—The Longer Range

In general, the range of an airplane is an empirical estimation that depends on the kerosene consumption rate (specific fuel consumption) of the engines. For example, consider a B787 airplane equipped with 2 GEnX engines carrying 130 ton kerosene. Assume the takeoff and landing consumes 20 tons kerosene, and only 110 ton is left for navigation. The sfc of GEnX during cruising is 14.5 g/kN.s [1]. Navigation just uses 1/5 thrust of the total 600 kN. Then, the estimated time span to use up the 110 ton kerosene is ~ 18 h. Under a navigation speed of 900 kmh, the range is estimated to be 16,000 km (Table 11.1).

For the navigation speed, an optimized value of 900 kmh is chosen in accordance with the best aviation efficiency η:

Table 11.1 Navigation data of B787 with its GEnX engines

B787 (130 t kerosene)		2 × GEnX (sfc = 15 g/kN.s)	Flight		
Weight of fuel for navigation	Navigation thrust	Fuel comsumption	Speed	Time	Range
110,000 kg	120 kN	6264 kg/h	900 kmh	18 h	15,805 km
(Note: takeoff uses 20 ton)		(= thrust * sfc * 3600* 0.001)			

Source https://en.wikipedia.org/wiki/Thrust-specific_fuel_consumption

$$V_{\text{optimized}} = \frac{(\text{*energy/time})}{(\text{navigation thrust})} \tag{11.1}$$

For example, the most optimal thrust is just 1/5–1/4 of the total GEnX thrust of 600 kN during the navigation journey. For more detailed theories on aircraft range/speed performance, please refer to [2–4].

The range of an airplane basically depends on the total carried energy (electricity or fossil fuel) and the output rate (speed or power) during navigation. Ignoring the energy needed for takeoff and landing, a rough estimation of the range follows the formula:

$$R = E\eta \frac{1}{g} \frac{L}{D} \frac{m_b}{m_a} \tag{11.2}$$

That is:

$$\begin{aligned} \text{Range(km)} = 3.6\ &\text{E(energy density, } \frac{\text{Wh}}{\text{kg}}) \\ &\times \eta(\text{Eta, 0.9 for electric, 0.3 for kerosene}) \\ &\times \frac{1}{\text{g}(9.8\,\text{ms}^{-2})} \\ &\times \text{L/D}\left(\frac{\text{lift}}{\text{drag}}\text{ratio} \approx 30\right) \\ &\times (m_b(\text{battery,fuel})/m_a(\text{plane + engine + fuel + battery + payload}) \end{aligned} \tag{11.3}$$

where E is the energy density, η is the power conversion coefficient, L/D is the lift-drag ratio, and m_b m_a is the weight of the fuel/airplane. The above formula is fairly corrected to evaluate the range of the airplane; for example, for the Boeing 787, we take the energy density E ranging from 4600 to 12,000 Wh/kg and the weight of the airplane from its takeoff to its landing (250–120 tons), and the longest range is from 9600 to 48,000 km, which covers its official reported longest range 16,000 km.

The above formula is used to analyze the range of the three types of aircrafts as shown in Table 11.2. Among them, RR's spirit of innovation (SOI) and Boeing's 787 are based on existing data. The SOI aircraft is purely electric, while Boeing 787 is a kerosene-fuel aircraft but adopts the most electric ingredients in history. The actual SOI range is not as high as expected. The possible reason is that in this test flight, the input data are not very accurate. SOI electric aircraft is not a very mature technology, and battery parameters, propeller efficiency, and lift-drag ratio have not been optimized (for example, propeller efficiency may not reach 70% and lift-drag ratio may not reach 29). Boeing 787 represents the highest level of the current aircraft with a mature technology and good data repeatability. The lift-drag ratio of the aircraft is 29, and the engine efficiency is 36%. According to these official data, the estimated range is very close to the expected value (16,000 km) from Boeing, which verifies the accuracy of the above range calculation formula. In this estimation, our proposed triangle hybrid aircraft is equipped with 50 kg fuel and 50 kg battery. The fuel is used for navigation, and the battery is used for vertical takeoff and landing. In addition, the maximum range of an electric aircraft is also estimated. Assuming that the weight of the aircraft is only the battery, the maximum efficiency is 0.9, and the lift-drag ratio is 29, the maximum range is 9.4 times the battery density [Wh/kg]. With the current best Li battery level of 150 Wh/kg, the maximum range of an electric aircraft is 1500 km. In general, the 150 Wh/kg E-density of the Li battery versus 4600 Wh/kg of kerosene is not comparable, and the energy density is the key for the navigation range. Of course, the kerosene-fuel airplane has a maximum limit of the range as well (9.4 * 4600 = 43,240 km compared to the current B787's 16,000 km range).

Table 11.2 Airplane range evaluation

Airplane	RR SoI	B787	3 × 3 D hybrid	
Parameters	Electric	Kerosene	Kerosene to navigate	Libattery for VTOL
	Existing cases			
E-density	168 Wh/kg	4600 Wh/kg	4600 Wh/kg	155 Wh/kg
P-density	400 W/kg	4000 W/kg		400 W/kg
P_VTOL	Na	Na		20 kW
mb (battery)	450 kg	0 kg	50 kg	50 kg
E_VTOL	Na	Na		8 kWh
mf (fuel)	0 kg	110,000 kg	50 kg	
E_navigate	76 kWh	506,000 kWh	230 kWh	
ma (engine + craft)	730 kg	100,000 kg	510 kg	
mp (payload)	80 kg	20,000 kg	80 kg	
MGTOw	1260 kg	230,000 kg	690 kg	
Eta	0.63	0.36	0.36	
Range	577 km	15,848 km	1452 km	VTOL
Note	Eta = 0.9 * 0.7	L/D = 29	60 kg engine, 300 kg LTG, 50 kg fuel, 50 kg battery, 150 kg plane	

Pure electric aircraft has a limited range due to the limitation of the energy density of the current Li battery technology. Therefore, hybrid electric airplanes are the best option for future electric aviation. Boeing 787 represents the best airplane design (aero-dynamic performance, lightweight material, etc.) and the best aero-engine adoption (GEnX, best aero-efficiency, best sfc, etc.). We are now looking forward to the best electric propulsions. In this book, we propose a new electric propulsion technology, the RDF jet, hoping to shed some light for future electric aero-engine. Since the RDF jet does not transfer kinetic energy through the central shaft, higher propulsion conversion can be expected. The rim drive motor uses a direct force to apply to the fan blade, which has larger torque and higher driving efficiency. As a roadmap anticipation of our Δ e-airplane in Table 11.2, the following parameters (but not limited to) are assumed: lift-to-drag ratio 29, the best electric motor efficiency 95%, the best gas generator efficiency 36%, lightweight building material density 0.5 g/cm^3, the sfc of the hybrid generator 15 g/kN.s, the best energy density 4600 Wh/kg, and the best engine/motor's power over weight ratio 7–9 kW/kg.

11.1.3 Aircraft Velocity—The Optimized Speed

For a certain airplane, there exists an optimal navigation speed. As shown in Fig. 11.3, the optimized speed is different for the best mileage at different heights. For example, the cruising speed of large aircraft (10,000 km altitude) such as Boeing 787 is 0.85 Mach (903 km/h), and the speed of medium plane (3000 km altitude) such as ARJ21 is 0.785 Mach (828 km/h).

There exist two optimal navigation speeds either for the best efficiency or for the longest range, correlating to different powers. The shape of the relationship between the power and speed of an aircraft is similar to a parabola (Fig. 11.4). However, the position of the minimum power does not exactly correspond to the maximum lift-drag ratio. The power, range, and speed are related to the lift-drag ratio, altitude, and weight, and the best range and speed follow different mechanisms. Ignoring the

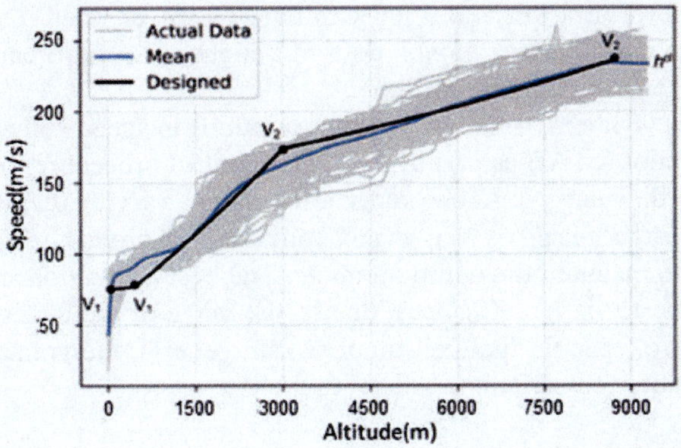

Fig. 11.3 Optimized navigation flight speed at different heights [5]

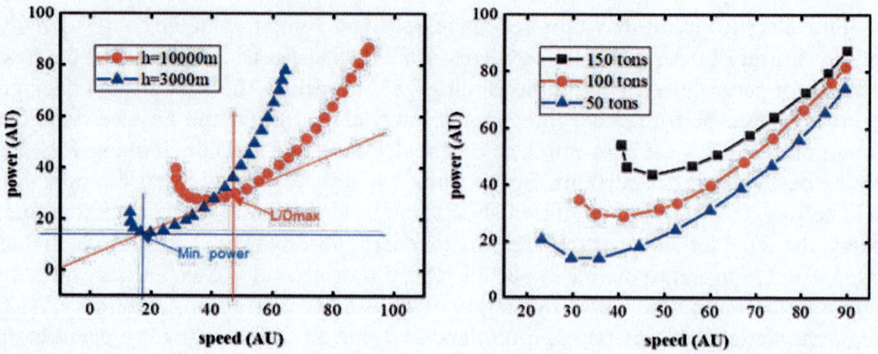

Fig. 11.4 Impact of different navigation speeds on the needed power at different altitudes and weights

takeoff and landing of the aircraft, during the flight journey, the range and speed follow different integrals. The range is the distance *S*, which is the integral of the velocity and speed, and the speed is for the efficiency, i.e., to achieve the minimum energy E during the whole journey:

$$S = \int V(t)dt \tag{11.4}$$

$$E = \int TdS = \int \frac{1}{2}\rho V(t)^2 C_L \times V(t)dt \tag{11.5}$$

As seen, they follow different relationships with the traveling time. The best flight distance follows the best L/D_{MAX}, i.e., the position of the tangential point between power and speed. On the other hand, the lowest energy consumption is at the lowest power position, that is, where the tangential ratio between power and speed is 0, as shown in Fig. 11.4. The L/D_{MAX} is also equivalent to the C_L/C_D ratio (typical value of C_L, C_D, C_L/C_D, and angle of attack is 2.21, 0.5, 0.02, and 25o, respectively [5]), which is highly dependable to the airplane design. B787 thus far has the best level of airplane design with dual GEnX engines. The C_L/C_D ratio is ~ 30. With a multi-electric engine design, a higher *L*/*D* ratio can be expected with improved aero-dynamics tailoring.

For a long-term flight, the variations in speed and power must be adjusted by the pilot (or AI) as part of the cruise control procedure to maintain the best aero-flight efficiency [6]. A long-range aircraft has a varying fuel weight which is a considerable part of the gross weight, and cruise control procedures must adjust the optimal power to maintain the optimum range. The best navigation speed has to accommodate the best aviation efficiency, that is, the range or the lowest energy consumption rate or sfc (specific fuel consumption). In general, the greater speed (powerfulness) does not necessarily correspond to the lowest sfc.

11.1.4 Thrust and Power—The Capability

An aircraft's capability mainly depends on the engine's powerfulness; that is, the maximum thrust of the engine can provide. When taking off from the ground, the engine is at full load of power and maximum thrust. For example, A1 uses the Boeing 747 plane as a prototype with a maximum takeoff weight of 396.9 tons. The maximum takeoff thrust of four CF6 engines can reach 1052 kN. The exhaust temperature and the engine speed reach the maximum values in the takeoff stage. Because the maximum capacity is used, the duration cannot exceed 5 min. After takeoff, less power is needed, allowing the engine to operate at lower level thrust for the best navigation efficiency (range and sfc). During flight, the engine may just use 1/5 of the full thrust for propulsion. For Boeing 747 plane, the NGTOW is 400 t, the *L*/*D* ratio is 30, and the needed navigation thrust is 134 kN, equivalent to just 1/8 of the maximum total thrust of 1052 kN of four CF6 engines. That is to say, only 13% of the total capacity is used. The analogy of just using 13% of the total power for navigation is to use a small portion of total power output to listen to a HiFi symphony from a high-end soundtrack system. When playing the music in a superpower audio system with much smaller output power, the sound quality is deep, rich, and energetic. The same principle applies for the A1 plane, which only uses 1/10–1/5 of the engine capacity during navigation with roomy and more optimized flight operation.

The maximum speed of the aircraft refers to the speed that can be reached or allowed when the aircraft engine operates at the maximum power or thrust. Obviously, the higher the thrust is, the higher the speed, and the greater the energy consumption as well. So, the maximum flight speed only represents the ability of an aircraft. The aviation should adopt the optimal cruise speed; that is, it is necessary to properly hold the thrust for maximum flight efficiency of the aircraft. Of course, special occasions are exceptions such that a fighter can only pursue its maximum thrust to ensure the maximum speed in combat.

Looking back the airplane history, in earlier times, the engine and propulsion are two separated units, i.e., the power is from the diesel engine and the thrust is from the propeller. Therefore, to quantify the engine's capacity, we use the horse power (HP or KW) to quantify the capability of an engine. Since the advent of gas turbine engines, we started to use thrust (kN) instead of power (kW) to measure the engine's capability; it is more convenient to use thrust to estimate the performance of an aircraft, because the power and propulsion are integrated in one in gas turbine machines. Now, for electric propulsion, we have to go back to use the power to characterize the engine's capability again, since most of the time, the propulsion and energy are two separate units which are different from the gas turbine machine with a consistent flow of the power/thrust. Therefore, it is necessary to correlate the thrust versus power in different aero-engine scenarios. Table 11.3 lists such a relationship between thrust and power. When the aircraft is at navigation, the output power is related to the flight speed. For example, when the flight speed of an ordinary commercial airliner is 900 kmh, then 1 kg of thrust corresponds to 2.5 kW (i.e., 10 kN vs. 2.5 kW) [7].

Table 11.3 Thrust versus power

		Velocity (km/h or m/s)		Thrust	P	kW per 10N (per kg thrust)	Notes
P vs T	flight	300kmh	83ms	120kN	10.00MW	0.83kW/kg	
		360kmh	100ms	120kN	12.00MW	1.00kW/kg	P=TV1
		450kmh	125ms	120kN	15.00MW	1.25kW/kg	e.g., 10 N thrust = 2.5
		600kmh	167ms	120kN	20.00MW	1.67kW/kg	kW power when
		900kmh	250ms	120kN	30.00MW	2.50kW/kg	V1=900kmh
		1200kmh	333ms	120kN	40.00MW	3.33kW/kg	V1: navigation speed
B787 GEnX	Takeoff_all	270kmh	75ms	600kN	45.00MW		P= P1+P2
	P1	270kmh	75ms	400kN	30.00MW		P1=FV2, P2=fV2
	P2	270kmh	75ms	200kN	15.00MW		
							F:acceleration
							f:friction
	f= mgk =	200*10*0.1	=	200kN			V2:takeoff speed
	Note:	navigation thrust = 120kN, takeoff thrust = 600kN					

During takeoff, the output power is the maximum for acceleration and to overcome ground friction. With two GEnX engines, the total thrust is 600 kN, corresponding to 45 MW power, which are used for acceleration (30 MW) and friction overcoming (15 MW), respectively. During navigation, the power consumption is the product of thrust and velocity. Although the navigation uses only 1/5 of the thrust, with the 900 kmh navigation speed, the power needed is also high, ~ 30 MW, with the cruising sfc ~15 g/kN.s.

Although we estimate the relation of thrust and power above, their relationship is more complicated in reality. Sometimes, we use the terms "power" and "thrust" interchangeably when discussing climb performance. This erroneously implies that these two terms are synonymous. In fact they belong to two different physics. It is therefore important to distinguish power versus thrust. Thrust is a force or pressure exerted on an object. Thrust is measured in pounds (lb) or newtons (N). Power, however, is a measurement of the rate of performing work or transferring energy (kinetic energy and potential energy). Power is typically measured in horsepower (hp) or kilowatts (kW). We can think of power as the motion (kinetic energy) a force (thrust) creates when exerted on an object over a period of time. Figure 11.5 shows the comparative relationship of the typical propulsions: gas turbine and propeller. Compared to the power versus speed characteristics, the gas turbine engine is certainly more advantageous than the propeller, especially at higher speeds. For the propeller machine, both the thrust and the power decrease with increasing speed. The AOC and ROC represent the angle of climb and the rate of climb [6]. AOC is seeking the shortest horizontal distance to climb as high as possible. ROC is seeking for the shortest time to reach that height. For example, when using the turbine machines, in order for the best AOL, one should use the thrust at the L/D_{max}; in order to achieve the best ROC, the power needed is higher than the L/D_{max} point.

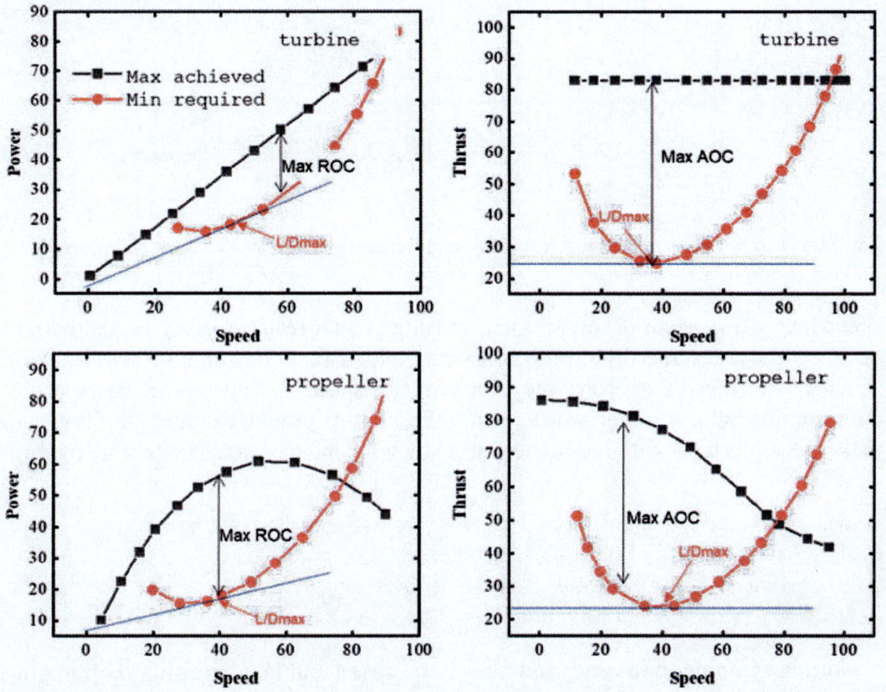

Fig. 11.5 Power, thrust versus speed of turbine and propeller

11.1.5 Lift Versus Weight—The Efficiency

As we describe above, the best efficiency of an airplane depends on the best design of an airplane, which mainly depends on the lift-weight ratio. When an airplane navigates at a constant speed, the thrust and drag forces, the gravity, and the lift are balanced. Taking a simple flying plate as an example, the relationship between the lift and velocity is shown in Fig. 11.6. The lift force is dependent on the attack angle α and will reach its maximum value when α is close to 20°. The flying plate will maintain at a certain height with a certain speed with the attacking angle of 20°. In reality, the aircraft uses airfoils instead of just a plain plate, but the floating principle is similar. An airfoil is a shaped structure taking full advantage of the lifting effect in accordance with Bernoulli's principle, with two surface profiles of the rounded end facing forward and the trailing edge which is narrow and tapered.

The speed of an airplane is also the balance between thrust and resistance. When the thrust is greater than the resistance, the aircraft will accelerate. As the speed increases, the resistance will also increase (~ V^2). When the resistance is equal to the thrust, the acceleration is zero, and the aircraft reaches the cruise speed. When the airplane cruises at higher power, it consumes more fuel. Therefore, the maximum speed is not necessarily the optimal cruise speed for the best efficiency. In fact, as the

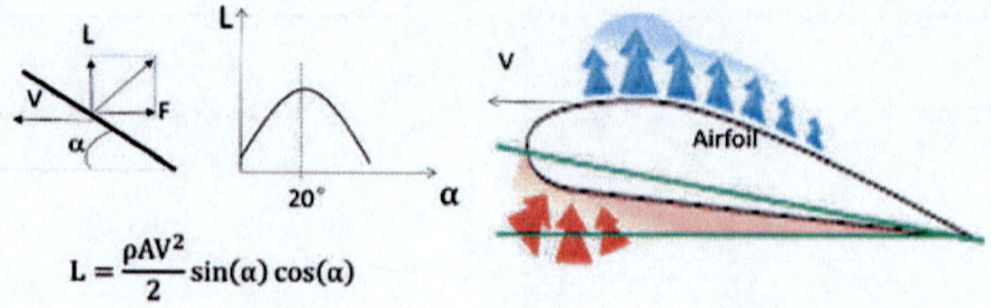

Fig. 11.6 Lift of a flying plate with a speed *V* and Bernoulli' lift of an airfoil for an airplane

speed increases, the air is compressed, causing greater resistance. At the same flight distance, the transonic flight time is shorter, but its fuel consumption will increase at least 1.5 times! Therefore, the best cruising speed should match the best fuel consumption rate. In other words, it is necessary to properly control the thrust to make the aircraft fly with the highest efficiency for the best sfc. The best navigation state is:

Maximum lift drag ratio + optimum cruise speed

During ramping, one needs to balance the speed and height during the ramping process, i.e., to balance the two energies:

$$\text{Kenetic Energy} = \frac{1}{2} \times m \times V^2 \quad \text{for speed}$$
$$\text{Potential Eergy} = m \times g \times h \quad \text{for height}$$

There are two methodologies in the climbing process: (1) to achieve maximum altitude in minimum horizontal distance with the best L/D_{max} and (2) to achieve a maximum flying distance for best efficiency. As shown in Fig. 11.7, two airplanes use 30 s to climb but with different climbing strategies. The first aircraft raises its angle of attack to climb 1000 m altitude in a shortest horizontal distance of 3000 m with a total traveling distance of 3162 m. Another airplane uses the same 30 s to fly 6000 m. In this scenario, it takes 3867 m horizontal distance to reach 1000 m height.

Except for the factors of air density, altitude, and speed, the best *L*/*D* ratio highly relies on the design of the wing airfoil (surface/shape) and overall airplane architecture to achieve the maximum lift with the minimum resistance. For example, the lift-drag ratio of the Boeing 787 shall be the best of its kind ~ 30; that is, a 200 ton aircraft needs 17 tons of thrust during the navigation. On the other hand, the lifting force of an airplane is from the resistant force of the aircraft, which consumes power. From this perspective, the smaller the resistance is, the better. That is, the smaller the weight of the aircraft, the better. Therefore, using lightweight materials to build airplane is always preferred; lightweight materials are always a key airplane technology which will be elaborated at the end of this chapter.

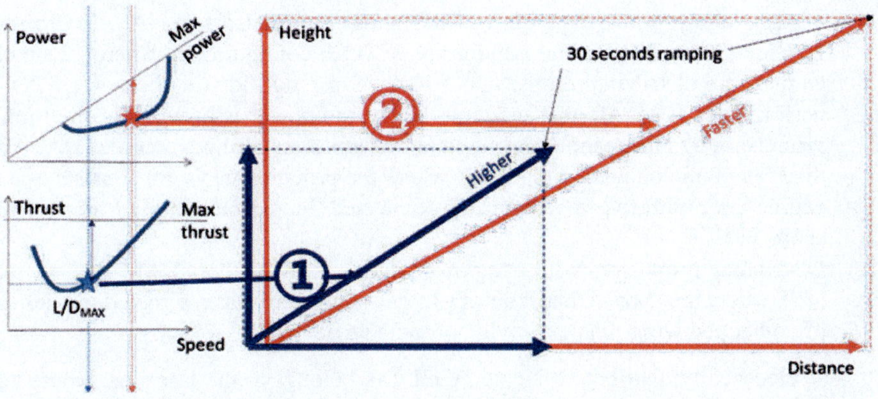

Fig. 11.7 Two strategies for airplane ramping: ① shortest distance to reach the height or ② quickest time to leave the airport

11.1.6 *Distributed Electric Propulsion*

The Concept of DEP

The DEP concept enables a higher overall efficiency of an airplane [8, 9]. Multiple electric thrusters can be placed flexibly and synergistically at various locations of an airplane for the best integrated aero-dynamic performance compared with just dual/four gas turbine engines equipped for Boeing 787. For example, careful integration of electric thrusters for boundary layer suction can improve propulsion efficiency, and proper positioning and configuration of the distributed propulsors can alleviate the vortex system on the lifting surface and mitigate the trailing vortex to increase dynamic pressure across blown surfaces for increased lift performance [10, 11]. Much lower noise signatures for the airport can be expected by using DEP since no thermal or less combustion is involved during takeoff/landing. DEP offers the capability to achieve revolutionary engine/aircraft integration to fit various disruptive aviation scenarios such as AAM.

The DEP concept originates from DP - distributed propulsion [12]. The traditional DP concept was awkwardly implemented either mechanically or via the thermal/airflow channels. While a formal definition of a DP system has not yet been established, in general, the distributed thrusts enhance system-level efficiency, capabilities, or performance as follows:

(a) Jet flaps provide spanwise thrust for high lift or cruise applications, such as those used in the Hunting Aircraft (BAE Systems, London, United Kingdom) and in a distributed propulsion system for a stealth fighter F-117 aircraft (Lockheed Martin, Bethesda, Maryland).
(b) Array of small individual engines synergistically coupled into the airframe, such that the propulsive units serve an integral role in enabling enhancements

of tight aero-propulsive coupling with multiple propulsors in NASA's Cruise-Efficient Short Takeoff and Landing (CESTOL) configuration, where 12 small engines are distributed on the upper surface of a hybrid wing body for STOL performance [13]. Another example is that the power is provided to multiple propulsors by one or more mechanical transmission methods, such as NASA's dual fan on the blended wing body, where the propulsion system consists of an engine core with two mechanically connected ducted fans through gears and shafts [14].

(c) Distributed propulsors with one power source, such as the ADAM III concept [15], where the "hot" exhaust air of two gas generators was redirected to a series of embedded wing fans to provide propulsive power.

The electrically distributed DP, i.e., the DEP, belongs to the last type, where an electrical energy source is used to provide power for multiple electric propulsors. The electrically aided distributed proportion is much more advanced because the power is transmitted to the individual electric energy via electric wires, which is far more convenient, faster responding, and much less energy loss during power transmission. Electronic engines can be distributed at any places on the airplane without considering any mechanical, thermal, or airflow constraints. For example, the triangular pattern of electronic propulsion can be arranged evenly below the airplane on the two sides of the wings and one at the front to form a stable propulsion for VTOL.

Scale Independence

Scale independence is one of the distinct features of the DEP. The power-to-weight ratio and the efficiency of the airplane/motors/controllers are indifferent whether a 100 kW power is dispersed into ten 10 kW shares or just two 50 kW drives. In addition, the cable weight/space from the power supply to the motor are relatively small. The compactness of the RDF jet also facilitates the use of the decentralized propulsion. The scale independence concept does not apply to piston engines or turbine engines. These engines use fossil fuels as a power source, and when reduced in size, their power (thrust)-to-weight ratio, efficiency, and reliability will be greatly affected with greater additional costs [16]. Only the DEP enables free distributed propulsion devices over the whole body at no cost to achieve integration advantages.

The Challenges

Distributed electric propulsion is regarded as a disruptive technology in the aviation industry. It involves multiple disciplines to bring substantial improvement of aircraft aero-dynamics. The scale independence of electric power systems provides a new degree of freedom for aircraft design. It helps to reduce the life cycle cost of aircraft and enhance the competitiveness of the aircraft market. The integrated design technology can excavate and utilize the coupling effect between different disciplines of DEP aircraft to a certain extent to reduce the structural weight, improve the flight performance and environmentally friendly characteristics of the aircraft, and improve the capability and robustness of the aircraft control system. The core factors that affect aircraft marketization are battery energy density, power density of motors and electronic control, and the efficiency of electric power systems mainly composed of these factors.

On the other hand, the main challenge lies in its excessive degree of multi-disciplinary of design freedom, and one must arrange their research focus one at a time. Although DEP is very advantageous, it is much more difficult to analyze these highly coupled disciplines and new physics-based methods to capture this complex interaction. Simplification of the complex system is necessary at the beginning stage of its R&D work with limited manpower.

Another bottleneck is that most of the current DEP work is mainly theoretical instead of experimental and infield tests. One key reason is that the current electric drives is neither optimized nor rotatable. Although small-scale unmanned aero-machines have successfully adopted multi-propeller propulsion in the UAS market, the orientation operation of larger-scale multi-propeller aircraft for VTOL aircraft is not that easy. Just from commonsense, it is not convenient to perform the directional operations of big propellers from vertical to horizontal for the VTOL and for the navigation. From this perspective, a new conceptual thruster is necessary. The RDF jet we proposed is a compact engine that possesses this distinct advantage of free-scale directional operations. The contributions of the RDF jet to the DEP are:

(a) Circumvent the tough engineering in traditional gas turbine engines, such as larger engines and high temperatures/pressures.
(b) Compact engine with rim driven inner fan blade instead of outstretching propeller.
(c) Small and lightweight enabling multiple and free distributed power, redundancy, and flexibility for VTOL/horizontal navigation.

Analysis of the Recent DEP Work

Huff et al. compared the noise signature of the DEP with gas turbine airplanes. The noise generated by the motor is much lower than that generated by compressor, combustion chamber, and turbine components. Compared with the noise of regional jet or single-channel passenger aircraft, the noise generated by the electric propulsion aircraft only by the motor system is 8–20 dB lower than the fan noise of regional jet aircraft and lower than that of single-channel commercial transport class aircraft [17]. NASA proposed [18] distributed hybrid electric propulsion used for large- and medium-sized passenger aircraft. The gas turbine burning fuel drives the generator to generate power, and then, the high-efficiency power transmission system transmits it to the drive motor of the fan or propeller to generate thrust. The system used for the horizontal takeoff and landing mode aircraft concept is also called distributed turbine electric propulsion. In the aero-dynamic design of hybrid electric propulsion future passenger aircraft, the coupling between distributed thrusters and aircraft bodies is used to improve the aircraft aero-dynamic characteristics. Schiltgen etc. [19] studied the aero-dynamic propulsion interaction and thermal system integration of TEDP aircraft systems with a confirmed aircraft configuration. Nguyen of NASA Ames Research Center proposed a new wing shape of distributed propulsion aircraft. The distributed propulsion installed on the wing can improve the aero-dynamic efficiency

(a) (b) (c)

Fig. 11.8 Joby S2 aircraft (figure courtesy: Joby Aviation, Santa Cruz, California), Lightning Strike aircraft (figure courtesy: Aurora Flight Sciences, Manassas, Virginia), and Airbus Vahana VTOL aircraft (figure courtesy: Airbus, Leiden, Netherlands)

by using the new wing shape. The preliminary conceptual analysis shows that for distributed propulsion aircraft, the proposed wing shape scheme can improve the lift-drag ratio [20]. Figure 11.8 illustrates a few showcases from NASA, the Aurora Flight Sciences, and Joby Aviation.

The feature/drawbacks of these showcases are as follows: they still use the old propeller concept, and there are very few detailed infield tests' data. In short, experimental work is highly needed, which requires consistent funding sponsors with persistent team efforts for many years of perseverant endeavors.

In the following, we illustrate a few showcases by using our RDF jets together with the technology details:

- Delivery airdrones for Amazon.
- The three-passenger VTOLer for rescue.
- The triangular 20 passenger hybrid aero-van.
- The RDF jets aided the STOL for Boeing 787.

A delta-shaped triangular delivery drone can be formulated offering a smoother aero-dynamic flow and provides the best occupancy. The edge is 7 m long, and the height is 2 m. MGTOW is below 8 tons. Nine RDF jets are distributed evenly beneath the Δ aircraft on three rotatable wings. The total weight of these engines is 9×60 kg $= 540$ kg. The total thrust is 9×900 kg $= 8.1$ tons. The wings are capable of flipping from 0 to 180°, allowing the airplane for vertical and forward propulsions as well as backward deceleration when landing. This small aircraft is suitable for city to city within 1000 km domestic flight. The flying height is ~ 1000 m above the ground (6 °C cooler with fresh air), offering a comfortable fresh air journey.

11.2 The Showcases—Delta VTOLer and STOL for B787

There are two categories of showcases by using the RDF jets: ① the VTOLer—independent airplanes with VTOL features powered by either pure electricity or hybrid power of kerosene/battery and ② the STOL—to leverage the weight for shorter takeoff distance by RDF jets powered by the batteries/supercapacitors.

11.2.1 VTOL: Δ Aircraft as Delivery Drone

A delta-shaped airplane is suggested with three rows of rotatable wings attached with nine RDF jets. The advantages of such a shape offer a smoother aero-dynamics flow and provide roomy occupancy space. Lightweight, good solidarity, easy retraction of the landing gear, and plenty of space for engine/fuel allocations are its distinct features. The delta shape has the advantage of a large wing area and low wing load (W/S[1]), together with good horizontal maneuverability, large sweep angle, and low resistance. The delta wing allows the leading edge to sweep back while the trailing edge is basically flat. The triangular wholesome wing is stable and has a larger attacking angle for the best lift-to-drag or thrust-to-weight ratio. During navigation, the thrust is equal to the drag force and lift equals gravity. The lift-to-drag ratio is

$$K = \frac{L}{D} = \frac{C_L}{C_D} \tag{11.6}$$

Lift can be expressed by air density, velocity, wing area, attaching angles, and lift coefficient:

$$L = \frac{1}{2}\rho V^2 S C_L \sin(\alpha) \tag{11.7}$$

When an aircraft flies at different angles of attack α, the windward area is different, which leads to different lift-drag ratios. The largest lift-drag ratio gives rise to the longest range at the best wing area and angle of attack configuration with the most optimized navigation speed. For example, for B787, the long-range speed navigation speed is 938 kmh.

Three delta (Δ-shaped RDF jet-driven aircraft is described below, fitting the first two scenarios, the small and medium size electric/hybrid aircraft for rescue, delivery drone, 20-passenger van crafts—all capable of VTOL operation.

20-seater VTOL Hybrid van Aircraft

A 20 passenger van-type air vehicle is proposed. This aircraft is equipped with nine RDF jet engines powered by 3D HKSC for VTOL and lightweight gas generator for the navigation/recharging. A streamlined airplane body is designed for the best Bernoulli lift-drag ratio. The nine RDF jets are arranged in three rows of rotating

[1] Typical wing load.

	Type	MTOW (kg)	Wing area (m^2)	kg/m^2
Airbus A380	Turbofan	575,000	845	680
Spitfire	Propeller	3039	22.48	135
Birds	Animal	0.2	0.01	10

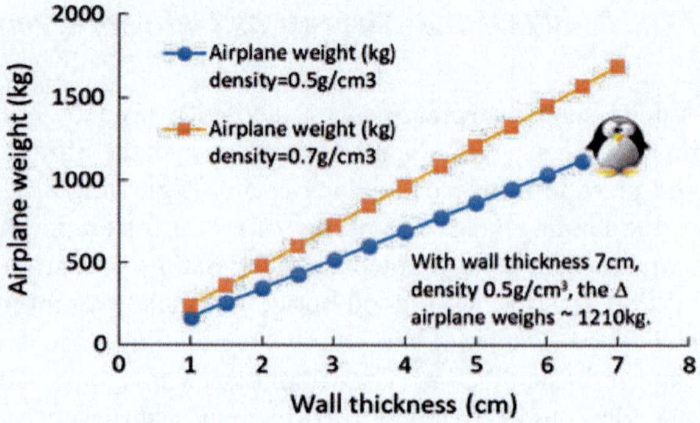

Fig. 11.9 Weight of the 20-passenger aircraft with framed FRC building materials of 0.5 and 0.7 g/cm^3

wings from the front to the back in a 2 3 4 order to fit the triangular shape of the aircraft. Gas turbine generators can be either Honeywell's 127 kg and 1 MW generator or other lightweight generators modified based on the small GE's G85 engine as prototypes. Two 3D HK SC packages are located on the two sides of the rear rotating wings. The spacing of each RDF jet beneath the airplane is distributed evenly with a spacing of approximately 1.5 m.

Details of the weight/size evaluation are as follows:

- *Airplane*

Lightweight materials are used to build the airplane. Assuming that the shell is made of 3D-nested fiber-reinforced composites (FRCs) [21] with a density of 500 kg/m^3 and a shell thickness of 7 cm, this 2-m-tall triangular airplane weighs approximately 1210 kg (the weight of the plane versus wall thickness is shown in Fig. 11.9). Together with the other weight, the MGTOW is about 5 tons.

- *Engine*

One Φ50 RDF jet weighs 60 kg, and nine jets weigh 540 kg. The size of the RDF jet is a 50 × 50 cm cylinder. Each offers up to 9000 kN thrust. The total max thrust is 8.1 ton, which is enough to lift a 5 ton MGTOG for VTOL.

- *Electricity*

Hybrid power is used, with a kerosene gas turbine generator to support long-term navigation and a 3D high-K supercapacitor for VTOL operation.

(1) We use a 90 kg 3D high-K supercapacitor (Chap. 10) to support nine RDF jets for vertical takeoff and landing purposes. Ten kilograms of 3D HK SC are embedded into each RDF jet. This 10 kg supercapacitor package occupies ~ 4340 cm^3 in volume, which can fit in the empty space of 20 × 15 × 15 = 4500 cm^3 in Φ50 × 50 RDF jet. With nine RDF jet engines of a 90-kg supercapacitor (energy density of 70 Wh/kg, power density of 10 kW/kg), the total energy is 5.4 kWh and the maximum power is 900 kW. Assume 1 kW lifts 9 kg weight, the nine RDF jets can lift 8.1 ton weight aircraft. With 5.4 kWh total energy, the 3D HC

SC can last this peak power for 5.4 kWh/900 kW = 22 s. In the meantime of the takeoff, LG can be started to recharge the supercapacitor to leverage the need on load for VTOL. The main use of SC is for VTOL and will be recharged after takeoff by a lightweight gas turbine electric generator started after takeoff.

(2) Light weight gas turbine generators such as Honeywell's 1-Megawatt (MW) Turbo-generator, with 127 kg in weight and 35 $\times$ 60 cm in size, can be used to support the RDF engines as navigation power supplier. Carrying 500 kg kerosene fuel with 4600 kWh/kg energy density and sfc of 0.2–0.3 kg/kW.h, the expected range of the plane is ~ 300 km. The total weight is 626 kg including generator plus the kerosene, the total energy is 4600 * 500 = 2300 kWh. The net energy density is 3674 Wh/kg [22].

- *Payload*

The airplane is designed to carry 20 passengers.The total payload is ~ 2 tons.

Adding the weight of the aircraft, engines, power/fuels, and payload, the total weight is ~ 5 tons. If each RDF jet provides 900 kg thrust, then nine engines can offer 8100 kg thrust. On the other hand, the maximum power is 900 kW, equivalent to 8100 kg lifting force for VTOL. Therefore, both the power and the thrust capability provide enough support to achieve the VTOL function of this aircraft.

Using the above data, we evaluate the range of this 20-seater van airplane. The main energy of navigation comes from the electricity of Honeywell gas generator using kerosene fuel. With the range estimation in Eq. (11.3), 1700 km can be achieved by carrying 500 kg kerosene fuels (Table 11.4), which is over the distance from Beijing to Shanghai (1100 km). This e-airplane has the VTOL capability powered by 3D HK SC, which can be recharged using the hybrid electric generator during navigation and used it for vertical landing purpose. The relevant information such as the weight of kerosene fuel, airplanes, engines and the payload, specific energy, and power density of both the supercapacitor and generator are also listed in the table.

This van aircraft is suitable for city to city of ~ 1000 km domestic flight. The flying journey is fairly comfortable at ~ 1000 m above the ground (6 °C cooler with fresh air). Optionally, this aircraft can also be used as "massive aero-cargo-delivery" (~2 tons payload) from city to city.

7-m Roomy Drone Delivery Aircraft

The post office was a governmental service in the early days which made a very important contributions in human civilization. The postcard delivers the intimate touching moment among relatives and friends. Postman delivered the postcard or a parcel that we are expecting from a far friend. Then, the express service from UPS EMS and DHL in the USA and Shentong, Zhongtong, Yuantong, SF, and Yunda, in China appears. And, the booming development of the delivery business, namely, the logistics becomes an IoT action – the Internet of Things—the things delivered based on the orders from internet. This is a huge marketplace since this boosts the national communications and economies by using the "Thing" as a medium. The Chinese translation of the word "logistics" is more vivid—"the cargo moving flow", which

Table 11.4 Range and VTOL estimation based on the relevant airplane/engine configurations

Airplane	20-seater Δ hybrid	
Parameters	Kerosene to navigate	3D HK SC for VTOL
E-density	3674 Wh/kg	70 Wh/kg
P-density		10 kW/kg
mb (battery)		90 kg
P_VTOL		900 kW
E_VTOL		6 kWh
VTOL lift		8100 kg
mf (fuel)	500 kg	
E_navigate	1837 kWh	
ma (engine)	540 kg	
Thrust (engine)	8100 kg	
ma (aircraft)	1210 kg	
ma (generator)	127 kg	
mp (payload)	2000 kg	
MGTOw	4477 kg	
Eta	0.36	
Range	1736 km	VTOL
Note	540 kg engine, 127 kg LTG, 500 kg fuel, 100 kg 3D HK SC, 1210 kg plane, 2000 kg payload	

describe the true meaning of the Thing in IoT. Indeed, people cannot just satisfy the verbal/visual communication only as it used to be (such as in the old movie "You Got a Mail" (a 1998 American romantic comedy-drama film)). Physical exchange is necessary to make the communication more wholesome. Electric drone delivery has a vast playground in the "cargo moving flow" business, both as a general public service or routined localized package delivery activities, which basically covers every corner of the country. This type of service is usually accomplished by helicopters, trucks, and trains. The drawback is its economics and its technological inconvenience. For example, a chopper is usually 30 m in diameter with a large outstretching long propeller, and its delivery expense is high.

A delta-shaped UAS delivery plane is formulated capable of vertical takeoff and landing. This drone is a mid-size delivery airplane that can be used as a tool for Amazon in the USA or Taobao in China. The size is 3–7 m-long edge and a height of 1.5–2 m, allowing more occupying room to contain more parcels for city-to-city delivery services. Distributed engines are arranged beneath this triangular aircraft. There could be two sizes of the delivery UAS planes equipped with 3 or 9, $\phi30 \times 30$ cm or $\Phi50 \times 50$ cm RDF jets (four options) distributed evenly beneath the airplane. A rough estimation of the weight is as follows:

$$\text{Weight} = \text{Density} * \left(\frac{1}{2}\left(L\left(\frac{\sqrt{3}}{2}L\right)\right) - \frac{1}{2}(L - \text{wall})\frac{\sqrt{3}}{2}(L - \text{wall})\right) * 2 \quad (11.8)$$

The building materials of the airplane body are lightweight 3D-nested materials from FRC or FRP composites. The density of such a framed material is 0.5 g/cm^3. Hence, the 7 m-wide/1.5 m-tall drone weighs 500 kg. Each Φ30 and Φ50 RDF jet weighs 20 kg and 60 kg, respectively (see estimation details in Chap. 9). The total weight is 60–540 kg. The weight of the electric power—supercapacitor, kerosene fuel, and electric generator—are similar as above. Optionally, the gas turbine electric generator can be from the modified J85[2] or CF700 engines.[3] Lightweight GE's J85 gas turbine engine was originally designed in 1954 and is still expected to power US aircraft until at least 2040. Such an engine is just 0.45 meterswide and 1.1 m long, with 1.5 tons of thrust with less than 200 kg weight (8:1 thrust/weight ratio). Another prototype option for the lightweight generator is GE's CF700 Turbofan with 330 kg engine weight providing 2672 kW power in power/weight ratios of 8.62 kW/kg (thrust 20.24 kN with specific fuel consumption: 68.3 kg/kN·h). The dimensions of the engine is 1918 mm in length and 838 mm in diameter.

Table 11.5 compares the range of the 7 m drone plane on different fuel load/payload configurations. The maximum package weight is 4 tons to keep the MGTOW within 5.4 tons. Delivery drones have quite a few payload/range configurations to fit in different markets. For example, the more fuels carried, the longer the range; the less the cargos, the longer the range. Optionally, a tiny catering drone can be formulated with just three RDF jets to support a 500 kg weight which can be purely operated electrically by battery or SCs. All the above delivery drones are capable of vertical takeoff and landing (VTOL).

3 x 3 Rescue VTOLer or Emergency Drone

A tiny 3 × 3 rescue VTOLer can be built with

- Three small RDF jet engines situated at the three corners.
- Three seaters occupancy for pilot, doctor, and patient.
- Three meter-long ΔVTOLer capable to take off in narrow and stringent site.

Such a tiny VTOLer can be used as a rescue plane or for emergency delivery or catering. Each engine is a Φ30 × 30 cm cylinder with a separation of ~ 2 m. One RDF jet engine weighs 20 kg, generating at least up to 6000 N thrust. Three RDF jets provide a maximum vertical support up to 1.8 tons. Three engines, one at the front and two at the back, form a stable triangular pattern, a very stable support for vertical takeoff and landing operations. The whole airplane weighs less than 700 kg. The three electric aero-engines can be powered by Li batteries and/or supercapacitors.

---

[2] *The J85, GE's "Little Tough Guy"*, https://blog.geaviation.com/technology/long-live-the-j85-ges-little-tough-guy/.

[3] *Model CF700. GE AE Website.* http://geae.com/engines/corporate/cf700.html, *accessed 2 Nov 2009.*

Table 11.5 Relationship of navigation range, payload, and fuels of the 7-m delivery drone

7-meter Δ droner							
E (kerosene) (Wh/kg)	mn (kerosene)	mv (3DHKSC) (kg)	ma (engine + craft) (kg)	mp (payload) (kg)	**MGTOW** (kg)	**Range** (km)	Eta
4600	100	100	1000	2000	3200	24	0.3
4600	*300*	100	1000	2000	3400	68	0.3
4600	*500*	100	1000	2000	3600	107	0.3
4600	*700*	100	1000	2000	3800	142	0.3
4600	*900*	100	1000	2000	4000	174	0.3
4600	*1100*	100	1000	2000	4200	202	0.3
4600	*1300*	100	1000	2000	4400	228	0.3
4600	*1500*	100	1000	2000	4600	252	0.3
4600	*1700*	100	1000	2000	4800	274	0.3
4600	*1900*	100	1000	2000	5000	294	0.3
4600	*2100*	100	1000	2000	5200	312	0.3
4600	510	100	1000	*500*	1600	187	0.3
4600	510	100	1000	*1000*	2100	151	0.3
4600	510	100	1000	*1500*	2600	127	0.3
4600	510	100	1000	*2000*	3100	109	0.3
4600g	510	100	1000	*2500*	3600	96	0.3
4600	510	100	1000	*3000*	4100	85	0.3
4600	510	100	1000	*3500*	4600	77	0.3
4600	510	100	1000	*4000*	5100	70	0.3

The batteries/SCs are placed in the middle bottom of the aircraft. Optionally, one lightweight generator (for example, a 127 kg Honeywell Turbogenerator) can be equipped to generate electricity using kerosene for long-range/endurance flight.

Table 11.6 compares the amount of battery versus the power needed for vertical takeoff. The minimum weight of the Li battery to lift a 700 kg VTOLer is ~ 460 kg (assuming its power density is 0.400 kW/kg), providing a 187 kW minimum power to lift 700 kg weight, based on the helicopter data—the 3.75 kg/kW lift-to-weight ratio. Only when the battery weight is over 460 kg, the maximum power from the battery can support the VTOL for 700 kg-weight plane. In additiona to this 460 kg margin, the more battery renders more mileage or more payload.

As can be seen, although the 460 kg Li ion battery can support the vertical takeoff of 700 kg VTOLer plane, it only leaves 199 kg weight for the airplane, engines, and payload. Assuming the 3 × 3 VTOLer is built with lightweight composite materials as before. This Δ-shaped VTOL aircraft with a side length of 3 m, height of 1 m, and wall thickness of 5 cm is built weighs 130 kg as estimated in Table 11.7, and the weight of 3 RDF jets is 60 kg. The total weight is 189 kg, and only 10 kg payload

Table 11.6 Minimum power for the VTOL and the necessary battery carried

Mgtow (kg)	Weight (payload) (kg)	Weight (battery)(kg)	Power provided(W)	Power needed (kW/kg)
700	699	1	0	186.67
700	649	51	20	186.67
700	599	101	40	186.67
700	349	351	140	186.67
700	299	401	160	186.67
700	249	451	180	186.67
700	199	501	200	186.67
700	149	551	220	186.67
700	99	601	240	186.67
700	49	651	260	186.67

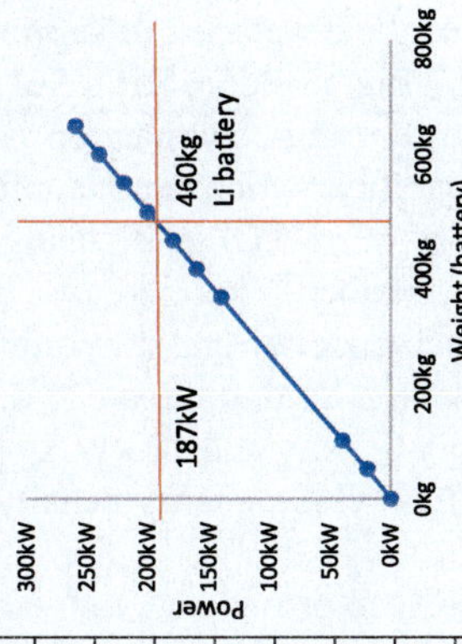

is left. Obviously, the ordinary battery is not good enough for this 3 × 3 rescue VTOLer. The 3D HK SC must be developed to support this VTOLer's operation.

Table 11.8 compares B787 big airplane, RR's SoI electric plane versus our 3 × 3 VTOLer. The distinct feature in this comparison is both the vertical takeoff and the horizontal navigating capability, which is not applicable for either B787 or RR's SoI plane. Technical details of our VTOLer by using the RDF jet engine powered by 3D HK SC and LTG are described. There are two options of this 3 × 3 VTOLer: purely electric driven by supercapacitor and a hybrid power of 3D HK SC and the lightweight generator. For the first kind, there are two types of SC used: high power density type (70 Wh/kg energy density and 10 kW/kg power density) for VTOL and a high energy density type (700 Wh/kg energy density and 1 kW/kg power density) for navigation.

The first type of 3D HK SC can provide enough power and thrust for the vertical takeoff. It can provide ~ 50 s 480 kW output power for vertical takeoff costing 7 kWh energy. Once takeing off, the second type supercapacitor offers the navigation energy for horizontal propulsion. With the total of 140 kWh energy, the estimated mileage is 180 km. Optionally, the second type of 3D HK SC can be replaced by Honeywell's lightweight generator plus 73 kg kerosene. The advantage of the hybrid option is that the generator can recharge the supercapacitor for vertical landing and its range is also much larger. It should be noted that this estimation is the most optimal one and a lot of technical challenges shall be conquered for RDF jet, 3D HKSC, and the airplane design.

Table 11.7 Weight of the aircraft versus the wall thickness and material density

1 m-height, 3 m-edge, ϕ30 RDF			
Wall thickness (m)	Airplane weight (kg)		
	Density = 0.5 g/cm^3	Density = 0.7 g/cm^3	
0.010	26.056	36.479	
0.015	39.052	54.673	RDF Jet 30L × 30D 19.193
0.020	52.026	72.836	
0.025	64.978	90.969	
0.030	77.908	109.072	
0.035	90.817	127.144	
0.040	103.704	145.186	
0.045	116.569	163.197	3 RDF jets 60 kg
0.050	129.413	181.178	
0.055	142.234	199.128	
0.060	155.034	217.048	Total 189 kg
0.065	167.812	234.937	
0.070	180.569	252.796	

Table 11.8 Comparison of the size, scale, of B787, SoI, and our 3 × 3 VTOLer

Para		Craft			
		Boeing 787	RR SoI	3 × 3 Jetcopter	
				Hybrid	Electric
Power/energy	VTOL	Na	Na	100 kg 3D HK SC, 70 Wh/kg, 10 kW/kg,	100 kg 3D HK SC, 70 Wh/kg, 10 kW/kg,
	Nav	130 ton kerosene, 4600 Wh/kg	475 kg Li battery, 155Wh/kg, 400 W/kg	127 kg Honeywell generator, 73 kg kerosene, 3000Wh/kg	200 kg 3D HK SC, 700 Wh/kg, 1 kW/kg,
Engine/thrust	kW		399 kW	480 kW	
	kN	600 kN (60 ton)		18kN (1800 kg)	
Energy (VTOL)	last 52.5 s	na	na	7 kWh	7 kWh
Energy (navigation)	related to the "range"	552,000 kWh	74 kWh	219 kWh	140 kWh
		(Kerosene)	(Li battery)	(Kerosene)	(3D HK SC)
Eta		35.60%	90%	90%	90%
Weight (MGTOW)		250000 kg	1250 kg	700 kg	
		(120 ton kerosene + 130 ton B787 + GEnXs)	(Airplane_engine + battery + 2men)	(190 kg airplane + engines + 3 × 70 kg man + 300 kg (100 kg Honeywell + 100 kg SC + 100 kg kerosene))	
Range	R ≈ eta * *E* * mb/mgtow	16000 km	53 km	282 km	180 km
Dimensions		60 × 60 × 17 m	7 × 8 × 2 m	3 × 3 x 1 m	3 × 3 × 1 m
Passenger		300	2	3	3
VTOL		No	No	Yes	Yes
Time		2011.9	2021.11	202x	202x
Pict				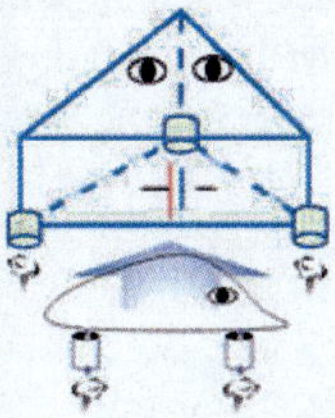	

(continued)

Table 11.8 (continued)

Para	Craft			
	Boeing 787	RR SoI	3 × 3 Jetcopter	
			Hybrid	Electric
	B787	Spirit of Innovation	3 × 3 rescue VTOLer	

The main feature of the 3 × 3 rescue VTOLer is the rotable jets and small VTOL area. The VTOL can be achieved in a small area as compared to helicopter, and no airport is needed. In the initial state, three electric engines sit vertically to project to the ground to lift up the aircraft. At a certain altitude (~ 100 m), the engines are gradually turned horizontally to provide horizontal thrust to move the airplane forward. The co-efforts of the RDF jets and the lift-drag ratio of the aircraft maintain the balance of the thrust and drag and the weight and lift.

Optionally, this three-seater triangular rescue aircraft can be used as emergency drone or for catering delivery. This VTOL UAS drone can deliver the payload with various weights in different ranges. Different configurations of the 3D HK SC and LTGs, airplane, and load can be formulated for different ranges (for example, a lighter airplane may leverage the power needed for the VTOL and more room for payload).

11.2.2 STOL: Shorter Takeoff for Airliner

RDF jets can also be used to aid the STOL for large commercial aircraft. The first principle of physics tells us that the most efferent way of takeoff is vertical lifting off (VTOL) of the ground and then starting horizontal propulsion, which skips the frictional energy loss during acceleration, especially for heavy commercial aircraft. For a 100-ton big airliner, it is impractical for its VTOL yet the RDF jets can leverage its MGTOW for shorter distance takeoff (STOL), saving a lot of airport resources and takeoff energy loss (friction loss). Nine RDF jet engines can be placed in a 3 × 3 triangular configuration under the front and the two wings providing a stable vertical 8-ton vertical thrust for easier takeoff (Fig. 11.10). Unlike the outstretching propeller, RDF is an enclosed cylindrical jet engine with rim driven inner fan blades, allowing an easy adjustment in various angles, i.e., the RDF can not only help the vertical takeoff but also be turned parallel for horizontal propulsion afterward, decelerating the aircraft when landing, or helping the air vehicle for directional maneuvering. The electrified propulsion will help the STOL and save energy and therefore shall be addressed for the next generation intelligent aviation.

Fig. 11.10 Configuration of 9 RDF jet engines located under the two wings and front to aid the STOL of a large aircraft

11.3 The Technologies—Rotatable Wing and VTOL Operation

The main integration technology of the engine and airplane is rotatable wings and the allocation methodology. The other two relevant technologies are the advanced unmanned control and lightweight building materials.

11.3.1 Rotating Algorithm of RDF Jets

The rotating wings are located beneath the airplane and operate as follows:

- Mount the RDF jet on the rotatable wing.
- Mount the rotatable wing beneath the aircraft.
- Rotate clockwise for vertical propulsion.
- Rotate counterclockwise for horizontal propulsion.

Rotating wing operation for VTOL aircraft includes lifting and flying of triangle VTOL aircraft. In general, there are five stages similar to a conventional flight: vertical takeoff, ramping, navigation, deceleration, and vertical landing (Fig. 11.11). The differences are as follows: the takeoff and landing are vertical, and no airport is needed. The navigation height is approximately 1000 m, the air is fresh, and the air conditioning is moderate.

At different stages of the flight, the different orientations and thrust levels are in accordance with the specific need in that scenario. For example, during the landing process, engines on the second row of rotating wings will turn to 180 degrees to provide the forward thrust to decelerate, while the other RDF jets gradually change the orientation from horizontal to 45° for a gradual descending. The five steps are:

- The first step is vertical takeoff.
- The second step is ramping by gradually rotating the wing to parallel.
- The third is parallel flight balanced by lift-to-drag and rotating wing.

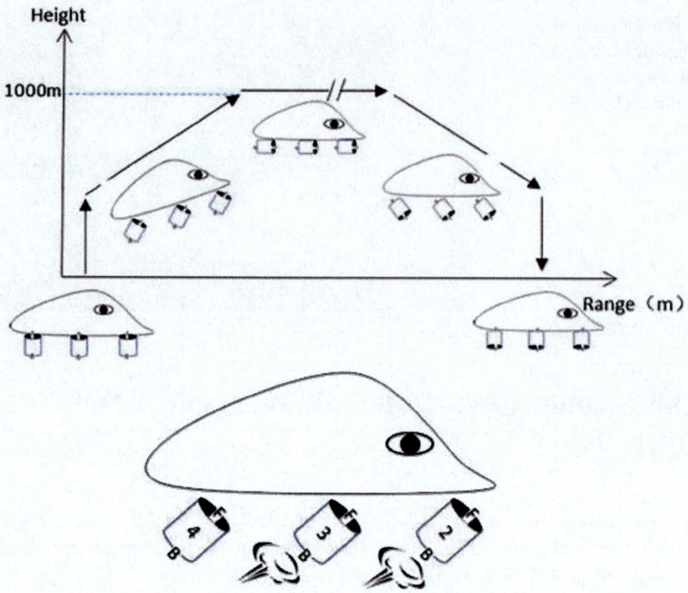

Fig. 11.11 Five stages of the electric plane

- The fourth is descending.
- The fifth is the landing.

We are explaining these five stages in details one by one.

(1) Vertical takeoff

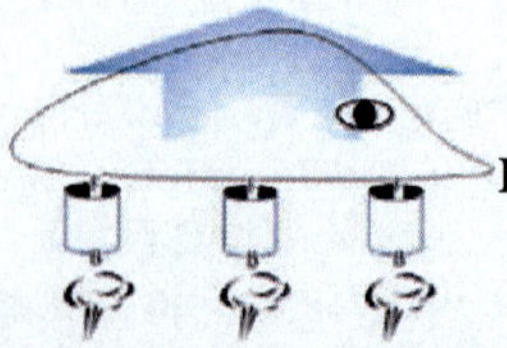

During vertical takeoff, all the RDF jets face the ground (the front on top and back on the bottom) to provide the vertical jet thrust. The engines are powered by 3D HK supercapacitors. Both the engines and the supercapacitor output the maximum strength to conquer the weight of the airplane. After taking off, the angle of the engines gradually rotates from 90 to 60° and then to 45° for ramping.

(2) Ramping

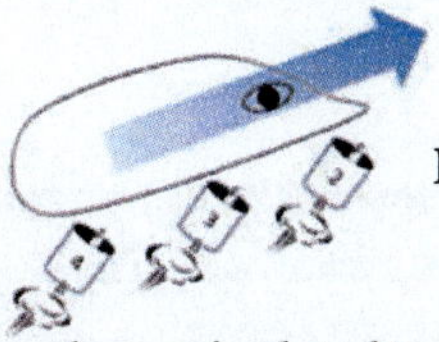

During the ramping stage, the orientation of the RDF jets is cleverly manipulated to adjust the horizontal and vertical angles for the best ramping efficiency. During the ramping process, the front and back rows of RDF jets can gradually turn to 45°, while the center row can still keep vertical projection providing the lift for the airplane. As the horizontal speed of the airplane increases by horizontal propulsion, the lift-to-weight will help the airplane hover in the air and leverage the need for vertical thrust from the RDF jets. At the ramping stage, the hybrid generator can be started providing power for RDF jets and recharging the supercapacitor. The airplane gradually ramps to a height of 1000 m with a steady navigation speed.

(3) Navigation

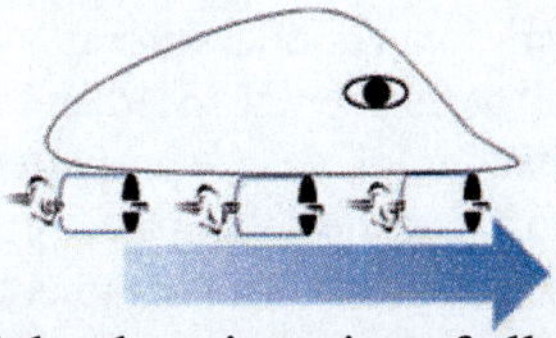

Once the triangular airplane reaches its navigation height, the orientation of all the RDF jets shall be adjusted to an appropriate angle for the best horizontal/vertical propulsion efficiency. Once the airplane is in the navigation state, the airplane will remain stable and at a certain altitude, i.e., 1000 m. This stable attitude is achieved by the co-effort of the RDF jet vertical jet and the lift-to-weight ratio of the whole airplane according to the streamlined curves during flight, similar to conventional airplanes. Of course, the airplane streamlines shall be delicately designed for the best *L/D* ratio with the help of CFD and test flights.

(4) Descending

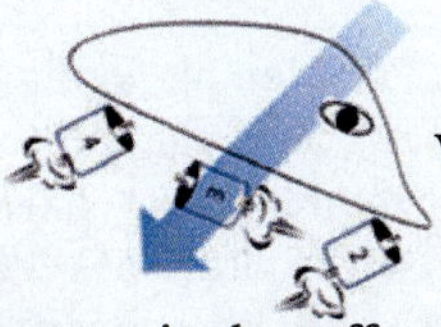

When the airplane is near the destination, the gas turbine generator is shut off, and the deceleration mode starts. At first, the engines in the middle row start facing forward to provide the decelerating thrust, while the first and third row engines start to turn from horizontal projection to vertical. These co-effects make the altitude of the airplane gradually decrease as well as the horizontal speed. With the help of the decelerating thrust from the second row RDF jets, the decelerating range is much shorter than that of the conventional landing approach.

(5) Vertical landing

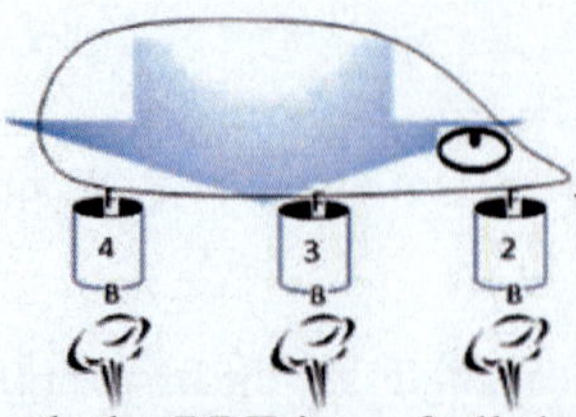

When the triangular airplane reaches zero horizontal speed, the RDF jets of all three rows face the ground to provide moderate thrust to buffer the airplanes gradually dropping down to the ground. Proper sensors are needed to detect the distance from the ground, allowing the foldable landing gear expanded to make the triangular airplane sit steadily on the floor.

11.3.2 VTOL Prerequisites

The VTOL capability requires much more efforts than conventional aircraft or a helicopter since it requires both power and thrust, both for horizontal speed and for vertical strength. Comparing to the current propulsion technologies, the propeller is just suitable for horizontal flight but not for vertical takeoff, and the helicopter is good at vertical takeoff only and less compatible with others (no speed). The commercial gas turbine engine is the most powerful yet not rotatable to shift the horizontal thrust to vertical VTOL power.

There are four prerequisites for a VTOLer:

(1) Thrust F > mg, vertical thrust must be greater than the weight.
(2) Power P > kg/kW, power-to-weigh ratio must be high, and 1 kW power must lift up at least 3.75 kg weight.
(3) Energy E > 1 min, enough energy to sustain enough time to lift up the aircraft to a certain height, i.e., 100 m.
(4) Adjustable V + H, flexible vertical and horizontal thrust, the rotatable wings.

None of the existing aircraft possesses all these features. For example, ordinary propeller-driven airplanes cannot take off vertically (no VTOL), although they can horizontally navigate at high speeds. The propeller thrust of Spitfire is just ~ 6.6 kN (660 kg) with its 1500 hp power, not being able to lift up itself (Spitfire Mk Vb's takeoff weight: 3039 kg) [23]. However, 1500 hp horsepower (1100 kW) can lift up a 4 tons if the plane is designed as a helicopter with long and wide propellers. But a chopper cannot fly fast. With RDF jets, we combine the above four features in one Δ VTOLer, capable of both vertical takeoff and enough horizontal speed, which is different from the propeller plane, helicopter, and gas turbine aero-crafts.

There is a "2 + 3" for VTOL aircraft: the two prerequisites are power and thrust, and the three components are engine (rim driven motor), propulsion (turbofan), and power (3D HK SC), as shown in the interactive matrix in Fig. 11.12. Both thrust and power must be considered in the three agents, engine, battery, and propeller, to achieve a VTOLer target as well as proper navigation range/speed. A rotating

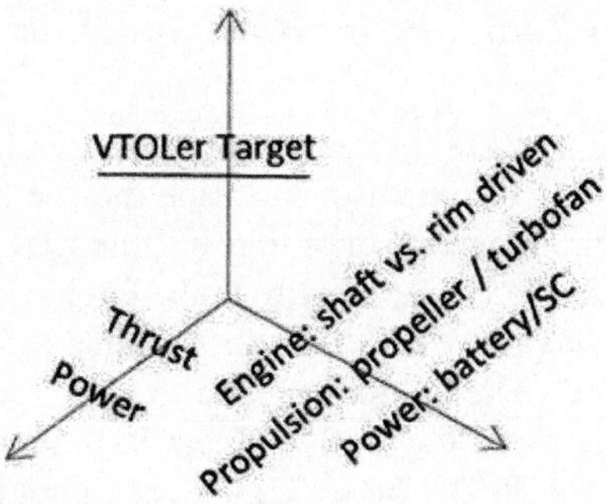

Fig. 11.12 VTOLer must possess both vertical takeoff and horizontal navigating capabilities with the best combination of the thrust and power provided by engine, propulsion, and power

wing is also needed to freely adjust the horizontal/vertical propulsion to provide both horizontal thrust for navigation and vertical lift strength to overcome gravity. The power of the battery/SC/LTG should be strong enough to drive the engine, and the engine must be powerful enough to drive the Taichi fan for enough thrust.

The thrust/power and energy for direct takeoff and landing of our Δ airplane are described as follows:

Takeoff power: To raise the aircraft with a speed of 1 m/s, the power needed is $P =$ FV = mgv = 700 kg * 9.8 * 1 = 6900 W. One power is from the battery, and the other is from the engine:

(1) *The battery*

The power density of the Li battery is 400 Wh/kg, and the maximum power of the 50 kg battery is 400 W/kg * 50 kg = 20 kW, meeting the minimum requirement of 6.9 kW.

(2) *The engine*

The general power versus weight ratio of the electric engine is greater than 1. For example, the mature Tesla car has a power/weight ratio of 2.5 kW/kg. Taking the conservative value of 2 kW/kg, 20 kg Φ30 RDF jet can provide 40 kW of power, which also meets the minimum requirement of 6.9 kW.

The vertical takeoff requires both power and thrust. In terms of lift-up strength, three RDF engines can provide thrust of 600–1800 kg, which is also greater than the weight of the aircraft of 700 kg. Therefore, both the power and thrust from the battery and the engine are good enough for vertical takeoff.

Takeoff energy. The battery/SC power must last long enough to lift the plane to the navigation height. The energy to raise a 700 kg to 1000 m is $E =$ mgh = 700 kg * 9.8 * 1000 m = 690 kJ = 1920 Wh (1 Wh = 3.6 kJ). The energy of a 50 kg lithium battery with an energy density of 155 Wh/kg is 7500 Wh, which is also greater than the needed 2 kWh. Therefore, the energy of the 50 kg battery is sufficient for the VTOL. After takeoff, the remaining energy of 5.5 kWh of the battery can be used for navigation. During this period, the RDF jets need to change from the vertical to the parallel direction to use its lift-drag ratio to maintain the altitude and the speed.

11.3.3 Propulsion and Energy in One Shell

The RDF jet can be integrated with 3D HK SC in the compact shell. There is an empty compartment inside the center rear space, which we can place the battery of the supercapacitor to drive this RDF jet. In this way, an independent RDF jet unit is formed combining both the engine and power in one.

Table 11.9 shows the detailed estimation of the weight and volume of the 3D HK SC. The volume needed to hold 10 kg 3D HK SC package is 4342 cm^3. The rear space of the $\Phi 50 \times 50$ cm RDF jet is one-third of the total volume, which is 0.33 * π * 25^2 * 50 = 32,381 cm^3, enough space to store 10 kg SC. Such an SC can be recharged in air by a lightweight gas turbine generator after takeoff.

This compact RDF jet/power integration takes advantage of the following:

- *Rim driven replaces the shaft drive. Archimedes once said, "Give me a pivot, I can prize up the earth". The rim drive is equivalent to lifting the earth with a lever from the long end. The torque is larger compared to drive the larger fan by a shafte. The ducted fan plus the shrunken tail nuzzle takes advantage of the large BPR GEnX engine to use the duct fan to create more than 80% of the total thrust of the gas turbine engine. The enclosed compact structure of the RDF jet allows easy operation of the VTOL, good concealment and low noise.*
- *The high-energy and high-power density supercapacitor capacity. A 3D HK supercapacitor can be packed or integrated inside the compact RDF jet structure to form an independent propulsion unit (a 30 × 30 cm RDF jet can fly by itself). Such a feature can only be feasible with the 3D HK SC (1 kg supercapacitor provides 700 Wh energy and 1 kW max power) since the current battery is just too heavy or too big to fit in such a niche space in the RDF jet. The energy and power density are too low for the battery (1 kg Li battery provides 150 Wh energy and 400 W max power, year 2022) to provide enough startup ability of the RDF jet for takeoff.*

Table 11.9 Weight and volume estimation on the 3D high-K supercapacitor

Weight, volume, numbers of 12″ SC wafer	
Volume of one Si wafer (m^3) (Φ12″, 200 μm thick)	1.41E − 05
Weight of one Si wafer (kg)	3.29E − 02
Weight of 100-layer SC on the wafer (kg)	2.48E − 03
Total weight of 12′ Siliocn wafer and SC (kg)	3.54E − 02
Volume of 100-layer SC on the wafer (kg)	1.24E − 06
How many wafers per 10 kg SC (#)	2.82E + 02
Total volume of 12′ Siliocn wafer and SC (m^3)	1.54E − 05
The volume of 10 kg 3D HK SC (m^3)	4.34E − 03
The volume of 10 kg 3D HK SC (cm^3)	4342

11.3.4 *Power Electronics and UAS*

Power electronic technology is needed from IEEE to apply electric power to electric engines. Robotic aero-control is needed for unmanned aero-system operations. Both involve distinct expertise of IEEE in both hardware and software integrations.

Power Electronics – IGBT + SiC

To transfer the power from the battery to the engine in an EV, high voltage, high current, and high power output plus their control are indispensable, as shown in Fig. 11.13. The IGBT transistor plays a key role in this control circuit, and SiC is the best material to build the IGBT for powerful and efficient electric circuitry. Reference [24] presents a thorough review of the development of SiC IGBT in the past 30 years, including structure design and performance and challenges of switching characteristics but purely from the IEEE's perspective.

(1) *IGBT*

Unlike the conventional MOSFET, which operates as a switch in low-power manipulations, the IGBT operates at much stronger electric power under high voltage and current. The device must be able to stand for the high voltage, passing enough current without incurring too much heat (power). Insulated gate bipolar transistors (IGBTs) have gained significant importance since their introduction in 1988 on the power control unit (PCU) in the (Hybrid) electric vehicle (H) EV applications market. The popular voltage is 600–3000 V IGBTs with up to currents of 2400 A. The IGBT is the power device of choice for the power control unit (PCU), which typically consists of one to two inverters plus a boost converter. The working frequency of the IGBT is mainly for power/voltage switching, which is relatively low (200 kHz).

(2) *SiC*

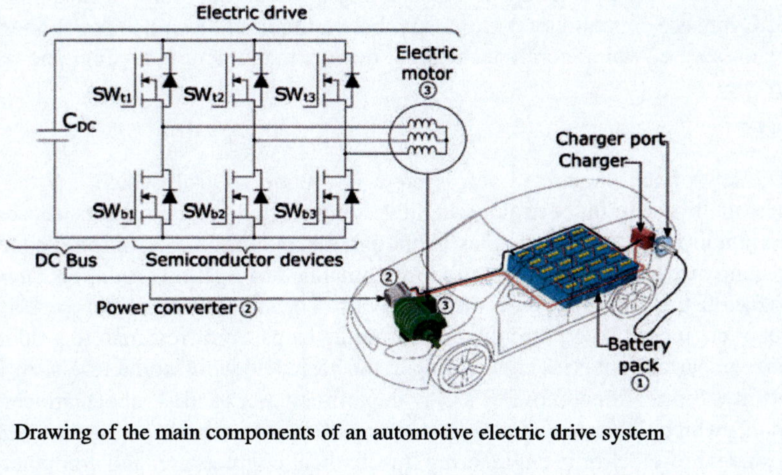

Fig. 11.13 Drawing of the main components of an automotive electric drive system

Silicon carbide device has 10 times higher voltage than silicon device, since its energy band gap is 2.8 times as compared to silicon, and its insulation breakdown field strength is as high as 3.2 MV/cm. Its thermal conductivity is 3.3 times that of silicon, which is 49 W/cm k. SiC power devices can work normally at a high temperature of 400 °C which can control a large current using a small device. The operating voltage is also much higher. The market price is about 5–6 times that of silicon material though. Silicon carbide MOSFET is not only suitable for a wide voltage range from 600 V to 10 kV, but also can break through the limitations of the existing electric vehicle motor design due to device performance, which is the focus of research and development.

(3) *SiC + IGBT*

SiC IGBT represents the top level of technologies of high-power/high-voltage electronic control electronics. There are two key technologies: SiC material (wide bandgap semiconductor silicon carbide) and IGBT (insulated gate bipolar transistor).

IGBT and SiC are not only mature technologies but also two fast ongoing developing fields involving many innovations. IGBT electronic control technology is not a bottleneck to limit the electrified propulsion compared to the EDF jet engine and the Delta airplanes. Infineon EV IGBT portfolios cover a wide range of voltages and powers from 100 W to 1 MW for automotive power modules and provide the ideal solution for converting the DC current from the electric battery into the AC current for electric motors in hybrid and electric vehicles. Through our leading technologies with very low energy conversion losses, our automotive power modules contribute to increasing the energy efficiency and battery range of electric vehicles. Intel is the world's largest semiconductor chip manufacturer founded in 1968. Samsung, Hynix, NXP and Philips, St Group, Texas Instruments (TI) AMD, and Micron are also the world renounced chip companies for years. Among them, Infineon (No. 11) is the world's major power device such as IGBT supplier and automotive semiconductor manufacturer. Infineon Technology Co., Ltd. was established on April 1, 1999 from the former Siemens group. In April 2020, Infineon had completed the acquisition of Cypress Semiconductor Company, becoming a wholesome semiconductor IDM manufacture, which combines design, device, manufacture, testing, and packaging in one.

UAS

The key of the unmanned aero-system is robotic control, which is a fairly brand new multi-disciplinary engineering involving electronics, computers, sensors, Wi-Fi communicating, and mechanics. Although the automatic control robotic technology is quite matured, proper adaptation for unmanned navigation needs special personnel to handle this expertise. How the robot control be started? How to become a qualified robot engineer? To go through these training steps from freshman to graduate helps us to undertake robotics engineering at the basic and wholesome levels. As an initial effort, a former director of DJI's UAV department proposed a "robot engineer learning plan", which illustrates a training roadmap of an aero-robotic engineer. He used to be a student of electronic engineering, mechanical engineering, and computers in four years of college life and two years of graduate school. His 6 years of true experience from freshman to graduate, which is basically his own learning curves, are more pragmatic for a robotic beginner to become an expert.

(1) *Overall*

The core of robotics is to deal with the interactions of the physical world. These interactions with the physical world are divided into three categories: sensors and processing algorithms (lidar, binocular vision, fusion algorithm), dynamics control of multirigid body system (dynamics and contact force control of industrial robot), and autonomous robot movement (locomotion, wheeled, foot, flying and other mobile robots). It is suggested that students who are interested in robotics should focus on these topics.

(2) *Freshmen*

Calculus and linear algebra are the two key courses for freshmen to learn and master. Linear algebra is especially important. Generally, excellent engineers and scientists should learn linear algebra at least five times in their five stages of robotic careers, including freshmen, convex optimization, linear systems, and machine learning. Two books are highly suggested: Linear Algebra Done Right and Linear Algebra Done Wrong. Another good resource is MIT's open class: linear algebra online. After learning linear algebra, a natural follow-up is to use a computer to perform matrix and vector multiplication. You must start learning programming as early as possible in your freshman year. The start learning programming is Python. Others such as e c/c++, Linux, HTML, and JavaScript are also necessary in the first year in college.

(3) *Sophomores*

In the sophomore year, one should pay special attention to the cultivation of practical ability. The best way to learn these skills is to participate in the training and program design of the robot community in your school. In your second year, one should invest more time in embedded systems while expanding the other "hand-made" capabilities. For example, we need to learn to use SolidWorks to draw basic mechanical schematics, to acquire the basic digital circuit knowledge for digital-to-analog conversion, to use 51, AVR, STM32 microcomputer, UART, SPI, I2C and other protocols for data receiving/sending, to use STM32 board, basic electric motor operation such rotation and driving, PID principle to debug the movement of four-wheel robot chassis, to implement basic sensors such as gyroscope, code disk, infrared, IP and router network configuration, microelectronic welding to unmanned system, etc. The amount of skills required by robot engineers is more than three times that of engineers in the IT industry. The foundation of these skills should be accumulated at the beginning of the sophomore year.

(4) *Junior*

Junior students must cultivate their ability to build a simple but complete robot independently, such as a Robocon-level robot, a rob masters' chariot, or a four rotor aircraft. Robocon-level robots involve many mechanical design, MCU development, motor drive development, code disk and ultrasonic sensor reading, chassis kinematics calculation, PID debugging, task scheduling logic debugging, and actuator control. The mechanical part of robomasters' chariot is simpler, but it also needs knowledge of pan tilt control, missile launching system control, power control, etc.; in addition, you can also learn visual recognition and automatic strike without doing these parts, so you need to start to study OpenCV.

(5) *Senior*

At the beginning of your senior year, you should start to investigate your future career on robotics since there are few different directions and start to learn some advanced technologies and theories, such as ROS, Simulink, gazebo, and Vrep.

The other task in the senior year is the internship project and thesis. You should use this opportunity to learn how to practice the project and how to present your result. It is very important for engineers to introduce and show their achievements through the ppts at any stage of their career. The ability to analyze and present may be a bottleneck limiting factor in your robot engineer career due to its multi-disciplinary nature. Only when you can clearly express your achievements to your team, can you get feedback from others and improve the overall cohesion of the team through communication. In this way, you can get more recognition in the team and contribute to the improvement of the overall work efficiency of the team.

(6) *First year of graduate*

When you are a graduate student, your goal is relatively clear, and you should have one distinct research direction. You need to define a scope in order to make one robot in two years' span of the graduate study, and you need to find a good topic to write a paper to publish. Exploring a hot topic is key for graduate studies. Automatic navigation and driving have been two hot spots in the last two years. On the one hand, automotive automation is the trend; on the other hand, the emergence of multi-rotor aircraft has created a great deal of demand for automatic flight. In addition to the robot visual localization algorithm, one also needs to learn other sensors and how to integrate these sensors with the visual localization algorithm.

(7) *Second year graduate student*

In the 2nd year of the graduate, you should have read at least 10 books and more than 50 papers and accumulated abundant programing. When you see a robot, you can decisively analyze what sensors, actuators, and computing platforms it uses and how much power its actuator can withstand. When you see a new algorithm, you can roughly judge its execution process and where it should be optimized. When you face a new knowledge you have not learned, you can analyze its connection with your previous learning and attach it to you knowledge tree. At this time, you can start to learn some more advanced technologies and tools, such as using FPGA and GPU to optimize algorithms, machine reinforcement learning.

The areas covered in robotics control are included according to the course classification method of robotics from Carnegie Mellon University, the world's top Robotics Education Institution. Robotics has four core areas:

(1) Perception: vision sensor, image sensor, tactile and force sensor, inertial navigation, sensor fusions.
(2) Cognition: artificial intelligence, knowledge expression, planning, task scheduling, machine learning, etc.
(3) Behavior: kinematics, dynamics, control, manipulation and locomotion, etc.
(4) Fundamentals of mathematics: optimal estimation, differential geometry, computational geometry, operations research, etc.

In general, autonomous control technologies are not a bottleneck to limit electrified propulsion. Most of its technologies can be transplanted for UAS.

11.3.5 Lightweight Aero-materials

Aero-materials include engine materials and airplane building materials. In Chap. 7, we discussed engine building materials to improve the power-to-weight ratio. In Chap. 10, we discussed the weight factor in batteries. Here, we discuss the building materials for airplane, which need it to be lightweight with enough strength.

Three-dimensional-nested FRC (fiber-reinforced composite [25]) and CMC (ceramic matrix composite [26]) materials are two favorite candidates to build the frame and shell of an airplane. FRC is a fiber-reinforced resin matrix composite based on organic polymers and is widely used in the aviation, automobile, and marine industries. The service temperature of resin matrix composites generally does not exceed 350 °C. CMC is a carbon matrix composite reinforced by carbon fiber and its fabric. It has the advantages of lower density (< 2.0 g/cm^3), high strength, and high temperature tolerance. Three-dimensional braided FRC and CMC have the advantages of low density (< 1.0 g/cm^3), high strength, high specific modulus, high thermal conductivity, and high dimensional stability. Fan blades built from the above FRC/CMC are already on the market. The main representatives are the commercial composite fan blades of GE90's series engine for B777, GEnX engine for B787, and LEAP-X engine for COMAC C919.

Lightweight 3D-framed CMC materials of the T300, T700, and T1000 series can be used to build the RDF jet shell and the Delta airplane. There are several categories of carbon fiber materials with different maximum stress tolerances of 6.35 GPa, 7.0 GPa, and 7.5 GPa compared to the high-strength titanium alloy $Ti_{13}V_{11}Cr_3Al$ with a tensile strength between 1100 and 1400 MPa at room temperature. Tensile strength refers to the maximum plastic deformation stress that a material can withstand. In other words, it is the maximum force that a material can bear before it breaks. Three-dimensional-framed CMC are both lightweight and high-strength/toughness materials. The density of solid CMC is 1.81 g/cm^3 with a fiber diameter of ~ 5–7 μm, and the famed lightweight CMC has a density of 0.5–0.7 g/cm^3.

Three-dimensional braided FRC, FRP, and CMC materials possess a very light density (0.3–0.7) and reinforced mechanical strength. The weight of a Δ-shaped VTOL aircraft with a side length of 3 m, height of 2 m, and wall thickness of 3 cm

Table 11.10 Weight estimation of airplane with different lightweight materials

Wall thickness (m)	Weight (kg)	
	Density = 0.5 g/cm^3	Density = 0.7 g/cm^3
0.010	29.950	41.930
0.015	44.888	62.843
0.020	59.800	83.720
0.025	74.687	104.562
0.030	89.550	125.370
0.035	104.388	146.143
0.040	119.200	166.880
0.045	133.988	187.583
0.050	148.750	208.250
0.055	163.488	228.883
0.060	178.200	249.480
0.065	192.887	270.042
0.070	207.550	290.570

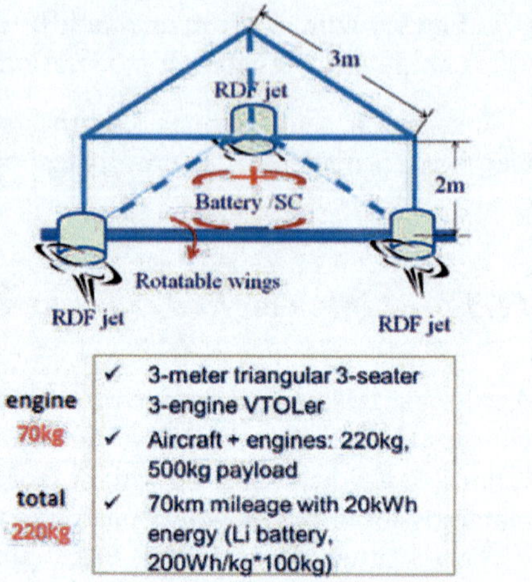

built with 3D frame resin-based composites and CMC materials is estimated in Table 11.10, and the weight of the RDF jet was estimated in Chap. 9 to be 70 kg. A total dry weight of 220 kg of the aircraft with three RDF jet engines is expected. This 3 × 3 m VTOL small aircraft can function as a rescue aircraft with adjustable horizontal and vertical wings to fit in various flexible infield scenarios.

Regarding to the density of the framed materials, we have measured two lightweight samples as in Fig. 11.14, one is from CMC, and the other one is from FRC. The density of the 3D-framed carbon-based ceramic matrix composites is lighter than the fiber-reinforced composite material and is preferred as the building materials for e-engine and e-airplanes. Of course, the mechanical strength of the 3D-framed material must fulfil the need of the aviation standard.

Fig. 11.14 Density measurement of the framed FRC and CMC materials

	3D framed CMC	3D framed FRC	
sample m/g	20	26	①
empty box m/g	53	53	
box + metal m/g	116	116	
box+ metal+ full water m/g	666	666	
box+ metal+ sample+ full water m/g	633	637	②
volume V/cm3	53	55	
density g/cm3	**0.377**	**0.473**	

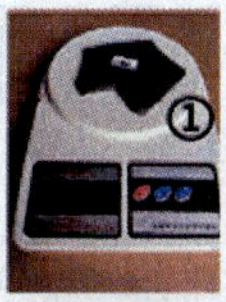

References

1. Thrust-specific fuel consumption: https://en.wikipedia.org/wiki/Thrust-specific_fuel_consumption
2. Eshelby, M.: Aircraft performance: theory and practice. American Institute of Aeronautics and Astronautics, Inc. (2000)
3. Jonas, J.: Jet airplane range considerations. J. Aeronaut. Sci. **14**(2), 124–128 (1947)
4. Cavcar, M., Cavcar, A.: Optimum range and endurance of a piston propeller aircraft with cambered wing. J. Aircr. **42**(1), 212–217 (2005)
5. Lyu, Y., Liem, R.P.: Flight performance analysis with data-driven mission parameterization: mapping flight operational data to aircraft performance analysis. Transp. Eng. **2**, 100035 (2020). ISSN 2666-691X. https://doi.org/10.1016/j.treng.2020.100035
6. FAA: Pilot's Handbook of Aeronautical Knowledge. https://www.faa.gov/
7. Sir Stanley Hooker: Not Much of an Engineer. The Crowood Press (2011). ISBN: 978-1853102851
8. Farokhi, S.: Future propulsion systems and energy sources in sustainable aviation (2019)
9. Davies, K., Norman, P., Jones, C., et al:. A review of turboelectric distributed propulsion technologies for N+3 aircraft electrical systems. In: Power Engineering Conference. IEEE (2014)
10. Hoogreef, M., Vos, R., Vries, R.D., et al.: Conceptual assessment of hybrid electric aircraft with distributed propulsion and boosted turbofans. In: AIAA Scitech 2019 Forum (2019)
11. Gan, Z.F.T., Mukherjee, B., Theron, J.P., et al.: A new distributed electric propulsion aircraft simulation tool for coupled flight dynamics, free wake, and acoustic predictions. In: VFS International 77th Annual Forum & Technology Display (2021)
12. Kirner, R., Raffaelli, L., Rolt, A., et al.: An assessment of distributed propulsion: advanced propulsion system architectures for conventional aircraft configurations. Aerosp. Sci. Technol. **46**, 42–50 (2015)
13. Kim, H.D., Berton, J.J., Jones, S.M.: Low noise cruise efficient short take-off and landing transport vehicle study. AIAA Paper 2006–7738, Sept 2006. https://doi.org/10.2514/6.2006-7738
14. Gohardani, A.S. (ed.): Distributed Propulsion Technology, pp. 173–184. Nova Science Publishers Inc., New York (2014)
15. Winborn Jr., B.R.: The ADAM III V/STOL concept. AIAA Paper 69–201, Feb 1969. https://doi.org/10.2514/6.1969-201
16. Moore, M.D.: Misconceptions of electric aircraft and their emerging aviation markets. In: 52nd Aerospace Sciences Meeting, p. 0535 (2014)

17. Huff, D.L., Henderson, B.S., Envia, E.: Motor noise for electric powered aircraft. In: 22nd AIAA/CEAS Aeroacoustics Conference, p. 2882 (2016)
18. Yoon, A., Arastu, F., Lohan, D., et al.: Direct-drive electric motor for STARC-ABL tail-cone propulsor. In: 2019 AIAA/IEEE Electric Aircraft Technologies Symposium (EATS). IEEE, pp. 1–10 (2019)
19. Schiltgen, B.T, Freeman, J.: Aeropropulsive interaction and thermal system integration within the ECO-150: a turboelectric distributed propulsion airliner with conventional electric machines. In: 16th AIAA Aviation Technology, Integration, and Operations Conference, p. 4064 (2016)
20. Nguyen, N.T., Reynolds, K., Ting, E., et al.: Distributed propulsion aircraft with aeroelastic wing shaping control for improved aerodynamic efficiency. J. Aircr. **55**(3), 1122–1140 (2018)
21. Kuruvilla, J., Oksman, K., Gejo, G., Wilson, R., Appukuttan, S. (eds.): Fiber Reinforced Composites: Constituents, Compatibility. Woodhead Publishing, Perspectives and Applications (2021)
22. Honeywell lightweight gas generator: https://www.honeywell.com/
23. Supermarine Spitfire: https://en.wikipedia.org/wiki/Supermarine_Spitfire. Spitfire Mk Vb specifications: max takeoff weight: 3039 kg, max power 1470 hp (1100 kW), maximum speed: 600 km/h, Range: 771 km, max thrust F = 6.6 kN (P = FV = 167 m/s*F = 1,100,000 W))
24. Han, L., Liang, L., Kang, Y., Qiu, Y.: A review of SiC IGBT: models, fabrications, characteristics, and applications. IEEE Trans. Power Electron. **36**(2), 2080–2093 (2021). https://doi.org/10.1109/TPEL.2020.3005940
25. Chen, H., Zhang, L.M., Jia, G.Y., et al.: The preparation and characterization of 3D-silica fiber reinforced silica composites. Key Eng. Mater. **249**, 159–162 (2003)
26. Jamra, F.R.M.: a brief review on ceramic matrix composites, it is attributes and it is utility in future generation gas turbine. IJIRST Int. J. Innovative Res. Sci. Technol. **1**(6), (2014)

Afterword

The motivation of this book is to apply the AI to AIAA in order to make it smarter and more intelligent—it is the basic relationship between IEEE and AIAA. It should be pointed out that although IEEE plays a main role in AI, there are many other contributors. IEEE resembles human brain who masters the other actions from other disciplines such as like ASME and MRS. The brain itself is not enough, we also requires the muscles and limbs. We must integrate electric/electrical engineering with other disciplines such as materials science, such as machinery/dynamics/material science, in order to achieve the ultimate goal of an intelligent AIAA. In this interaction, a leader—a dream keeper, is needed who can persistently coordinate the whole cooperative progress without deviating the original target. Not only this leader have the executive power and perseverance, but also has the authority to execute the key AIAA/IEEE actions/plans.

From the technologies perspectives, we have presented two distinct spots in this book which can be invested upon immediately, the two points of interests to make AIAA smarter, one is to use electronics, the other is to use electricity.

(1) The three levels by using electronic technology to make AIAA more intelligent are sensors, data processing and reactions/actuation.

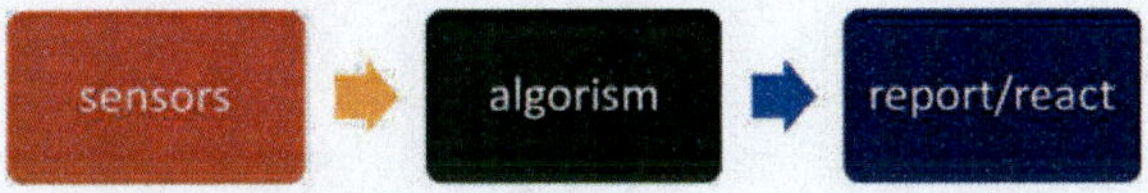

The three steps of AI

Among the three steps, sensor technology is the most basic foundation. It is the most essential part especially right now both from its necessity and the feasibility. Sensors are like the bases of a skyscraper, although we see it is tall and glorious. we should at first lay down a solid foundation. Before we cook a delicious meal, at least we need to get the rice ready. We should collect enough

© The Editor(s) (if applicable) and The Author(s), under exclusive license to Springer Nature Singapore Pte Ltd. 2023

F. L. Duan, *When AIAA Meets IEEE*,
https://doi.org/10.1007/978-981-19-8394-8

sensing data in all dimensions and then we can do the data mining and algorithms to process them to make the right judgment and then the correct reaction. The building the sensor is the first step and at currently time is still very lacking especially in the AIAA area. The difference between aero-sensors and others lies in that it's relatively more difficult to make and also these sensors work in a relatively tougher environment such as high temperatures up to a 1000 °C. Smart method such as MEMS and AM are two basic approaches to fit the purpose. MEMS is good at build multiple sensors in batch, AM is good at making the sensor onsite and on tough locations. The curved surface patterning and the tough working environment are the two main challenges to for AIAA's sensors.

(2) Electrified aviation is another way for a smarter AIAA by the aid of IEEE. There are three main cleverness. One is the smart electric engine / propulsion. The other is the integrated design of the e-airplane with multiple e-engines. The third one is the mobile electric power grid incorporating the most advanced achievements in electric power including battery, supercapacitor and lightweight gas generator. There are six major areas of the electrification on aviation as shown in the chart:

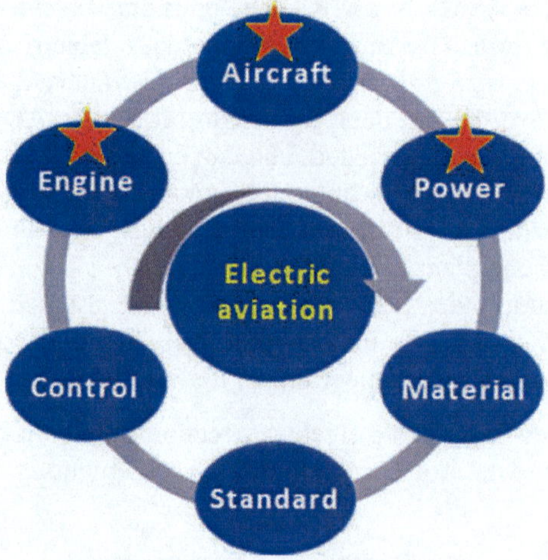

The composition of e-aviation

Typically, these six compositions basically cover all the aspects of e-eviation, where at the current moment the three star areas are the key technologies which need to be paid more attention to. The engine, power and aircraft are the three main topics in this book from Chaps. 9, 10 and11. We present the RDF jet as a new type of e-engine which is most suitable for e-propulsion with high thrust yet light weight/compact size. To provide this amount of propulsion, more effective electric power is needed. Therefore we present the hybrid electric power supplier for both long term navigation range and VTOL operation by lightweight gas turbine generator plus the 3D high K super capacitor. For the most efficient aero-dynamic performance we design a Δ shaped streamlined airplane which equipped multiple electric engines with rotatable wings. For the rest of the remaining three parts, they are either fairly mature technologies or belong to the various international FAA organizations/managements which are not covered in this book.

With the obvious advantage of green electricity from the new sustainable energy, from the Sun, the Wind and the Water, it is already expected that the electrical cars will be the mainstream transportation toys on the street within the next 5 years, although fossil thermal engines viehcles will still remain as a legendary human achievement on the road. Likewise, driven by the strong needs of smarter and greener AIAA and pushed by the great achievements of the twenty-first century on both AIAA and IEEE, we also anticipate the e-airplane will dominate on the air traffic in 10 years, either as a flying car or a car copter, the same as e-cars as the main road viehcles with the two incoming advancements of AI- and e-aviation technologies.

Glossary

3A Refer to three automations: industrial automation, office automation, home automation, but different from the meaning of AI, the artificial intelligence. AI has much more complex thinking and judgment based on the multiple sensors and sensor fusion.

3D HK SC A three dimensional high dielectric super capacitor Which is suitable for the to provide the high energy density as well as the high power density for the electric engine and electric aviation, made by the MEMS Micro fabrication technique together with a super high dielectric constant (50,000) material.

3D printing An equivalent term to additive manufacturing which constructs a three-dimensional object from a CAD model or a digital 3D model by a variety of processes in which material is deposited, joined or solidified under computer control.

3D thinking The three-dimensional quadrants (urgency, significance and the good timing) in a task management of a project. To get the something done, one needs to control the proper balance of the three factors. The Chinese call it chance, privilege, feasibility that a good deed shall happen at a specific time and a specific place with specific people. If such timing is not yet ready, you may either wait or manage to create that opportunity. Trying to do it forcibly may just get things even worse.

AAM The abbreviation of advanced air mobility, also called advanced aerial mobility, refers to the "adoption of electric and hybrid propulsion aid an aircraft for versatile air operations."Advancing aerial mobility involves the emergence of transformative and disruptive new airborne technology to transport people and things to locations not traditionally served by current modes of air transportation, including both rural and the more challenging and complex urban environments.

Additive manufacturing The construction or manufacture of an 3D object by deposited, joined or solidified materials under computer control with material being added together (such as plastics, liquids or powder grains being fused), typically layer by layer.

© The Editor(s) (if applicable) and The Author(s), under exclusive license to Springer Nature Singapore Pte Ltd. 2023

F. L. Duan, *When AIAA Meets IEEE*,
https://doi.org/10.1007/978-981-19-8394-8

Aero-engines The power component of an aircraft propulsion system, also called aircraft engines. There are few typical types: piston engines, gas turbines and electric motors. In history, the airplane and aero engine are two independent units. Nowadays with the advent of the multi electric engines and rotatable wings, the integrated design of airplane with multiple engines becomes possible to achieve more advanced aerodynamic performance.

AI Artificial intelligence (AI) is originally defined as the intelligence demonstrated by machines as opposed to the natural intelligence displayed by humans. Initially AI is the simulation of human intelligence by machines that are programmed to think like humans and mimic their actions. Nowadays AI has the extended meanings of the smarter brain, the IoT and the intelligent activities/machines that do a more effective job or even replace humans to fulfil a task.

AIAA Refers to American Institute of Aeronautics and Astronautics. AIAA (pronounced A-I-double-A) the world's largest aerospace organization in the aerospace community for more than 80 years.

Amazon An American multi-national technology company that focuses on e-commerce and online shopping.

Battery An electric device which stores the electricity inside a cell or a system. Nowadays most often people refer the battery as rechargeable and portable electric storage unit such as Li battery and fuel cell battery which can be carried by a portable computer, iPhone, EV car or by an electric airplane.

Boeing 787 One of the world most advanced, American-built wide-body jet airliners developed and manufactured by Boeing Commercial Airplanes. The most distinguished features of the Boeing 787 is its most electrified components in the airplane, and also the lightweight composites materials to build the turbo fan for the most fuel efficient aviation. The first B787 airplane was delivered in September 2011.

BYD A major full-electric automobile and Li rechargeable batteries manufacturer in Shenzhen China. BYD Co. Ltd (Chinese: 比亚迪) is also called "Build Your Dreams".

CATL Contemporary Amperex Technology Co. Limited (Chinese: 宁德时代), abbreviated as CATL, is a Chinese battery manufacturer and technology company founded in 2011 that specializes in the manufacturing of lithium-ion batteries for electric vehicles and energy storage systems, as well as battery management systems. With a market share of 32.6% in 2021, CATL is the biggest lithium-ion battery manufacturer for EVs in the world.

COMSOL A finite element analysis, solver, and simulation software package for various physics/chemistry/biology and engineering applications, especially coupled phenomena and multiphysics.

Delivery drone An unmanned aerial vehicle used to transport packages, medical supplies, food, or other goods. Delivery drones are typically autonomous with robotics control. Delivery drones are the indispensable tools to transport the online shopping goods such as Amazon in US and Taobao in China.

Delta-shaped airplane An aircraft with a delta shaped wing in the form of a triangle. It is named for its similarity in shape to the Greek uppercase letter delta (Δ). The delta form has unique aerodynamic characteristics and structural advantages such as more space occupancy and more lift-to-drag ratio. Δ-shaped is most suitable shape for the multi-electric propulsion aircraft with rotatable wing both for vertical take off and horizontal navigation.

DEP A distributed electric propulsion with an arrangement in which the propulsive and related air flows are distributed over the aerodynamic surfaces of an aircraft. The purpose is to improve the craft's aerodynamic, propulsive and/or structural efficiency over an equivalent conventional design. Anticipated benefits include improved fuel efficiency, emissions, noise, landing field length and handling. DEP comprises multiple small fans or propellers driven by electric motors. Typically, each individual thruster is direct driven by its own relatively small and lightweight electric motor. The electrical power may be provided by any suitable source.

DJI A Chinese technology company headquartered in Shenzhen, Guangdong (Chinese: 大疆) who manufactures commercial unmanned aerial vehicles (drones) for aerial photography and videography. It's revolutionary coaxial twin rotor design of T40 agricultural unmanned aircraft brings 40 kg spraying and 50 kg seeding load. It is equipped with dual atomization spraying system, smart map system, active phased array radar and binocular visual perception system, integrating flight defense and aerial survey, which can easily lead you to precision agriculture. One T40 costs ~ CMY4000 (year 2022).

E-aviation Electrified aviation including the VTOLer and STOL

E-engine An electrical machine that converts electrical energy into mechanical energy. e-engine is a derivative from an electric motor, which operates through the interaction between the motor's magnetic field and electric current in a wire winding to generate force in the form of torque applied on the motor's shaft. In this book, a new electric engine is proposed which is rim driven instead of the shaft-driven with its distinctive advantage of the bigger torque.

E-motor An electrical machine that converts electrical energy into mechanical energy or vice versa. Electric engine and electric generator are two derivatives from an electric motor, which operate through the interaction between the motor's magnetic field and electric current in a wire winding to generate force in the form of torque applied on the motor's shaft. An electric generator is mechanically identical to an electric engine, but operates with a reversed flow of power, converting mechanical energy into electrical energy. Electric engines can be powered by direct current (DC) sources, such as from batteries, or rectifiers, or by alternating current (AC) sources, such as a power grid, inverters or electrical generators. Most of the electric generators are either the diesel generator of the gas turbine generator.

Energy density The amount of energy stored in a given media per unit volume or per unit weight. The synonyms are volumetric energy density, energy per unit mass, specific energy, gravimetric energy density. There are different types of reactions to release different types of energies in the media, which can be a super capacitor, battery, fossil fuel, kerosene, hydrogen tank. The types of reactions can be: nuclear, chemical, electrochemical, and electrical.

Engineering man The word "engineering" does not just mean "to build a skyscraper". Engineering = "science, life and dream", which is the cross combination of science, technology and sociology. Engineering = IQ + EQ, including team and responsibility chain, which is the great difference between pure artists/sientists / craft masters. We need a smart man, we also need the united power. Together we can make a difference. The core concept of Engineering is IPO (Input → Process → Output): ① O - A feasible and necessary right question; ② P - the right theory or method; ③ I - input variables and which variable is the key factor, and how to optimize the combination of variables.

E-plane An electric airplane which is driven by the electric engine either to the propeller or to the duct fan. The electric power to drive this electric engine could come from the battery super capacitors or the generators, that is, an e-airplane can be either pure electric or hybrid.

E-power An equivalent term as electric power. e-power is from either /all electric generators, batteries, and super capacitors. Specifically for the aviation need, electric power is also called mobile electric power which requires the three resources possess both the higher power for VTOL capability and higher energy density for longer range.

EV An electric car, battery electric car, or all-electric car, is an automobile that is propelled by one or more electric motors, using only energy stored in batteries. Compared to internal combustion engine (ICE) vehicles, electric cars are quieter, have no exhaust emissions, and lower emissions overall. Many countries have established government incentives for plug-in electric vehicles, tax credits, subsidies, and other non-monetary incentives while several countries have legislated to phase-out sales of fossil fuel cars, to reduce air pollution and limit climate change. Tesla and BYD are world's two popular brands of the EV cars.

Gas turbine Also called a combustion turbine, is a type of continuous flow internal combustion engine in which burning of an air–fuel mixture produces hot gases that spin a turbine to produce power. The three main parts (the core) in the direction of flow include a rotating gas compressor, a combustor, and a compressor-driving turbine.

GEnX An advanced dual rotor, axial flow, high-bypass turbofan jet engine in production by GE Aviation for the Boeing 787. GEnX adopted the most advanced technology including lightweight composite materials, big BPR ratio, and overall superior aerodynamic and aircraft propulsion performance.

Green Also referred to as environmentally friendly, eco-friendly, nature-friendly. Green is a sustainability and marketing term referring to the good deeds (machines, energies, activities, associations) on reducing the CO_2 and other harmful gases emissions and using more sustainable energy resources other than the fossil fuels or coals.

IEEE Institute of Electrical and Electronics Engineers. IEEE (pronounced I-triple-E) is a professional association for electronic engineering and electrical engineering (and associated disciplines). As of 2018, it is the world's largest association of technical professionals with more than 423,000 members in over 160 countries around the world. Its objectives are the educational and technical advancement of electrical and electronic engineering, telecommunications, computer engineering and similar disciplines.

IGBT Insulated Gate Bipolar Transistor, a power semiconductor device of energy conversion and transmission, an indispensable device which is used in the electric vehicles connecting the electric power to the electric engines, to transfer and to control the electric power output from batteries/hybrid generators towards the e-engines.

IoT The Internet of things, which may consist of two meanings: (1) the internet info + things, i.e., The physical things can be exchanged via the internet. With aid of the text, video and audio message from internet, the physical object can be exchanged via the drones deliveries like Amazon in US and Taobao in China. (2) describes physical objects with sensors, processing ability, software, and other technologies that connect and exchange data with other devices and systems over the Internet or other communications networks.

IPO IPO = Input – Process – Output, IPO is the technical route to fulfil a project, a topic and a task. I, P and O are the three major components of a project, which is to set the specific goal ("O", the output), then determine the input variable "I" that affects the output "O", by going through a process "P", which is the proper mechanism/model such as simulation or experiment. IPO forms the direction of an engineering endeavor, i.e., from Input → Process → Output. Although "O" is the last in the sequence and the final result of the whole engineering process, it may be initiated first, that is, what we call "being able to ask the right question is half the solution". The ability to define problem, and to ask executable question is a very important ability. To determine a specific target "O" is to ask the right question. This "O" is not only necessary but also feasible. To some extent, this "O" is the innovation, the new idea, the core of IPO. The right "O" initiates and directs the development flow of a project, a scientific topic or as task.

Jingdong JD.com, Inc. is a Chinese e-commerce company headquartered in Beijing

Li battery Most of the time is equivalent term of Li-ion battery. Strictly speaking, Li battery include Li-ion battery, LiS and other lithium-based batteries.

lift-to-drag ratio The amount of lift generated by a wing or airfoil compared to its drag. The lift/drag ratio (L/D) is determined by dividing the lift coefficient by the drag coefficient, CL/CD. A ratio of L/D indicates airfoil efficiency.

Li-ion battery Or lithium-ion battery, a rechargeable battery composed of cells in which lithium ions move back and forth from negative to positive electrodes during the discharging and charging process.

LTG Lightweight turbine generator, which is an advanced electric generator using the gas turbine principle with lightweight material with moderate weight (less than 1 ton) and high power output (500 k to 10 MW), compact size (diameter < 0.5 m, length < 1.5 m).

MEMS Microelectromechanical systems, also written as micro-electro-mechanical systems constitute the technology of microscopic devices, made up of components between 1 μm to a millimeter in size. They usually consist of a central unit that processes data (an integrated circuit chip such as microprocessor) and several components that interact with the surroundings (such as micro sensors). Because of the large surface area to volume ratio of MEMS, forces produced by ambient electromagnetism (e.g., electrostatic charges and magnetic moments), and fluid dynamics (e.g., surface tension and viscosity) are more important design considerations than with larger scale mechanical devices. MEMS technology is distinguished from molecular nanotechnology or molecular electronics in that the latter two must also consider surface chemistry.

Micro/Nano The science, technology and engineering that belong to the range of 10^{-8} to 10^{-3} m

Mindset An established set of attitudes, esp. regarded as typical of a particular group's frame of mind, attitude, disposition. A firmly established mindset could create a powerful incentive to adopt or accept prior behaviors, choices, or tools, sometimes referred to as cognitive inertia, or "groupthink."

NEV A new word proposed as opposed to the EV (electric vehicles) since more and more electricity to drive the EVs (including EV cars, ships and airplanes) may not come purely from the fossil fuel but from the more sustainable energy resources like solar, windy and tidal, i.e., the new energies. The green resources are basically "free energy" which does not consume any natural sources.

Payload The carrying capacity of an aircraft in weight which may include cargo, passengers, flight crew, munitions, scientific instruments or experiments, or other equipment.

PI Also called a dream keeper, refers to the holder of an independent grant or the leader of a research for the grant project, usually in the sciences, technologies and engineering, such as a laboratory study, a clinical trial, or an VC-sponsored startup industrial entrepreneur. It is used widely for the person who makes final decisions and supervise funding and expenditures on a given research project.

Power density The amount of power (time rate of energy transfer) per unit volume or per unit weight, in energy transformers including power supply unit such as batteries, fuel cells, generators, and power users such as engines, motors, jets, thrusters, etc. Also called specific power, power-to-weight ratio, in the unit of W/kg, W/Liter.

Power-to-weight ratio The horsepower of the engine divided by the weight. Power-to-weight ratio (PWR, also called specific power, or power-to-mass ratio) is used to compare the performance of any engine or power source. It is also used as a measurement of performance of a vehicle as a whole, with the engine's power output being divided by the weight (or mass) of the vehicle. Power-to-weight is often quoted by manufacturers at the peak value, but the actual value may vary in use and variations will affect performance.

Pt-Dot-TC A unique insitu thermocouple device proposed in this book, which is very easy to fabricate and can be used for the onsite accurate temperature calibration and the measurement.

RDF jet A rim driven fan jet thruster for electric aircraft specially for VTOL feature. It is composed of a rim driven motor, the Taichi fan blade, and the duct jet. Among them, rim driven motor is good at larger torque, Taichi fan blade is good at more air intake, and duct and shrunken outlet is for higher jet speed. The RDF jet is a small, light and flexible rotatable e-propulsion engine for various aircrafts with VTOL/STOL features.

Rim-driven A novel type of electric magnetic drive unit that does not use a central hub for transmission of the driving torque. Conventional hub centric propellers typically use a shaft driven by a turbine, a diesel engine or an electric motor. The blades of the rim-driven thruster, by contrast, are mounted on an outer ring rather than a central hub. The ring constitutes the rotor of an electric motor and sits within a surrounding stator, which is also ring-shaped and creates the necessary torque.

Rotatable wing Refers to the wing capable of rotating from vertical to horizontal direction. Such wing attaches the multiple electric engines and since these e-engines are small, compact and light, it is relatively easier to rotate them with various angles with the rotatable wing. When rotating engines in various angles, the associated aircraft is capable to move from vertical to horizontal propulsions or vice versa.

SC Super capacitor, also called an ultra capacitor, is a high-capacity capacitor with a capacitance value (10–100 times more energy) much higher than conventional electrolytic capacitors. Unlike ordinary capacitor, the so-called super capacitor does not use the conventional solid dielectric, but rather, it involves electrochemical oxidation–reduction (redox) in the charging/discharging process, which possesses intrinsic limitation on the high specific power output density. But comparing with batteries, it can accept and deliver charge much faster and possesses much more charging/recharging cycles. In this book a new 3D high K dielectric super capacitor is invented which operates in different ways as the conventional SCs.

Sensor A device that detects events or changes in its environment for the purpose of sensing a physical phenomenon and convert it into electric signals to other electronics, frequently a computer processor. Sensor is one of three key e-devices i.e. the RLC (passive device of resistor, inductor and capacitor), the transistor (3-terminal active device) and sensor (converting device of other physical signals to electric signals). Sensors are always used with other electronics.

SFC Also called thrust-specific fuel consumption (TSFC), is the fuel efficiency of an engine with respect to thrust output. sfc (in the unit such as g/(s·kN)) is the fuel consumption per unit of thrust (kilonewtons, or kN) or per unit of Walt (kilowatts, or kW). It is thus thrust or power dependent, meaning that the src is dependable on the output thrust or power. In general, the higher thrust of power, the higher src.

Spitfire British single-seat fighter aircraft used by the Royal Air Force and other Allied countries before, during, and after World War II, which made a great contribution to win the War in many countries.

STOL Short takeoff or landing, which is mainly achieved by leveraging the weight of the aircraft. In general, the takeoff distance shorten X% with X% weight leverage.

Tai Chi The Tai Chi logo ☯ is an interactive chart of the positive (Yang((Chinese: 阳)) and negative (Yin(Chinese: 阴)) energy flows, where the Yin and Yang are both from the Tai Chi. Tai Chi (Chinese: 太极) also called taiji ("Supreme Ultimate") appears in both Taoist and Confucian philosophy. These two ancient Saints believe that everything in the world consists of two parts: Yin and Yang, such as positive and negative, good or bad, etc.. This Chinese philosophical theorem basically believe: (1) positive and negative energy flows coexist inside any item, (2) the Yin and Yang energy flows towards each other, (3) there is a Yin within the Yang, and there is a Yang energy inside the Yin. The constant positive and negative energy flows form the Life. The term taiji is a Chinese cosmological concept for the flux of Yin and Yang energy flow.

Taobao Taobao (Chinese: 淘宝) is a Chinese online shopping platform. It is headquartered in Hangzhou and is owned by Alibaba. It is one of the most-visited website globally in 2021 in China. Taobao.com was registered on April 21, 2003 by Alibaba Cloud Computing (Beijing) Co., Ltd and becomes popular since 2016 with over 1 billion product listings as of 2016 the combined transaction volume of 3 trillion CMY in 2017. Taobao provides a platform for small businesses and individual entrepreneurs to open online stores that mainly cater to consumers in Chinese-speaking regions (Mainland China, Hong Kong, Macau and Taiwan) and abroad, which is made payable by online accounts. Its stores usually offer an express delivery service. Sellers are able to post goods for sale either through a fixed price or an auction. Auctions make up a small percentage of transactions, whereas the majority of the products are new merchandise sold at fixed prices. Taobao users usually read feedback and compare items from multiple shops. Taobao's popular payment platform is Alibaba's Alipay.

Tesla An American multinational electric vehicle and clean energy company. Tesla, Inc. is one of a pioneering EV companies who bring the revolutionary leading automobile with the obvious advantages of quieter and no exhaust emission. EV car will phase-out the fossil fuel car eventually within a few years with government incentives for plug-in electric vehicles, tax credits, subsidies, and other non-monetary services.

TFTC Thin film thermocouple, an advanced temperature measuring sensor which can be attached to any surface of a machine part. Compared to conventional sensors, TFTC sensors are smarter which can be built with the MEMS process in small size and can be built in batch, together with many other advantages.

The 3rd way An innovative engineering approach to find the 3rd path which is neither yes nor no to avoid the dilemma but can cleverly resolve the issue. This term has different meaning as its political term.

UAS An unmanned aircraft system (UAS), which includes the UAV together with the ground-based controller and communications. The UAV is the key of UAS, which is an unmanned aerial vehicle (UAV), commonly known as a drone, is an aircraft without any human pilot, crew, or passengers on board.

VAATE The Versatile Affordable Advanced Turbine Engines (VAATE) program provides the framework for addressing future turbine engine needs. VAATE is the first a AIAA plan which proposes the intelligence engine blue print.

VTOL Refers to vertical take-off and landing, an aircraft that can hover, take off and land vertically without relying on a runway. Specifically, the technology behind VTOL is the capability of a vertical thrust provided by the electric propulsion. Therefore, the VTOL is most often coined as eVTOL. An ideal VTOLer is an aircraft with multiple electric engines which is capable to provide the thrust for both the vertical takeoff and landing as well as the horizontal navigation with rotatable wings.

天时地利与人和 A Chinese proverb that says "favorable time and place together with the right people are the three basic elements of a success".

Index

A

Additive manufacturing, 38, 41, 43, 67, 367

Advanced Air Mobility (AAM), 8, 17, 168–170, 173, 174, 198, 224, 233, 261, 271, 276, 280, 325, 335, 367

Aero-engine, 3, 4, 6, 7, 10, 14, 17, 18, 36, 47, 54, 66, 69, 75–79, 81, 82, 84, 88, 90, 98, 100, 103, 113, 114, 137, 146, 150, 161, 168–170, 175, 188, 195, 204, 205, 209, 210, 212, 218, 220, 223, 224, 227–231, 233, 237, 255, 257, 259, 268–270, 274, 275, 278, 281, 305, 307, 323–325, 329, 331, 343, 368

Airbus, 6, 276, 325, 326, 338

Aircraft, 5, 6, 8, 9, 15–17, 69, 128, 168–178, 180, 182, 183, 186–188, 190, 191, 194–199, 203, 204, 209, 210, 212–216, 218, 219, 221–225, 228–230, 233–235, 256, 260, 264, 266, 270–272, 276–278, 280, 283, 284, 302, 305, 307, 309, 311, 323–325, 327–331, 333–341, 344, 346, 348, 349, 352, 353, 358–360, 364, 367–370, 372–375

Amazon, 25, 134, 169, 171, 175, 231, 258, 338, 342, 368, 371

American Institute of Aeronautics and Astronautics (AIAA), 3–6, 8–10, 12, 13, 15–18, 24, 25, 36, 40, 44, 47, 48, 66, 69–71, 81, 96, 98, 100, 103, 105, 119, 120, 127, 136, 143, 145–151, 156–162, 168, 192, 206, 209, 261, 275, 278, 282, 283, 363–365, 368, 375

Artificial Intelligence (AI), 23–32, 44, 66, 145, 156, 158, 330, 359, 363, 365, 367, 368

Aviation, 3, 4, 6, 10, 13, 16–18, 25, 30, 145, 146, 159, 160, 168–170, 172, 174, 175, 178, 182, 192, 194, 196, 205, 211–213, 215, 222, 224, 229–231, 233, 239, 243, 252, 253, 256, 260, 263–269, 271–276, 278, 280–284, 289, 296, 301, 302, 304, 305, 307, 314, 317, 324–326, 329–331, 335, 336, 338, 348, 359, 360, 364, 367–370

B

Battery, 4, 6, 13, 108, 150, 159, 172, 173, 175, 180, 184, 192, 194, 196–199, 201–204, 206, 225, 231, 233, 235, 238, 243, 257, 261, 264–267, 269–271, 274–276, 278–282, 284, 285, 287–296, 302–304, 309, 325, 328, 336, 338, 343, 344, 352–356, 359, 364, 368–373

Boeing 787, 191, 194, 195, 202, 205, 223, 228, 259, 260, 268, 269, 327–329, 334, 335, 338, 368, 370

C

CATL, 291, 368

Cloud

- cloud data, 25, 29
- cloud online, 25

© The Editor(s) (if applicable) and The Author(s), under exclusive license to Springer Nature Singapore Pte Ltd. 2023
F. L. Duan, *When AIAA Meets IEEE*,
https://doi.org/10.1007/978-981-19-8394-8

D

Device Under Test (DUT), 49, 108, 120–122, 124
Digital Multi Meter (DMM), 76, 107, 108
Distributed Electric Propulsion (DEP), 9, 168, 172, 194, 205, 234, 324, 335–337, 369
Drone delivery, 26, 174, 175, 229, 278, 341, 342

E

E-engine
 duct fan, 230, 234, 235, 370
 duct jet, 227
 hybrid, 172, 175, 197, 364, 370, 371
 propulsion, 6, 8, 9, 17, 146, 168, 205, 209, 230, 233, 364, 368, 373
 rim driven, 230, 369
 rotations per minute (RPM), 197
 thrust, 8, 175, 201, 203, 206, 213, 227, 230, 233, 235, 276, 348, 364, 375
 torque, 180, 181, 206, 235, 369
Electric Vehicle (EV) cars
 BYD (Build your dream), 14, 18, 159, 160, 172, 196, 197, 260, 274, 283, 287, 290–295, 368, 370
 Tesla, 159, 160, 172, 196, 197, 260, 261, 270, 274, 283, 290–294, 353, 370, 374
Electrified aviation
 e-aviation, 168, 192, 194, 196, 206, 365, 369
 e-engine, 206, 364, 369
 efficiency, 159, 281
 e-motor, 13, 160, 168, 205, 206, 240
 e-plane, 196, 305
 e-power, 13, 18, 145, 146, 168, 231, 305, 370
 eta, 271
Energy density, 198, 201, 203, 212, 230, 231, 264–266, 268–275, 280, 281, 284, 285, 289–291, 293, 295, 299, 301–303, 305, 309, 311, 312, 327–329, 336, 341, 346, 353, 367, 369, 370
E-plane
 carcopter, 194, 195
 delta-shaped airplane, 339, 368
 delta-wing aircraft, 339
 DJI, 172, 173, 194, 199–201, 203, 356, 369
 e-Fan, 6, 276
 Eviation Alice, 173, 196, 197
 flying car, 170, 173, 195, 365
 RR (Rolls-Royce) SOI (Spirit of Innovation), 196, 197
 skycar, 170, 195
 triangular aircraft, 342

F

Flight
 Fossil fuel
  ammonia fuel, 272, 273
  diesel, 228, 268, 270, 274, 304, 369
  gasoline, 185, 199, 206, 265, 272, 273, 302
  kerosene, 185, 230, 268, 270, 272, 369
Fuel Cell (FC)
 hydrogen, 184, 265, 273, 280, 284, 287, 294–298
 hydrogen tank, 272
 liquid hydrogen, 299
 methanol, 272, 295–297, 301

G

Gas turbine machines
 Bypass Ratio (BPR), 209, 210, 221, 223, 225, 227, 237, 283
 combustion, 212, 220, 221, 223, 233, 270
 compressor, 220, 221, 223, 225, 229
 cooling effect test, 162
 cooling performance test, 48, 54
 core, 220, 221
 eta, 223
 duct fan, 218, 224, 231, 257
 gas turbine engine, 212, 225, 227, 237, 283, 331
 gas turbine generator, 212, 224
 inlet, 220–222
 jet nozzle, 220
 Ni-Cr alloy, 259
 nozzle guide vane (NGV), 49, 54, 58, 140
 turbofan, 209, 212, 215, 217, 222, 224, 225, 227, 237, 256, 257
 thermal barrier coating (TBC), 48, 50–54, 75, 81, 82, 88
 turbine blade, 47–49, 54, 55, 58, 61, 68, 69, 71, 80, 81, 107, 109, 120, 125, 127, 132, 139, 141
 yttria-stabilized zirconia (YSZ), 75, 82, 84, 88, 107, 128
General Electric Company

GEnX, 226, 227, 237, 370
LM6000, 226, 266, 268, 269, 275, 305, 307
General Electric Company (GE)
CF700, 226, 275, 307, 343
J85, 305, 307, 343
Gravimetric energy density, 272, 273, 369
Green
airport pollution, 8, 191, 309
airport resources, 8, 190
CO, 191
CO_2, 183, 190
noise, 288, 309
NOx, 8, 186, 191, 295

I
Infield testing, 25, 89, 120, 142, 192, 194, 337, 338
Infrared thermography, 48–50, 61, 127, 142
Input Process Output (IPO), 35, 36, 146, 254, 370, 371
Institute of Electrical and Electronics Engineers (IEEE), 3, 4, 6, 9, 10, 12, 13, 15–18, 23–25, 32, 33, 47, 70, 81, 120, 143, 145–150, 156–162, 168, 192, 209, 261, 275, 278, 281, 355, 363, 364, 371
Insulated Gate Bipolar Transistor (IGBT), 13, 14, 236, 243, 260, 355, 356, 371
Intellectual quality (IQ) and Emotional Quality (EQ), 16, 147, 155, 157, 158
Intelligent engine, 66, 81
Internet of Things (IoT), 9, 10, 17, 23–27, 29, 30, 32, 66, 157, 174, 175, 183, 277, 341, 368, 371

J
Jingdong, 25, 371

L
Landing Takeoff cycle (LTO), 187, 188, 191
Li battery
18650 battery, 290, 294
4680 battery, 197
blade battery, 197, 283, 287, 291, 294, 295
LFP battery, 291, 292, 294
Li-ion battery, 159, 231, 264, 274, 276, 280, 283, 284, 290, 293–295, 371
Li(NiCoMn)O2 battery, 292
Lift-to-drag ratio, 203, 324, 325, 329, 339, 369, 371
Lift-to-weight ratio, 203, 344, 351
Lightweight aero-material, 359
Lightweight Turbine Generator (LTG), 17, 168, 172, 174, 229, 231, 266, 271, 274, 275, 277, 278, 281, 283, 284, 304–308, 317, 324, 346, 348, 353, 371
LTG
BTW (by the way) gas generator, 225, 229
rim driven gas generator, 308

M
MEMS
curved surfaces patterning, 40, 41, 43, 47, 100, 101, 364
CVD, 314
dry film, 68, 98, 105
hand-draw, 44, 95, 98–100
inductive coupling plasma (ICP), 41, 42, 67
Integrated circuits (IC), 10, 12, 23, 30, 41, 66–68, 313, 315, 372
lift-off, 39, 40, 43, 73, 93
mask-free, 98, 106, 107
on site fabrication, 98, 99
patterning, 40, 41, 43, 97, 101, 105
photolithography, 39, 43, 73, 93, 98, 281
photo resist (PR), 40, 67, 68, 98
physical vapor deposition (PVD), 39, 100, 280, 313, 314
polydimethylsiloxane (PDMS), 101
thin film, 39, 40, 89, 93, 96
Micro, 10, 12, 28, 38, 47
Middle way, 147, 148, 182
Mindset, 3, 4, 16, 18, 146, 150, 158, 168, 372
Multi-disciplinary, 337, 356, 358

N
Nano, 3, 35, 157, 372
Navigation, 171, 173, 178, 185, 187, 191, 198, 199, 210, 218, 230, 231, 239, 268, 274, 275, 277, 278, 284, 305, 317, 324, 326–332, 334, 337, 339–341, 346, 349, 351, 353, 356, 358, 359, 364, 369, 375
New Energy Vehicle (NEV), 14, 29, 265, 281, 290, 292, 294, 372

P

Pax, 6, 181, 182, 185, 259, 276, 278
Payload, 171, 175, 201, 204, 218, 325, 341, 343, 344, 348, 372
Permanent magnet, 13, 14, 205, 236, 238, 240, 242–245, 247, 249–252, 260, 283, 306, 308
Power density, 13, 14, 197–199, 203–205, 212, 230, 231, 260, 261, 264–271, 274, 275, 278, 281, 285, 288, 301–303, 308–312, 336, 340, 341, 344, 346, 353, 354, 367, 372
Power electronic, 355
Power-to-weight ratio, 13, 14, 127, 203, 205, 206, 225, 336, 359, 372
Principal investigator (PI), 158, 162
Project-driven, 6, 157
Propeller, 4, 146, 170, 173, 192, 194, 195, 197, 199, 201, 203, 209, 210, 212–219, 223–225, 227, 228, 230, 233–236, 243, 253, 255, 276, 278, 328, 331–333, 337, 338, 342, 348, 352, 369, 370, 373
PtAl, 94–96
Pt-Dotted thermocouple (Pt-Dot-TC), 121–124, 134, 135, 372

R

Range, 5, 27, 29, 47, 76, 78, 84, 86, 92, 103, 107, 108, 112, 114, 122, 123, 170–172, 174, 181, 186, 192, 195, 198, 199, 201, 203, 210, 213, 217, 221, 231, 235, 253, 259, 260, 266–268, 274–277, 288, 305, 313, 317, 324–330, 339, 341–344, 346, 348, 351, 352, 356, 364, 370, 372
Resistor Temperature Detector (RTD), 12, 59, 69
Rim driven
  rim driven motor (RDM), 242, 243, 249, 252, 259, 260, 352, 373
  rim driven propulsion (RDP), 255, 256, 280
Rim Driven Fan (RDF) jet, 6, 9, 153, 168, 172, 174–176, 187, 188, 191, 196, 203, 204, 209, 212, 216, 218, 223, 225, 226, 228, 230, 231, 233–235, 237–240, 243, 244, 249, 252, 254–257, 259–261, 270, 274, 276, 277, 281, 283, 284, 308, 311, 324, 329, 336–344, 346, 348–354, 359, 360, 364, 373
Robotics, 11, 23, 24, 27, 28, 31, 32, 205, 355–358, 368
Robustness
  high vibration test, 140
  impact tests, 136, 139, 141
  mechanical shocking test, 139
  peeling off, 90, 98, 129, 137
  shocking tests, 89, 136, 138–141
  thermal shock test, 100, 137
  vibration tests, 136, 141
R-type thermocouple, 52, 121, 124, 129

S

SAW
  Inter digital transducer (IDT), 79, 109–112, 114, 130
  Lithium-Niobate-based, 130
  LN, 79, 111, 112, 131, 134
  near-field antenna, 79, 114, 130, 132, 133
  piezoelectric, 78, 79, 113, 114, 130
  radio frequency (RF), 26
  wireless sensing, 49, 71, 78, 80, 100, 109, 110, 130, 131
SAW (surface acoustic wave) sensor, 77, 78, 114
Seebeck effect, 73
Sensor, 4, 6–8, 10, 12, 15–18, 23, 24, 26–41, 44, 47–49, 51, 54–56, 58–61, 66, 68–71, 73–90, 94, 96–98, 100, 101, 103–107, 110–112, 114, 119–121, 124–127, 129–131, 134, 136, 137, 139–143, 145–149, 158, 161, 168, 242, 309, 352, 356–359, 363, 364, 367, 371–374
Shock wave, 139, 215, 217, 227, 228, 233, 243, 253
Short Take Off or Landing (STOL), 168, 176, 177, 187, 190, 229–231, 235, 259, 260, 270, 275, 276, 280, 324, 325, 336, 338, 348, 349, 369, 373, 374
Silicon carbide (SiC), 13, 355, 356
Specific energy, 233, 264, 269, 272, 273, 301, 341, 369
Specific Fuel Consumption (SFC), 223, 268, 269, 307, 326, 329–332, 334, 341, 343, 373
Specific power, 194, 196, 204, 205, 233, 264, 268, 302, 304, 372, 373
Spitfire, 209, 215, 352, 373
Supercapacitor or Super capacitor (SC), 4, 146, 172, 175, 180, 192, 212, 225,

231, 234, 238, 243, 261, 264–267, 270, 271, 275, 277, 280, 284, 285, 294, 302, 303, 305, 308, 309, 311–317, 338, 340, 341, 343, 346, 350, 351, 354, 364

T

Tai Chi, 17, 233, 234, 237, 252, 253
Taobao, 134, 175, 231, 342, 368, 371, 374
Temperature measurement
  accuracy, 54, 58, 71, 120
  calibration, 49, 61, 120, 124, 127, 134
  cooling air, 55, 58
  dynamic test, 53
  temperature distribution test, 58
  temperature responsiveness, 76, 127
  transient temperature test, 126
Thermal stress, 47, 70, 75, 79, 81, 88, 90–92, 98, 120, 137, 138
Thermal voltage, 49, 53, 106, 107, 124, 125, 134, 136, 137, 142
Thermistor, 75, 76, 100, 107–109
Thermocouple (TC), 35, 49–57, 60–62, 68, 71, 73, 75
Thin Film ThermoCouple (TFTC), 4, 44, 47–53, 55–58, 60–62, 71, 73–76, 90, 94–100, 106, 107, 120–122, 124, 126–130, 136–138, 142, 143, 149, 162, 374
Thin Film ThermoCouple (TFTC) array, 49, 57, 60, 96, 125, 127
3D framed Ceramic Matrix Composite (CMC), 359
3D HK SC (three dimensional high dielectric super capacitor), 13, 17, 168, 172, 174, 212, 226, 231, 261, 271, 274, 275, 277, 284, 304, 309–314, 317, 324, 340, 341, 346, 348, 350, 352, 354, 367
3D printing, 38, 41, 43, 44, 261, 367
3D thinking, 18, 367
3rd way, The, 18, 153, 160, 161, 374

U

UAS, 13, 16, 17, 157, 168, 173, 194, 198–200, 203, 218, 225, 231, 233, 278, 324, 337, 342, 348, 355, 356, 359, 374

V

Versatile Affordable Advanced Turbine Engines (VAATE), 48, 66, 69, 70, 375
Volumetric energy density, 272, 273, 293, 297, 299, 301, 369
VTOL
  Advanced Air Mobility (AAM), 8, 168, 169, 173, 174, 198, 233, 276, 280
  eVTOL, 8, 176, 194, 375
  green, 17, 168, 187, 212, 231
  horizontal propulsion, 218, 233, 348
  rotatable wing, 171, 177, 194, 195, 198, 202, 218, 228, 234, 266, 349, 365, 375
  smart aviation, 176, 178, 181
  vertical propulsion, 174, 177, 195, 228, 353
VTOLer, 17, 169–172, 174, 175, 201–204, 210, 259, 325, 338, 343, 344, 346–348, 352, 353, 369, 375
  catering drone, 169
  delivery drone, 171, 175, 176, 338, 339
  delta airplane, 356, 359
  emergency drone, 343, 348

W

Weight-to-power ratio, 201
Wired-TC (thermocouple), 7, 39, 49, 54–58, 74, 97, 120, 123–125
天时地利与人和, 156, 375
太极, 374
工程导学, 146, 155